Computer Graphics – Systems and Applications

H. Hagen D. Roller (Eds.)

Geometric Modeling

Methods and Applications

With 140 Figures

Springer-Verlag
Berlin Heidelberg New York
London Paris Tokyo
Hong Kong Barcelona
Budapest

Prof. Dr. Hans Hagen
Fachbereich Informatik
Universität Kaiserslautern
Postfach 3049
W-6750 Kaiserslautern, FRG

Dr. Dieter Roller
Hewlett Packard GmbH
Postfach 1430
W-7030 Böblingen, FRG

ISBN-13:978-3-642-76406-6 e-ISBN-13:978-3-642-76404-2
DOI: 10.1007/978-3-642-76404-2

Library of Congress Cataloging-in-Publication Data

Geometric modeling: methods and applications/H. Hagen, D. Roller (eds.).
p. cm. – (Symbolic computation. Computer graphics – systems and applications)
Includes bibliographical references and index.
ISBN-13:978-3-642-76406-6
1. Computer graphics – Congresses. I. Hagen, H. (Hans), 1953- . II. Roller, D. (Dieter), 1951- . III. Series.
T385.G464 1991 06.6 – dc20 91-6723 CIP

Softcover reprint of the hardcover 1st edition 1991

45/3140-543210 – Printed on acid-free paper

Preface

This book is based on lectures presented at an international workshop on geometric modeling held at Hewlett Packard GmbH in Böblingen, FRG, in June 1990. International experts from academia and industry were selected to speak on the most interesting topics in geometric modeling. The resulting papers, published in this volume, give a state-of-the-art survey of the relevant problems and issues. The following topics are discussed:

– *Methods for constructing surfaces on surfaces:* four different solutions to the multidimensional problem of constructing an interpolant from surface data are provided.

– *Surfaces in solid modeling:* current results on the implementation of free-form solids in three well established solid models are reviewed.

– *Box splines and applications:* an introduction to box spline methods for the representation of surfaces is given. Basic properties of box splines are derived, and refinement and evaluation methods for box splines are presented in detail. Shape preserving properties, the construction of non-rectangular box spline surfaces, applications to surface modeling, and imbedding problems, are discussed.

– *Advanced computer graphics techniques for volume visualization:* the steps to be executed in the visualization process of volume data are described and tools are discussed that assist in handling this data.

– *Rational B-splines:* an introduction to the representation of curves and surfaces using rational B-splines is given, together with a critical evaluation of their potential for industrial application.

– *Scattered data interpolation and applications:* a tutorial introduction is given to the multivariate scattered data interpolation problem. Various types of data sets, strategies for dealing with them, and readily available algorithms, along with situations where they have been useful, are discussed.

– *Variational principles in curve and surface design:* some new techniques for the design of smooth curves and surfaces are presented, based on a calculus of variations approach.

– *Functional splines for modeling:* some implicit curves and surfaces, as used for interpolation, approximation, blending and filling, are introduced and interpreted as functional splines fulfilling geometric continuity conditions.

– *Numerically-controlled milling of CAD surface data:* the main steps in the extraction of machine-understandable information from surface data are discussed, with emphasis on the problem of collisions between tool and surface model.

– *Aspects of form feature modeling:* starting from the definition of a form feature as a region of interest on the surface of a part, examples are given and a more precise definition is formulated from the point of view of CAD modeling.

– *Advanced methods for parametric design:* parametric design is introduced and examples of application areas are given. Methods are presented for parametric modeling in CAD. Illustrative examples are given and the characteristics of the methods are summarized.

– *A tutorial introduction to blossoming, also called polarization:* this is a powerful new technique for analyzing Bézier and B-spline curves and surfaces. The basic concepts and uses are introduced and several implementation consequences are identified.

We would like to thank all participating speakers, and the audience, for what appears to have been a very successful workshop.

Kaiserslautern, Böblingen
January 1991

Hans Hagen
Dieter Roller

Contents

Methods for Constructing Surfaces on Surfaces

Robert E. Barnhill and Thomas A. Foley
Computer Science Department
Arizona State University
Tempe, Arizona 85287 U. S. A.

Abstract. Given data defined on a (domain) surface, we construct an interpolant, which is a "surface defined on a surface." We provide four different solutions to this multidimensional problem.

1. Introduction. We invented the phrase "surfaces defined on surfaces" several years ago (Barnhill 1985) to describe the construction of (possibly higher dimensional) surfaces defined on physical objects. There are many applications of this subject. For example, in a project with NASA-Ames, we were given wind tunnel measurements of pressures on an aircraft wing and asked to predict the pressure at an arbitrary location on the wing. The wing itself was a collection of bicubic patches and constituted the domain surface. We then constructed the pressure "surface" as a trivariate surface interpolating to the measured pressures at the given data sites on the wing. Many other physical phenomena are similar to this example, for instance, gravity anomalies of the earth, mineral deposits, ozone, rainfall and temperatures. Time can also enter the problem as an additional variable and visualized by means of animation. Surfaces on surfaces can arise in design if we think of the NASA pressures indicating a possible requirement for redesign of the wing.

Surfaces defined on surfaces is not the same as trivariate (or multivariate) interpolation. As illustrated by the NASA example, the geometry of the domain surface should be involved in the approximation: it would be unusual, although possible, to treat the problem simply as trivariate interpolation.

We have developed several methods for interpolating to data given on a surface and visualizing the results, in a sequence of papers (Barnhill, Piper & Stead 1985), (Barnhill, Makatura & Stead 1987), (Barnhill, Piper & Rescorla 1987), (Barnhill & Ou 1990) and (Foley, Lane Nielson, Franke & Hagen 1990). Four principal methods are utilized: a direct method, a distance-weighted interpolant, a triangular interpolant, and a domain mapping technique.

2. Direct Method (Barnhill, Piper & Rescorla, 1987). We assume that the domain surface is defined parametrically, for example, by means of piecewise bicubic interpolation. Although the method is general, this example will fix the ideas. We shall also speak of "pressures", for concreteness.

There are three main steps:

(1) *Geometry step*

Preprocess the surface domain so that the various parts of the domain surface are defined by piecewise bicubic interpolation.

(2) *Pressures at defining geometry sites*

Determine the pressures at the defining geometry sites by means of a scattered data interpolant.

(3) *Pressures at arbitrary locations on the domain surface*

Use a smooth interpolant to the pressures at the defining geometry sites to predict the pressure at an arbitrary location on the domain surface.

The details of these steps comprise the remainder of this section.

2.1. Representation of Wing. The wing geometry points are points in R^3 which define the surface of the wing. They have a gridded structure and are of the form

$$\mathbf{w^{i,j}}, \quad i = 0, \ldots, 14, \quad j = 0, \ldots, 16. \tag{2.1}$$

We estimate the derivative data needed for piecewise bicubic interpolation and call the overall interpolant $\mathbf{w}$:

$$\mathbf{w} : D = [0,14] \times [0,16] \rightarrow \mathbf{R}^3. \tag{2.2}$$

Thus the $\mathbf{w^{i,j}}$ are the images of the gridpoints in D under the mapping $\mathbf{w}$. That is, the top of the wing is the image of the rectangle D under a C^1 map which takes $\{(14, v) : 0 \leq v \leq 16\}$ onto the trailing edge and the other two sides of D onto single points with u-tangents with uv-derivatives zero along these degenerate edges, see Fig. 1. Piecewise bicubic interpolation is used instead of tensor product spline interpolation in order to localize the interpolant and to more nearly maintain the wing's convexity.

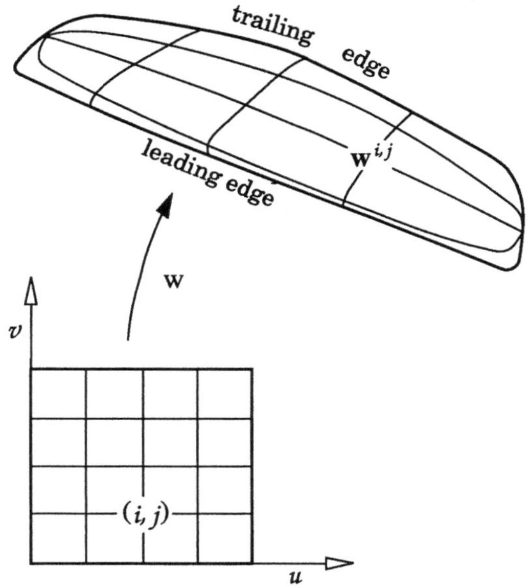

Fig. 1. Geometry of the wing: piecewise bicubic representation

2.2. Pressures at Wing Geometry Points. The given data are of the form $(\mathbf{x_k}; f_k)$ where the $\mathbf{x_k}$ are the three-dimensional data sites (plug taps) and the pressures f_k are measured. For each wing geometry point $\mathbf{w^{i,j}}$, a corresponding pressure $f^{i,j}$ is calculated as follows:

Construct the hyperbolic multiquadric (Hardy 1971) M of the form

$$M = M(\mathbf{x}) = \sum_{k=1}^{n} c_k \{d_k^2(\mathbf{x}) + R\}^{\frac{1}{2}}, \tag{2.3}$$

where $d_k(\mathbf{x})$ is the three-dimensional Euclidean distance from $\mathbf{x}$ to $\mathbf{x_k}$, R is specified positive number, the sum is over the n plug taps closest to $\mathbf{x}$, and the number c_k are the solution of the $n \times n$ interpolation system $M(\mathbf{x}_k) = f_k$. Thus the multiquadric corresponding to $\mathbf{w^{i,j}}$ is a function of $i, j : M = M_{i,j}$. The desired pressure at the wing geometry point is simply its image under its multiquadric M :

$$f^{i,j} = M_{i,j}(\mathbf{w^{i,j}}). \tag{2.4}$$

We observe that the plug taps need not be on the geometric wing constructed as the piecewise bicubic image of D. However, since the location of the plug taps enters our solution only via three-dimensional distances, this possible complication does not matter.

2.3. Pressures at Arbitrary Points on the Wing. We define the pressure at an arbitrary point $\mathbf{x}$ on the wing by $f = f(\mathbf{x}) = g(\mathbf{w}^{-1}(\mathbf{x}))$ where g is a C^1 piecewise bicubic interpolant to the pressures $f^{i,j}$ at the wing geometry points. However in practice, we need not consider inverse functions: we evaluate and display the piecewise bicubic interpolant g to the pressures $f^{i,j}$ along isoparametric lines in D.

3. Distance-weighted Interpolant (Barnhill, Piper & Rescorla 1987). This method utilizes a localized distance-weighted interpolant with Dirichlet tessellated domains of data dependence and geodesic distances. The interpolant is a convex combination of interpolating "nodal" functions.

Given an underlying function f with values f_i at nodes P_i, where $P_i \in S$, i=1,..,N, and S is the domain surface, define

$$f(P) = \sum_{i=1}^{N} W_i(P) q_i(P) / \sum_{i=1}^{N} W_i(P) \tag{3.1}$$

where the weights $W_i(P)$ are modified inverse distance functions:

$$W_i(P) = l_i(P)/d_i^2(P). \tag{3.2}$$

The function $d_i(P)$ is a 'distance' function that approximates the geodesic distance along the surface from a given point P to point P_i on the surface. The function $l_i(P)$ is a 'weight' function which has local support induced by a 'neighbor' structure. It measures the influence of P_i at the point P, and its value varies from 0 to 1. The function $q_i(P)$ is a 'nodal' function which interpolates to f at P_i and fits the value of f on a set of neighbouring nodes of P_i in a least squares sense. The construction of the $l_i(P)$, $d_i(P)$ and $q_i(P)$ are described in the following subsections.

3.1. The Tessellation. To construct weight functions $l_i(P)$ and nodal functions $q_i(P)$, we must know which domain data points are neighbors of P_i on the surface. In the general trivariate interpolation problem, these neighbors can be found from a Dirichlet tessellation of the given domain points in three dimensions. However, a problem arises when data sites are physically close together, but far apart on the surface, for example, points that are on opposite sides of a surface, but are nearby based on Euclidean distance. In this case, we insert barrier points to separate these data sites from being neighbors. The Dirichlet tessellation of the given domain points and the inserted barrier points is constructed. The general idea of creating barrier points is to first determine a barrier inside the domain surface that separates the parts of the surface. The barrier points are obtained by perpendicular projections of the given data sites onto the barrier. The type of possible barrier depends upon the shape of the domain surface, and can be a point (for a sphere), a finite line (for a torpedo-like surface) or a finite plane (for a thin plate surface). From the 3D Dirichlet tessellation, the neighbors of a given data site P_i fall into two classes : (1) barrier point neighbors and (2) non-barrier neighbors. In the construction of the weight function $l_i(P)$, the neighbors of P_i include both of the two classes, whereas in the construction of the nodal functions $q_i(P)$, only non-barrier neighbor points, which form the structure of the domain surface, are considered. Barrier points are illustrated in Fig. 2.

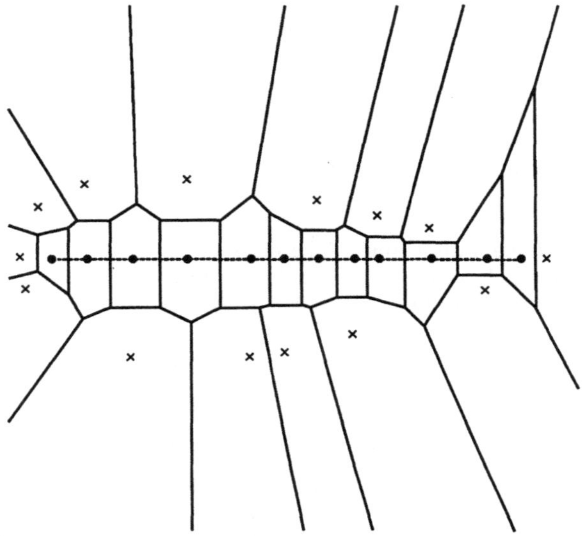

Fig. 2. The Dirichlet tessellation of data points and barrier points. Each barrier point is a perpendicular projection of a data point on that barrier (dash line).

3.2. The Weight Functions. The weight functions $l_i(P)$ measure the influence of P_i at the point P. The function $l_i(P)$'s support is a localized region about the point P_i and its value varies from 0 to 1. Specifically, $l_i(P)$ is defined as

$$l_i(P) = \prod_{j \in J} l_{ij}(P) \tag{3.3}$$

where J is the set of indices of all neighbors (barrier points and non-barrier points) of P_i. The function $l_{ij}(P)$ is defined as

$$l_{ij}(P) = \begin{cases} 1 & \text{if } t < 0; \\ H(t) & \text{if } 0 \le t \le 1; \\ 0 & t > 1 \end{cases} \tag{3.4}$$

where $H(t) = -t^2(3-2t)+1$ is a cubic Hermite basis function,

$$t = \frac{< P - P_i, V >}{< V, V >}, \tag{3.5}$$

and $V = P_j - P_i$. The contours of $l_{ij}(P)$ are planes perpendicular to the vector V. The contour plane $l_{ij}(P) = 1$ separates the surface into two regions $\widehat{R}_j$ and R_j. The region R_j forms the support of the l_{ij} and $R(i) = \cap_{j \in J} R_j$ forms the support of $l_i(P)$, i.e. $l_i(P) \neq 0$

if and only if $P \in R(i)$. Therefore the function $l_i(P)$ provides a local support weighting function whose support captures the neighbor structure of the Dirichlet tessellation. If two points, P_i and P_k, are separated by a barrier point, then the supports of their corresponding weighting functions are disjoint, that is $R(i) \cap R(k) = \emptyset$. Thus if $P \in R(i)$, then $l_k(P) = 0$ and vice versa. Hence we guarantee that the data point on one 'side' of the surface would not affect that on the opposite 'side'.

3.3. Distance Functions. The distance function $d_i(P)$ measures geodesic distance along the surface. In general, we do not have an explicit representation of domain surfaces. Therefore, it is difficult, if not impossible, to evaluate exact geodesic distance between points on the surface. A strategy for overcoming this problem is to approximate the surface locally by a sphere. Then geodesic distances along the surface are approximated along the spheres. The approximation method is as follows: For each data point P_i we construct a sphere which is a least square fit to the non-barrier neighbors of P_i and which also passes through P_i. The center of the sphere (Cx, Cy, Cz) is determined as follows: Let J denote the set of indices of all non-barrier neighbors of P_i. For each non-barrier neighbor P_j of P_i, we construct a plane π_{ij} which is the perpendicular bisector of the line segment P_iP_j . Let h_{ij} denote the distance from the center of the sphere (Cx, Cy, Cz) to the plane π_{ij}. We know $h_{ij} = 0$ if and only if P_i and P_j lie on the sphere. Thus, (Cx, Cy, Cz) can be determined by solving a least squares problem that minimizes the sum $\sum_{i \in J} h_{ij}^2$.

The distance function $d_i(P)$ from P_i to a point $P \in S$ is defined as follows: Project the point P onto the surface of the sphere associated with P_i. Then $d_i(P)$ is evaluated as the geodesic distance along the sphere from P_i to the projection of P. The distance is exact if the domain surface is a sphere. If the domain surface is other than a sphere, $d_i(P)$ provides a local approximation to geodesic distances.

3.4. The Nodal Functions. The nodal functions $q_i(P)$ are a local approximation to f in the neighbourhood of P_i that satisfy $q_i(P_i) = f_i$. These functions take into account the nature of the domain surface around the P_i. Consider the tangent plane, at one of the data points P_i, to the sphere passing through P_i. Let (u, v) denote the projection of P onto this tangent plane, where $(u, v) = g(P)$ denotes a local cartesian coordinate system in the plane with origin $g(P_i) = (0, 0)$. Let $(u_i, v_i) = g(P_i)$. The function value f_j at neighboring points P_j to P_i can now be associated with the image points $(u_j, v_j) = g(P_j)$. Then

$$q_i(P) = \widehat{q}_i(u, v) = f_i + c_1u + c_2v + c_3uv + c_4u^2 + c_5v^2 \tag{3.6}$$

is the bivariate quadratic in (u, v) which approximates the f_j at the domain points (u_j, v_j) in a least squares sense. Since quadratic polynomials have six degrees of freedom, we need at least five non-barrier neighbor points of P_i to determine the unknown coefficients. If the number of neighbors of P_i are less than five, additional points near P_i are introduced.

The local cartesian coordinate system (u, v) can be obtained in practice by a coordinate transformation from $P = (x, y, z)$ to $P' = (u, v, w)$, where the center of the sphere is transformed (0,0,0) and P_i is transformed to (0,0,1). This is accomplished by a translation followed by a rotation and a scaling.

4. A Triangular-based Method (Barnhill & Ou 1990). We now consider the problem of interpolating to function values f_i at the points $P_i \in S$, $i = 1, \ldots, N$ using a triangular-based method. Triangular interpolation methods for the specific case of a sphere have been discussed in (Lawson 1984), (Renka 1984) and (Nielson & Ramaraj 1987). Basically, the methods are extensions of planar methods for triangular interpolation (Barnhill, Birkhoff & Gordon 1973), (Herron 1979) and (Nielson 1979). This paper extends the spherical case to that of an arbitrary convex surface. The main steps are similar to those described in the above papers, that is:

(1) Triangulation of the domain, based upon the data points $\{P_i\}$,

(2) Estimation of the gradient vector at each data point P_i,

(3) Evaluation of the interpolation function f at arbitrary points P on S.

These three steps are described in detail in the following subsections.

4.1. Triangulation over a Closed Convex Surface. Methods for triangulation of scattered data points over a sphere are discussed in (Lawson 1984), (Renka 1984), (Ramaraj 1986) and (Nielson & Ramaraj 1987). Here, we consider the triangulation of scattered data points over an arbitrary convex surface.

Lawson points out (Lawson 1984) that a triangulation grid with N vertices which covers a sphere will have $2N-4$ triangles. This will also be true for any closed convex surface. In the case of spherical triangulation, the general step is first to form an initial triangle and then loop through the remaining $N-3$ points, adding one point and a triangle at a time. At each stage, the triangular grid is modified and then optimized by swapping the diagonal of quadrilaterals formed by two adjacent triangles, if necessary. There are three standard criteria for optimization: the max-min angle test (Nielson & Ramaraj 1987), Renka's analog of the circle test (Renka 1984) and Lawson's convex polygon test (Lawson 1984). Nielson points out that, empirically, these three criteria appear to lead to the same triangulation (Nielson & Ramaraj 1987). These criteria are not easy to implement in the general convex domain surface case, therefore, we consider an alternative method.

First, we introduce a property for the surface triangulation.

'One Sided Property' A triangle $V_1V_2V_3$ in a surface triangulation on a set of data points is said to have one sided property if all the other data points lie on, or on the same side of, the plane determined by the three points $V_1V_2V_3$. A surface triangulation is said to have the one sided property if every triangle in the triangular grid has the one sided property.

The procedure for a convex surface triangulation discussed here is to construct a triangular grid that has the one sided property. The first step is to find an initial triangle with the one sided property. This initial triangle is constructed so that the end points of one edge have the minimum distance among all the data points. The triangulation is then built up by adding one triangle at a time. Consider a current triangulation. For each boundary edge V_1V_2 we seek a point Q from the point set $\{P_i\}_{i=1}^{n} - \{V_1, V_2, V_3\}$ such that the triangle V_1V_2Q has the one sided property, where the point V_3 is the vertex of triangle $V_1V_2V_3$ in the current triangulation. After the point Q is found, the new triangle V_1V_2Q is added to the triangulation. The boundary set is updated by first deleting the edge V_1V_2. If V_1Q is in the current boundary set then it must be deleted from the boundary set. Otherwise V_1Q is added to the boundary. The edge V_2Q is treated analogously. The procedure continues until the list of edges in the boundary set is exhausted. The procedure is not difficult to implement and avoids the diagonal swapping of earlier methods.

The following theorem illustrates that, in the spherical case, the result of the new procedure will lead to the same triangulation as that based on the criteria of Renka's analog of the circle test. In the proof we follow the convention that a spherical triangulation is a triangular grid covering the whole sphere and every spherical triangle in the triangular grid is a proper spherical triangle. (A proper spherical triangle is one such that the lengths of all its edges are less than $\pi \times R$.)

Renka's analog of the circle test is the following: Consider two spherical triangles $V_1V_2V_3$ and $V_1V_3V_4$ labeled in counter-clockwise order. Consider a circle through the three vertices V_1, V_2, V_3. In the planar case, the edge V_1V_3 is swapped with V_2V_4 if V_4 lies interior to the circle. In the spherical case , if the projection of V_4 lies interior to the circle through V_1, V_2 and V_3, then the geodesic edge joining V_1V_3 is replaced by the geodesic edge joining V_2V_4.

Theorem. *If a spherical triangulation has the one sided property, then it is an optimal triangulation with respect to Renka's analog of the circle test.*

Proof: Suppose the theorem is false, that is, a spherical triangulation has the one sided property but there exists a spherical quadrilateral $V_1V_2V_3V_4$ such that its diagonal V_1V_3 needs to be swapped with V_2V_4 for optimization. Without loss of generality, let the equation of the sphere be $x^2 + y^2 + z^2 = 1$. Let the equation of the plane determined by $V_1V_2V_3$ be $z = h$ where $h < 1$.

Since $V_1V_2V_3$ is a proper triangle and V_1V_3 is not an optimal diagonal, then the projection of V_4 will be interior to a circle C which is the intersection of the sphere and the plane $z = h$. This means that the z coordinate of the point V_4 is greater than h and hence it lies in the upper part of the sphere. Since the triangulation has the one sided property, then all the data points lie on or above the plane $z = h$. Since the triangular grid covers the whole sphere, there exists some spherical triangle the edge of which will pass the lower part (below the plane $z = h$ of the sphere). Thus it is not a proper triangle and this is a contradiction.

4.2. Estimation of the Gradient Vector at Each Data Point. We make the assumption that the surface S is smooth so that for each point on the surface there exists a plane tangent to S.

A function f defined on S is differentiable (Lawson 1984) at a point $P \in S$ if and only if there exists a three-vector G satisfying

$$\lim_{\substack{|\Delta P| \to 0 \\ (\Delta P + P) \in S}} f(P + \Delta P) - f(P) - G^T \Delta P = \delta(|\Delta P|) \tag{4.1}$$

or

$$f(P + \Delta P) - f(P) = G^T \Delta P + \delta(|\Delta P|)$$

where $(P + \Delta P) \in S$ and $\delta \to 0$ as $|\Delta P| \to 0$

It can be shown that if a vector G satisfies (4.1) then so does any vector of the form $G_\lambda = G + \lambda N$, where λ is a real number and N is a unit vector orthogonal to the tangent plane T at P. The unique vector of this form, denoted by $G(P)$, which lies on the tangent plane T is called the 'gradient' of f at P.

Consider f as the restriction of a function F defined in an open 3 dimensional neighborhood of the point P. If F is differentiable at P and has gradient ∇F, then $G(P)$ is the orthogonal projection of ∇F onto the tangent plane T.

The gradients $G(P_i)$ at one of the surface data points P_i are then estimated by the following procedure.

(i) (Shape approximation) A sphere is generated that passes through P_i and approximates, in a least squares sense, the neighboring domain data points of P_i. The tangent plane of this sphere at P_i is taken as an approximation to the tangent plane of S at P_i.

(ii) (Coordinate transformation) A local coordinate system is constructed such that in this system, P_i has the coordinate (0,0,1) and the center of the sphere is the origin. The coordinate transformation is a translation followed by a rotation and scaling and hence will be defined by:

$$P' = MP + B,$$

where $P' = [u, v, w]^T$, $P = [x, y, z]^T$, $B = [x_0, y_0, z_0]^T$ and M is a 3×3 non-singular matrix.

(iii) (Function approximation) The neighboring domain data points of P_i are projected orthogonally onto the u-v plane of the local coordinate system passing through P_i and tangent to the sphere (w-components in the new coordinate system are ignored). The function value of each neighbor of P_i is associated with its image on the tangent plane. A least squares quadratic fit is now constructed which is constrained to interpolate f_i at (0,0).

$$q(u, v) = f_i + a_1 u + a_2 v + a_3 u^2 + a_4 uv + a_5 v^2 \tag{4.2}$$

Since there are five coefficients to be determined, at least 5 neighbor points of P_i should be taken for the fit.

(iv) (Gradient approximation) With $(a_1, a_2, 0)$ as the gradient vector at P_i of the approximation function, then an approximation of the gradient $G(P)$ is obtained by transforming $(a_1, a_2, 0)$ back to the original coordinate system.

$$G(P) = (M^T)^{-1}(a_1, a_2, 0)^T \tag{4.3}$$

4.3. Interpolation in a Single Triangle. The method described here requires the function values f_i and the gradients $G(P_i)$ (or estimated gradients) to be known for each data point P_i. Basically, an extension of a planar method for triangular interpolation (Barnhill, Birkhoff & Gordon 1973), (Herron 1979) and (Nielson 1979) based on univariate interpolation along line segments can be used. (Lawson 1984), (Renka 1984), (Nielson & Ramaraj 1987) have extended the planar methods to the case of surface of a sphere by replacing the univariate interpolation along line segments by interpolation with respect to arc length along geodesics. We extend the side vertex method of (Nielson 1979) below.

The method used here is based on the assumption that the shape of a convex surface S can be approximated by a sphere locally. The geodesic between two points on the convex surface S is approximated by the geodesic on the sphere.

To approximate the function value at a given point $Q \in S$, we first find the surface triangle $P_1P_2P_3$ containing the point Q. The procedure of how to find this triangle is discussed later.

We approximate the shape of S near the point Q by a sphere passing through the four points P_1, P_2, P_3 and Q. The interpolation is defined as the convex combination of three partial interpolants:

$$f(Q) = w_1 f_1(Q) + w_2 f_2(Q) + w_3 f_3(Q) \tag{4.4}$$

where

$$w_i = \begin{cases} b_{i+1}b_{i+2}/(b_{i+1}b_{i+2} + b_{i+1}b_i + b_ib_{i+2}) & \text{for } Q \neq P_{i+1} \text{ or } P_{i+2} \\ 0 & \text{for } Q = P_{i+1} \\ 1 & \text{for } Q = P_{i+2} \end{cases}$$

The b_i are barycentric coordinates of Q in the planar triangle $P_1P_2P_3$ and f_i are side-vertex interpolants each requiring two Hermite cubic interpolations and one linear interpolation. The subscripts are to be evaluated modulo 3 to one of the values 1,2, or 3. More precisely, $f_i(Q) = f(Q)$ where:

$$f(Q) = H_0(t)f_i + H_1(t)f(M) + H_2(t)|P_i - M|\frac{\partial S}{\partial[P_i - M]}(P_i) + H_3(t)|P_i - M|\frac{\partial S}{\partial[P_i - M]}(M)$$

$$f(M) = H_0(s)f_j + H_1(s)f_k + H_2(s)|P_j - P_k|\frac{\partial S}{\partial[P_j - P_k]}(P_j) + H_3(s)|P_j - P_k|\frac{\partial S}{\partial[P_j - P_k]}(P_k)$$

where M is the intersection of the two curves P_kP_j and P_iQ,

$$t = \frac{|Q - M|}{|Pi - M|}, s = \frac{|M - P_k|}{|P_i - P_k|}$$

$|P_1 - P_2|$ denotes the geodesic distance between the two points P_1 and P_2,

$$H_0(x) = x^2(3 - 3x)$$

$$H_1(x) = (1 - x)^2(2x + 1)$$

$$H_2(x) = x^2(x - 1)$$

$$H_3(x) = x(1 - x)^2$$

and $\frac{\partial S}{\partial[P_1-P_2]}(P_2)$ denotes the directional derivative at point P_2 along the geodesic joining P_1 and P_2.

These directional derivatives can be computed from the gradient values at the four points P_i, P_j, P_k and M. We have discussed how to estimate the gradients at the vertices of the grid. The gradient value at point M is approximated by linear interpolation of the gradients at these vertices.

The following method will determine whether the point $Q \in S$ is contained in a surface triangle $P_1P_2P_3$: Construct a sphere that passes through the four points P_1, P_2, P_3 and Q with center point C, then Q is contained in the surface triangle $P_1P_2P_3$ if and only if the following three determinants are all greater than or equal to zero:

$$\begin{aligned} &\det(Q-C, P_2-C, P_3-C) \geq 0 \\ &\det(P_1-C, Q-C, P_3-C) \geq 0 \\ &\det(P_1-C, P_2-C, Q-C) \geq 0. \end{aligned} \tag{4.5}$$

These criteria require that the order of the vertices of each triangle $P_iP_jP_k$ in the triangulation be pre-arranged in such a way that

$$det(P_i - O, P_j - O, P_k - O) \geq 0$$

where $O \in R^3$ is the pseudo-center of the data point set $\{P_i\}$, i.e. $O = (O_x, O_y, O_z)$, where

$$\begin{aligned} O_x &= (x_{max} - x_{min})/2 \\ O_y &= (y_{max} - y_{min})/2 \\ O_z &= (z_{max} - z_{min})/2. \end{aligned}$$

The values x_{max}, y_{max}, z_{max}, x_{min}, y_{min}, z_{min} are the maximum and minimum coordinate components of the positional data point set $\{P_i\}, i = 1, 2, .., N$.

The computational expense in generating spheres and applying the tests given by (4.5) for each triangle in the triangulation would be great, therefore, a simple method is used to trivially reject most of the triangles that do not contain Q: For a given triangle $P_iP_jP_k$ in the triangulation, if Q and the pseudo-center O lie on the same side of the plane determined by $P_iP_jP_k$, then Q is not contained in the triangle $P_iP_jP_k$. If triangle $P_iP_jP_k$ is not ruled out, then the tests given by (4.5) are used. In our experience, the number of triangles that potentially contain Q is significantly reduced by this simple method. For example, in all our test cases there were less than 5 remaining triangles that required tests (4.5). Note also that if all but one triangle is trivially rejected, then Q must be contained in that triangle, as will always be the case for convex surfaces.

5. Domain Mapping (Foley, Lane, Nielson, Franke & Hagen 1990). For the domain mapping method described in this section, we assume that we are given N arbitrary points P_i on a closed surface D and associated real values f_i. We address the problem of constructing a function $F(P)$, defined for all P on D, that satisfies $F(P_i) = f_i$, for $i = 1, \ldots, N$.

The domain mapping technique conceptually involves mapping the surface domain D to a sphere, solving a corresponding interpolation problem on the sphere, and then mapping back to D for a solution. The surface domain D need not be convex, but it assumed that D is topologically equivalent to a sphere. Because the domains may be defined in several different ways, we find it convenient to consider several different cases that commonly occur. A discussion of the implementation and graphical display techniques are given in (Foley, Lane, Nielson, Franke & Hagen 1990).

Any interpolant to scattered data on a spherical domain may be used in this method, and many examples are given that use the C^2 modified reciprocal multiquadric interpolant in (Foley 1989). That method and the spherical multiquadric method described in (Hardy & Goepfert 1975) and (Pottmann & Eck 1990) are global methods that are effective on many test cases and they are easily implemented. The spherical multiquadric method to scattered data on a spherical domain is of the form:

$$F(P) = \sum_{i=1}^{N} c_i B_i(P),$$

where $0 < R < 1$ is a user defined constant and

$$B_i(P) = (1 + R^2 - 2R < P, P_i >)^{1/2}.$$

The coefficients c_i are computed by solving the N by N linear system of equations $F(P_i) = f_i$, for $i = 1, \ldots, N$. Other interpolants that are based on a spherical triangulation of the data sites P_i are described in (Lawson 1984), (Renka 1984) and (Nielson & Ramaraj 1987). These interpolants are defined piecewise over the spherical triangulation in a manner similar to triangulation based methods on a planar domain.

With the notable exception of an implicitly defined surface, a closed surface D of genus zero is commonly defined by a mapping B from a simpler domain A onto D. Special attention is given to the cases where 1) A is a planar rectangle and B is a periodic parametric mapping, and 2) A is a sphere and B is a radial projection. Another case that is considered is where only discrete points on D are given and D is not known explicitly. This case allows the formation of interpolants on approximations to implicitly defined surfaces. Finally, the domain mapping approach can be applied to any domain which is a trivariate deformation of the previous cases, assuming that the known deformation is a one-to-one and onto mapping. Each of these cases are covered separately in the subsections that follow. The following assumptions are made about the mapping B from A onto D in the cases where A is a sphere or a rectangle. If A is a sphere, then the mapping B is also one-to-one. If A is a rectangle, then the mapping B is one-to-one except at the pre-images of two special points on D that we will call polar points. This is a common situation for a domain D defined by a parametric mapping of a rectangular domain A, where two edges of A are mapped to two points on D and the remaining portion of A is mapped in a one-to-one manner onto D.

The surface on surface interpolant $F(P)$ which satisfies $F(P_i) = f_i$ is constructed using the following steps.

Step 1 : For $i = 1, ..., N$, find a_i in A such that $B(a_i) = P_i$ in D.

Step 2 : Find a mapping E from A onto the surface of the unit sphere S so that for s in S and P in D, the mapping $P = BE^{-1}(s)$ is a one-to-one map from S onto D.

Step 3 : For $i = 1, ..., N$, compute points $s_i = E(a_i)$ on the unit sphere S.

Step 4 : Construct the scattered data interpolant $G(s)$ on the sphere S which satisfies $G(s_i) = f_i$, for $i = 1, ..., N$.

Step 5 : For P in D, find a point a in A such that $B(a) = P$ and define $F(P) = G(E(a))$.

Steps 1 and 5 are potentially difficult problems for a general domain D because they may involve the inversion of a nonlinear map B. However, for the evaluation of the interpolant F in Step 5 over all of D, as opposed to the evaluation at a single fixed point, the inversion can be avoided by evaluating $F(B(a)) = G(E(a))$ over all a in A. In fact, if the domain D is only known by the mapping $B(a)$, for a in A, then to actually know that a point P is in D, we must know some point a in A such that $B(a) = P$. In this case, no inversion of the mapping B is necessary and the term "find" in Steps 1 and 5 can be replaced by "let".

The continuity of the interpolant $F(P)$ on D depends on the continuity of the interpolant $G(s)$ on the sphere and on the mapping $BE^{-1}(s)$ from S onto D. For the examples presented in (Foley, Lane, Nielson, Franke & Hagen 1990), the domains D have C^0, C^1, C^2 and C^∞ continuity, while $G(s)$ is the C^2 method in (Foley 1989).

5.1. *D* is a mapping of a rectangle. A common situation is when D is a parametric surface defined by $B(u, v)$, where (u, v) is in the rectangle A given by $[umin, umax) \times [vmin, vmax]$. It is assumed that B is one-to-one except at the preimage of two polar points d_1 and d_2 in D. Without loss of generality, assume that $B(u, vmax) = d_1$ and $B(u, vmin) = d_2$ for all $umin <= u < umax$. A mapping E from A to the unit sphere S in Step 2 which establishes a one-to-one correspondence between S and D is given by

$$E(u, v) = (-cos(v')sin(u'), cos(v')cos(u'), sin(v')), \quad \text{where}$$

$$u' = \frac{u - umin}{umax - umin} * 2\pi \qquad \text{and} \qquad v' = \frac{v - vmin}{vmax - vmin} * \pi - \frac{\pi}{2}.$$

For all $umin <= u < umax$, we have $E(u, vmax) = (0, 0, 1)$ and $E(u, vmin) = (0, 0, -1)$. If $E^{-1}(s)$ is the set of points in A such that $E(a) = s$, then the composition $BE^{-1}(s)$ is a one-to-one map from S onto D.

5.2. *D* is a mapping of a sphere. Another useful method for defining a closed surface domain D is to project out radially from a sphere, for example, the surfaces in (Foley, Lane, Nielson & Ramaraj 1990). Let A be the surface of the sphere with center c and radius r. Suppose that the domain surface D is defined by

$$D = \{P = a + h(a) * (a - c)/r : \text{for} \quad a \in A\}, \tag{5.1}$$

where $h(a)$ is some non-negative function defined on A. For our general five step technique, simply define the mappings $B(a) = a + h(a) * (a - c)/r$ and $E(a) = (a - c)/r$. Furthermore, given a point P in D, the point a in A that satisfies $B(a) = P$ can easily be computed as the intersection of the ray $\overrightarrow{cP}$ with the sphere A.

5.3. D is only known at discrete points. There are instances where the domain D is not known precisely, it is known only that the sampled data points P_i are on some surface. If the set of points P_i have a "star-like" property with respect to some central point c, then a surface which passes through the P_i can be developed and this surface can be used as an approximation to the unknown surface domain D. If the points P_i are triangulated into a polyhedron, then it is assumed that there exists a central point c such that each ray $\overrightarrow{cP_i}$ intersects the polyhedron only at P_i. It is not assumed that the topology of the polyhedron is known, and no direct use is made of any triangulation of the points P_i. The approach is to find a surface that passes through the P_i, which is a radial projection of a sphere.

Supposing that a central point c has been found, let $r = min[||P_i - c||]$ for all i, and let A be the sphere with center c and radius r. Compute $h_i = ||P_i - c|| - r$ and let a_i be the intersection of ray $\overrightarrow{cP_i}$ with the sphere A. Define the mapping $E(a) = (a - c)/r$ from A to the unit sphere S and let $s_i = E(a_i)$. Construct the interpolant $H(s)$ over the unit sphere S that satisfies $H(s_i) = h_i$, for $i = 1, ..., N$. If the multiquadric methods in (Foley 1989) or (Pottmann & Eck 1990) are used, then save the LU decomposition of the coefficient matrix for the N by N linear system because it can be used again in computing $G(s)$ in Step 4 of the general process. Similarly, if a triangulation approach is used to compute $H(s)$, then save pertinent information that can be used in Step 4. Define the surface D by (5.1) using $h(a) = H(E(a))$.

With D now represented as a radially projected function defined over a sphere, the method in Section 5.2 can be applied and much of the computation needed to construct the interpolant $F(P)$ has been accomplished in the construction of the domain D. Steps 1, 2 and 3 of the general technique have already been computed. To compute the interpolant $G(s)$ in Step 4 that satisfies $G(s_i) = f_i$, the saved LU decomposition in the computation of $H(s)$ can be used because the two N by N linear systems of equations to be solved have the same coefficient matrix.

If more points are known to be on the unknown surface D, but there are no measured function values for these points, then all of the points on D can be used to form $H(s)$, which is used in the approximation of the surface D. However, the two linear systems of equations solved in computing $H(s)$ and $G(s)$ will have different coefficient matrices. This technique can be used if the star-like surface domain D is defined by an implicit equation of the form $d(x, y, z) = 0$.

5.4. Deformations of the preceding cases. Suppose that D is defined by a mapping B from A which can be handled by one the preceding cases. Let X be a one-to-one trivariate deformation that maps D onto D', such as one from (Barr 1984) or (Sederberg & Parry 1986), where D' does not intersect itself. Using the mapping $B'(a) = X(B(a))$ from A to D', the same general procedures apply.

The domain mapping technique can be extended to arbitrary topologies, provided that the scattered data interpolation problem can be solved on a topologically equivalent surface. For example, if a scattered data interpolant is developed for data on a torus, then the problem can be solved on surfaces of genus one.

Acknowledgements. This research was supported by the Department of Energy under grant DEFG0287ER25041 and by the National Science Foundation under grant DDM-8807747.

REFERENCES

1. Barnhill, R.E. (1977), *Representation and approxiamtion of surfaces*, in Mathematical Software III, J. R. Rice, ed., Academic Press, New York, 69-120.
2. Barnhill, R.E. (1985), *Surfaces in computer aided geometric design: A survey with new results*, Computer Aided Geometric Design 2, 1-17.
3. Barnhill, R.E, Birkhoff, G.B and Gordon, W.J. (1973), *Smooth Interpolation in triangles*, J. Approx. Theory 8, 114-128.
4. Barnhill, R.E., Makatura, G. T. and Stead, S. E. (1987), *A new look at higher dimensional surfaces through computer graphics*, in G.E. Farin, ed., 'Geometric Modeling', SIAM, Philadelphia, 123-129.
5. Barnhill, R.E. and Ou, H.S. (1990), *Surfaces defined on surfaces*, Computer Aided Geometric Design 7.
6. Barnhill, R.E., Piper, B.R. and Rescorla, K.L.(1987), *Interpolation to arbitrary data on a surface*, in G.E. Farin, ed., 'Geometric Modeling', SIAM, Philadelphia, 281-289.
7. Barnhill, R.E., Piper, B.R. and Stead, S. E. (1985), *Surface representation for the graphical display of structured data*, Computer Aided Geometric Design, 2 (1985), 185-187. A later form appears in The Visual Computer, 1 (1985), 108-111.
8. Barr, A.H. (1984), *Global and local deformations of solid primitives*, Computer Graphics 18, no. 3, 21-30.
9. Foley, T.A.(1989), *Interpolation to scattered data on a spherical domain*, in M. Cox and J. Mason, ed., 'Algorithms for Approximation II', Chapman and Hall, London.
10. Foley, T.A., Lane, D.A., Nielson, G.M., Franke, R. and Hagen H. (1990), *Interpolation of scattered data on closed surfaces*, Computer Aided Geometric Design 7.
11. Foley, T.A., Lane, D.A., Nielson, G.M. and Ramaraj, R. (1990), *Visualizing functions over a sphere*, I.E.E.E. Comp. Graphics and Applic. 10, no. 1, 32-40.
12. Hardy, R. L. (1971) *Multiquadratic equations of topography and other irregular surfaces*, J. Geophys. Res., 76, 1905-1915.
13. Hardy, R. L. and Goepfert, W.M. (1975), *Least squares prediction of gravity anomalies, geoidal undulations, and deflections of the vertical with multiquadric harmonic functions*, Geophys. Research Letters 2, 423-426.
14. Herron, G.J. (1979), *Triangular and Multisided Patch Schemes*, Ph.D. Thesis, Mathematics Department, University of Utah, Salt Lake City.
15. Lawson, C.L. (1984), *C^1 Surface interpolation for scattered data on a sphere*, Rocky Mountain J. Math. 14, 177-202.
16. Nielson, G.M. (1979), *The side-vertex method for interpolation in triangles*, Jour. Approx. Theory 25, 318-336.
17. Nielson, G.M. and Ramaraj, R. (1987), *Interpolation over a sphere based upon a minimum norm network*, Computer Aided Geometric Design 4, 41-57.
18. Pottmann, H. and Eck, M. (1990), *Modified multiquadric methods for scattered data interpolation over a sphere*, Computer Aided Geometric Design 7.
19. Ramaraj, R. (1986), *Interpolation and display of scattered data over a sphere*, Masters Thesis, Computer Science Department, Arizona State University, Tempe.
20. Renka, R.J. (1984), *Interpolation of data on the surface of a sphere*, ACM Trans. Math. Software, 417-436.
21. Sederberg, T.W. and Parry, S.R. (1986), *Free-form Deformation of Solid Geometric Models*, SIGGRAPH '86 Conference Proceedings, 151-160.
22. Shepard, D. (1968), *A two-dimensional interpolation function for irregularly-spaced data*, Proceedings of the 1968 ACM National Conference, 517-524.

Surfaces in Solid Modeling

Pere Brunet Alvar Vinacua
Department of Software
Polytechnic University of Catalonia

1 Introduction

The objective of geometric modeling is to provide efficient, flexible and powerful tools to represent three-dimensional objects and to manipulate such representations. Such manipulations consist in creating new objects, applying geometric transformations, editing the shape or performing other operations to generate new shapes, interrogating the model and rendering it.

A geometric modeling system is the implementation of an unambiguous model and the tools required to perform these manipulations [Req 80]. A representation is dubbed unambiguous if every valid internal representation corresponds to a single real object. Ambiguous systems are not really useful, then, since interrogations become impossible, as the internal representation may correspond to several real-world objects.

Schemes based upon the representation of surfaces by locally algebraic surfaces have been proposed in the last few years[Sed 85], [Dah 89], but most existing models are parametric, and describe the surface through a set of adequately stitched polynomial (or rational) patches [BFK 84]. The most frequently used mathematical models of the individual patches are B-splines and Bézier patches. In the first, the continuity between neighbors is parametric, in the latter it is geometric (meaning that the associated geometric entities —tangent planes, curvatures— are continuous).

As for the topology of the interconnections with the other patches, it has evolvedm simple rectangular meshes to very complex topologies with triangular Bézier patches, for instance. This has increased the flexibility of these schemes, making it possible to model closed smooth surfaces that enclose a volume, a crucial feature for their integration in solid models.

In fact, surface models can describe either open surfaces or closed sets of patches that enclose a finite volume. Solid models, on the other hand, represent closed regions of the space in a unambiguous way. Both models have evolved independently [Req 80], and whereas surface models are prepared to deal with free form surfaces of complex form, volume models are usually restricted to objects limited by plane or quadric algebraic surfaces (cylinders, cones, spheres, etc.) that are well suited for the classification tests (inside - outside) of points against solids.

The two best known schemes for the representation of solid objects are the boundary representation and the constructive solid geometry [Req 80]. In boundary representations, the solids are represented by their boundary which is in turn represented by a disjoint set of faces bordered by one or more circular rings of edges which intersect in vertices. The model stores both geometric and topological information describing the connection between neighbour geometric elements. Constructive solid geometry, in turn, represents solids by ordered binary trees. In it, non terminal nodes can represent boolean operations like union, intersection and difference while terminal nodes can be instances of primitive solids located conveniently in space. Both schemes have specific advantages and disadvantages: whereas boolean operations are computationally expensive in boundary representations, in CSG trees both rendering operations and geometric interrogations are rather complex. Moreover, both models are usually restricted to planar or quadric surfaces. On the other hand, octrees form one of the classical decomposition schemes [Req 80]; they are trees that represent solids by encoding the recursive subdivision of a finite universe. They usually approximate the surface of the solid [Mea 82] by a layer of small cubes of a minimum specified size and represent the enclosed volume through the set of black nodes.

Several authors have addressed the problem of including free-form surfaces in solid models. Varady and Pratt [VaP 84] for instance propose a set of solid-surface operations, and include a discussion on the existing solid modelers. Goldman [Gol 87] in turn discusses different types of surfaces and some strategies for introducing them into solid models. In this context, the present paper presents several possibilities for introducing free-form surfaces in the diferent solid models (boundary representations, constructive solid geometry, octree representations), focusing mainly on boundary and octree representations.

2 Boundary Representation

In order to define and model the surface of a complex object, the user can either first generate a wire-frame representation of a mesh of curves that represent the characteristics of the shape of the object, or supply a set of irregularly distributed points (with associated tangent planes) in the space which have to be interpolated. In the first case, the curves can in turn be obtained from direct digitalization of real data, through a 2D design process, or by curving the initial edges of a polyhedron that sketches the shape of the final object. In the second case, the data points can also come from the vertices of a polyhedral approximation of the solid; either a set of triangular patches can be defined based on a triangulation of the initial grid [Nie 87], [Dah 89], or the surface can be directly obtained by recursive subdivision of the polyhedron surface [DoS 78], [Nas 87], [Bru 88].

2.1 Transfinite Interpolation of a Curve Grid

Concerning the problem of constructing a continuous surface which interpolates (filling the holes) the grid of curves of the wireframe representation of a given solid, many approaches have appeared in the literature. The problem of the transfinite interpolation of general curves has been suitably solved through the Coons approach [Coo 69], [BBK 78]. However, these approaches either build a surface which is not composed of standard Bézier patches, or require a grid with strong limitations in the shape of the curves. In any case, the main problem of the surface interpolation is that the user is either faced with non-rectangular regions in the grid, or with vertices where a number of edges other than four meet.

[ChK 84] proposed a special rational patch, which derives from the family introduced by Gregory [Gre 74] and allows the smooth junction of Coons patches. Patches depend only on the curved edges of the interpolated face. On the other hand, triangular, five-sided and six-sided faces are admitted in addition to the usual four-sided patch faces. It must be observed that, although the bicubically blended Coons patch $S(u, v)$ interpolates four boundary curves and normal derivatives along them, normal derivatives cannot be specified independently, as in the corners they must fulfill

$$\partial^2 S(u,v)/\partial u \partial v = \partial^2 S(u,v)/\partial v \partial u$$

Both the [Gre 74] proposal and the Gregory patch proposed in [ChK 84] guarantee that the above compatibility condition is fulfilled, independently of the specific derivatives in the boundary curves. The Gregory patch is similar to a cubic Bézier patch, but the four central control points are doubled, thus giving a total number of 20 control points [ChK 84]. At the evaluation of the patch at a specific point (u, v), virtual central control points are used, computed as a linear interpolation of the pair of corresponding control points.

Gregory patches can be used for the interpolation of a mesh of four-sided faces. As it is difficult for the designer to specify the derivatives along the boundary curves, they can be estimated from tangent planes at the junctions of boundaries, by linear interpolation of the derivatives. [ChK 84] also consider the smooth connection of three, five and six-sided faces, by dividing them into a number of rectangular faces converging to a central point and generating internal curves.

Sarraga's approach [Sar 86], [Sar 89] uses Bézier patches for the G^1 interpolation of a grid of cubic Bézier curves. Bézier patches are preferred for compatibility reasons and simplicity of manipulation. After imposing coplanarity of the tangent and transverse derivatives of two adjacent patches, it is concluded that the coefficient of the tangent derivative must be a cubic polynomial along the common boundary, and the one corresponding to the transverse derivative must be quadratic; in this case, the computation of the twist Bézier points around a vertex is independent of the data in neighbor vertices. A transfinite interpolation scheme is obtained, which uses Bézier patches of order 6 x 6. Three special situations are discussed in detail: four-edge grid points in which two curves cross, three-edge vertices, and five-edge vertices in the grid. In all cases, patches are four-sided and algorithms exist for the automatic surface generation in both rational and non-rational cases. However, the main problem that remains with this approach is that four-edge vertices must be formed by two pairs of tangent curves. More recently, Peters [Pet 90] has derived a set of necessary and sufficient conditions on the mesh data that allow the Bézier interpolation. They are based mainly on the existence of a second fundamental form at the vertex such that it is fulfilled by all mesh curves converging to it.

2.2 Recursive Subdivision

A quite different scheme for the modeling of solids limited by complex surfaces is recursive subdivision. In this case, an initial polyhedron acts as a sketch of the final surface, which is automatically generated by the subdivision algorithm. The surface can be modified either by changing the shape of the initial polyhedron, or by acting upon certain subdivision parameters. Recursive subdivision schemes derive from the work of Chaikin [Cha 74], that proposed a method for deriving a curve from a polygon. Chaikin's algorithm successively refines the polygon by knocking out its corners, at 0.25 and 0.75 of the length of every edge. It is easy to show that Chaikin's curve, in the limit, is the quadratic B-spline generated by the initial control polygon. In a similar way, any polyhedron can be successively refined in order to generate a limit surface; in every step of the subdivision algorithm, new geometric elements (faces, edges and vertices) are incorporated to the polyhedral model, which approaches the limit surface.

Several algorithms based on these subdivision principles have been appeared, differing in the way in which the new geometric elements are obtained. In the Catmull and Clark approach [CaC 78], every four sided face is divided into four faces in every subdivision step. In rectangular grids, the linear relation between vertex coordinates at succesive steps can be completely determined, and a standard bicubic B-spline surface is generated in the limit. In n-sided topologies, however, the choice of the coefficients of the linear relation affects the shape of the limit surface, and it can be shown that it is not possible to obtain curvature continuity [DoS 78].

The Doo and Sabin biquadratic approach [DoS 78] is based, on the other hand, on a shrink of every polyhedron face. The algorithm of a single subdivision step consists of the following steps:

- For every face with n vertices $V_1 \ldots V_n$, compute the corresponding vertices of the associated F-face of the refined polygon

$$V_i' = \sum_1^n \delta_{ij} V_j$$

where

$$\delta_{ij} = \begin{cases} \frac{(n+5)}{4n}, & \text{if } i = j \\ \frac{1}{4n}(3 + 2\,cos(2\pi\,(i-j)/n)), & \text{if } i \neq j. \end{cases}$$

- For every edge of the old polyhedron, form a new four-sided E-face connecting the images of the edge endpoints on each of the faces sharing the edge, Fig. 1.
- For every vertex V of the old polyhedron, form a new V-face connecting the images of V on each of the n faces converging to V. It must be observed that from the first subdivision step, there are no vertices joining $n \neq 4$ edges, Fig. 1.

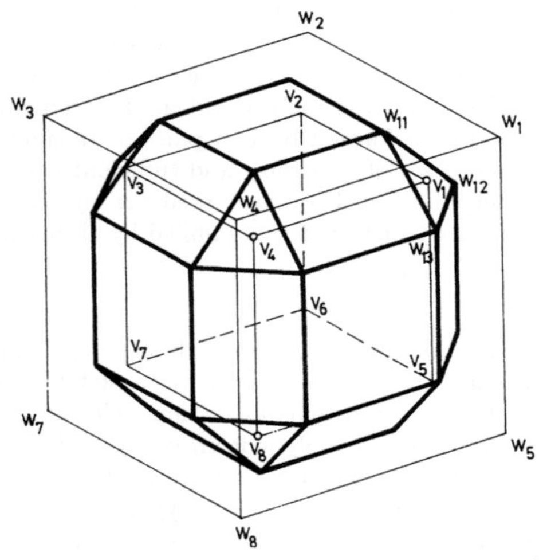

Fig. 1 A polyhedron, with vertices $V_1 \ldots V_8$; the auxiliary corresponding polyhedron (vertices $W_1 \ldots W_8$), and the polyhedron obtained after the first step of the subdivision process (vertices W_{ij})

In this approach, vertices with $n \neq 4$ faces in the initial polyhedron are converted to n-sided faces after the first subdivision step. They will reduce in successive steps and tend, in the limit, to their centroid (the same occurs, obviously, in non-four sided faces of the initial polyhedron). Except for these points, the rest of the polyhedron presents a rectangular topology and tends, in the limit, to a biquadratic B-spline. However, these (finitely many) extraordinary points are singular points of the surface; in their vicinity, the surface is not representable as a biquadratic B-spline. Doo and Sabin [DoS 78] have shown both that the limit surface always presents tangent plane continuity, and that the continuity behaviour of the surface near extraordinary points is regular and independent of n. See also [MiP 87] for a discussion on the recursive subdivision schemes, and the resulting continuity of the limit surfaces.

The biquadratic approach of Doo and Sabin has been extended [Nas 87] for the generation of a surface that interpolates the set of vertices of the initial polyhedron. In this approach, an auxiliary polyhedron with a set of vertices $W = (W_i)_1^N$ having the same topology as the initial polyhedron with vertices $V = (V_i)_1^N$ is first obtained, Fig. 1. The set W is computed by imposing that the limit surface interpolates V, which results in a linear set of equations. On the other hand, it is possible to incorporate tension parameters (shape handles) to every vertex $V = (V_i)_1^N$ of the initial polyhedron, in order to control the shape of the surface during the subdivision process [Bru 88]. They can be used to maximize the VC^2 condition, as it is shown in the example of Fig. 2.

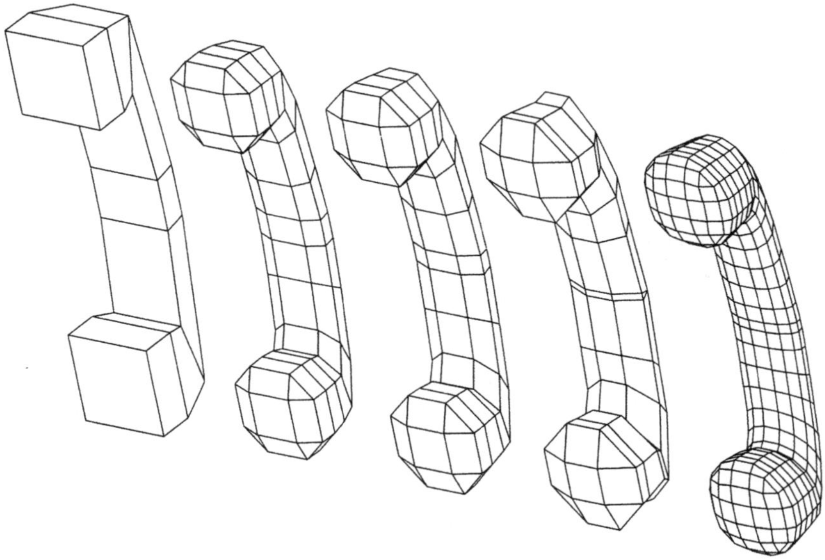

Fig. 2 Subdivision of an object: from left to right, the initial polyhedron, the polyhedron obtained after the first subdivision step using three different choices for shape handles, and the second subdivision using the intermediate values of shape handles

2.3 Triangular Parametric Patches

In 1987 Nielson proposed in [Nie 87] a transfinite triangular patch based upon the idea of Hermite interpolation along rays emanating from the vertices. It interpolates boundary curves and surface normals along those curves. Therefore, obtaining visual continuity of the first order along those edges becomes immediate. For higher orders of smoothness, the idea could be extended, using a higher degree interpolant on each ray.

The idea is attractive because it provides a fairly simple yet flexible triangular patch. To evaluate the patch at a point with baricentric coordinates b_1, b_2 and b_3, we ought to construct three Hermite interpolants along the three lines shown in Fig. 3 (say h_1, h_2 and

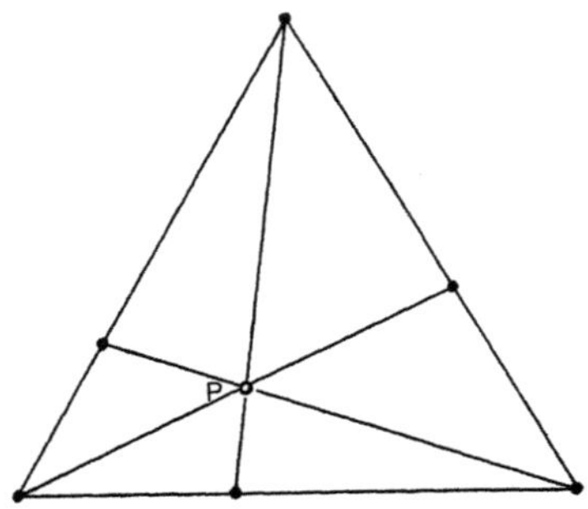

Fig. 3 The three interpolants for a given point

h_3). These interpolants are computed using the boundary data at the six points shown in the figure. The value of Nielson's patch at $P = (b_1, b_2, b_3)$ is then obtained by a weighted average of the three values $h_1(P)$, $h_2(P)$ and $h_3(P)$:

$$G_{\alpha,\beta}(b_1, b_2, b_3) = \frac{b_1^2 b_2^2 h_3(P) + b_1^2 b_3^2 h_2(P) + b_2^2 b_3^2 h_1(P)}{b_1^2 b_2^2 + b_1^2 b_3^2 + b_2^2 b_3^2}$$

yielding a rational interpolant to the data. The subindices α and β denote two free tension parameters (actually used in the computation of the h_i). They arise from the fact that the data at the endpoints for each of those Hermite interpolants does not contain first derivatives. It merely constrains these to be perpendicular to the given normals; Nielson adopts a certain direction for the derivative at each endpoint, but this still leaves the norm of the derivatives free, yielding those degrees of freedom.

Being transfinite, this patch is not of great usefulness in this form in many applications. However, one can build discrete interpolation schemes based on this one through discretization. In fact, using possibly the simplest applicable discretization, Nielson builds a six (vectorial) parameters VC^1 interpolant, where the user gives a triangulation of data points, and normals at the corners of each triangle (i.e. at each data point). The scheme then produces a first order visually continuous patch. In fact in this case simpler weight functions can be used, where all the squares of the b_i have been dropped. It reputedly fares well in modeling closed surfaces interpolating the data. The tension parameters α and β may be set to one, or may be decreased —flattening the surface down to the polyhedron formed by the data points— or increased —"inflating" the surface further.

To our knowledge this scheme has not yet been pursued in commercial solid modelers, but it shows promise of good performance and may be implemented in the near future.

Other schemes exist for solving this problem, that adapt to the specially stringent requirement that the surface may close onto itself (thus binding a solid). See for instance [Her 85] and the scheme proposed by Piper in [Pip 87], using polynomial triangular Bézier patches.

2.4 Piecewise Quadric Surfaces

Another recent proposal, suitable for modeling solids with free-form boundaries was made by Dahmen in [Dah 89]. He also starts with a triangulation where the vertices are the data points, and with a tangent plane (or normal) associated with each data point. So far the data are essentially the same as in Nielson's scheme. However Dahmen also requires that a line be associated with each triangle in the triangulation, satisfying certain special conditions (which may not always be feasible). If one can find such a set of lines, it is called a *transversal system.* He then constructs an interpolating surface formed by pieces of quadrics stitched together in a VC^1 fashion. Each piece of implicit surface is contained in a certain tetrahedron and there are twelve such patches for each triangle in the original triangulation.

These tetrahedra are built using the transversal system's lines and a Powell-Sabin-type split of the original triangles (see Fig. 4). Once this is done, the implicit patches are

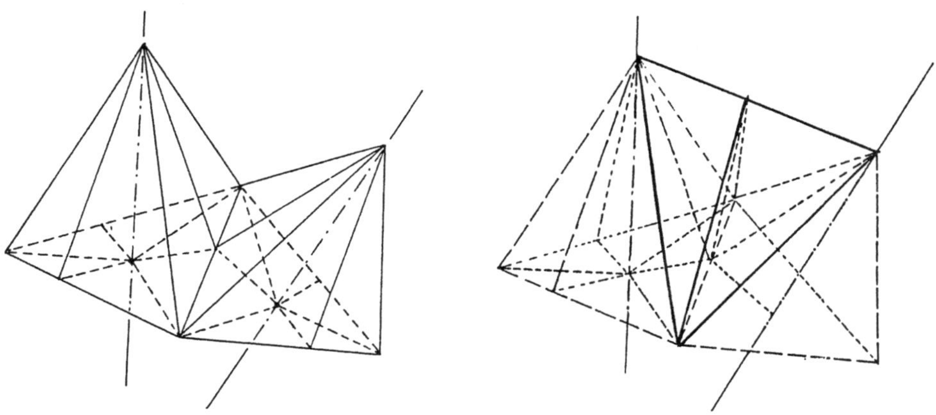

Fig. 4 Tetrahedra constructed from the original triangulation and the transversal system

determined in a Bézier form, by computing the corresponding weights (see also [Sed 85]).

Although the scheme is more involved, requiring the input of a transversal system, which may be hard to find, it may have advantages from the implicit representation. Intersections of an implicit patch with a straight line (or any other parametric entity) are easy to compute, and so is the classification of a point with respect to the solid, a fundamental operation in connection with solid models.

2.5 Non-interpolating Schemes

When the objective is to design certain solids with free-form boundaries, it may not be necessary —or even desirable in some cases— to use an interpolation scheme. One ought to think in terms of an operator sitting in front of a graphics workstation, and conceive a flexible enough interface for the user to generate and manipulate the model of an object he has imagined. One then faces a basic problem: the amount of information that the user has to specify to pin down his idea is so vast that it becomes overwhelming. The way out is then to provide a set of tools that is powerful enough to make the task feasible.

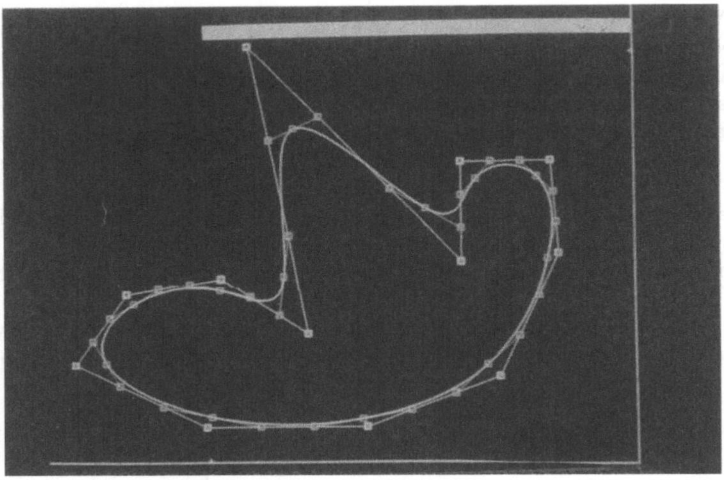

Fig. 5 A polygonal and the corresponding V2-spline

One approach to this problem is to provide a set of tools to instantiate objects in a certain family, and another set of tools to deform or otherwise edit the resulting shapes. A germ of this is constructed in [Bat 88]. There, a program is presented that is geared to the design of "almost-revolution" solids. The user starts by designing a curved shape on a plane. This is done initially using Farin's V2-splines [Far 82]. The user enters a series of points defining a polygonal (either open or closed), and the program constructs a series of Bézier curves with a shape similar to that of the polygonal itself (see Fig. 5). Before proceeding the user is given the opportunity of editing the polygonal he has entered, or of adding tension to the curve (breaking the second order continuity of the original curve but preserving the first order continuity), or even editing the Bézier control points directly (almost surely breaking even the first order continuity).

When he is satisfied with the shape of the curve, the program generates a solid by rotating that curve around an axis, adding covers if the curve and axis do not define a closed region. The rotation is only approximated, since polynomial patches are used. The user can then use a series of editing tools on the revolution object (so it may cease to be a solid of revolution), subdividing patches, or moving their (corner) control points in space.

When a control point is moved, a crease on the surface will appear. If that is not what is wanted, the user may request that the program attempts to perform the requested alteration preserving the smoothness by suitably modifying the neighboring patches. As is well known, this cannot always be accomplished, since there exist certain *compatibility conditions* as one walks around a corner [Wat 88], [Pet 90]. The program detects these situations and warns the user about the problem.

This approach allows for the relatively easy definition of objects which are far from being revolution solids. See for instance Fig. 6. Further, this setup lends itself to be ex-

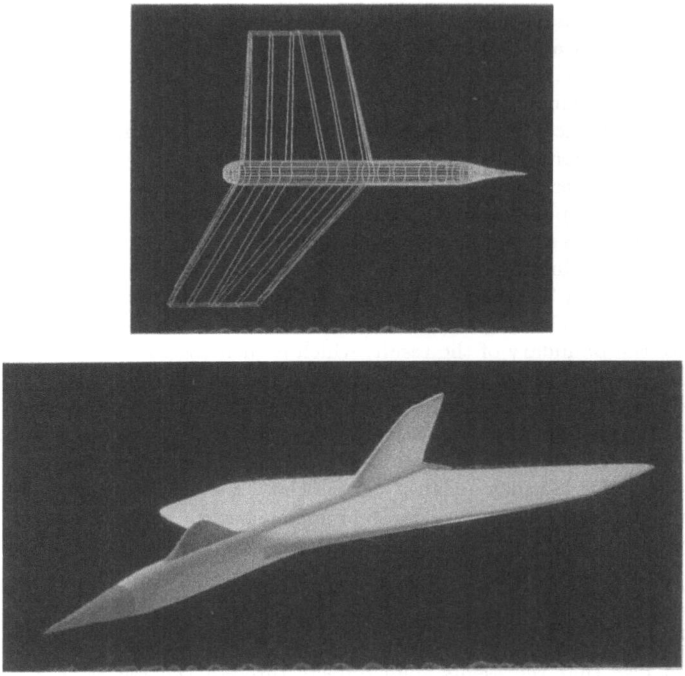

Fig. 6 A solid modelled by the program described in [Bat 88]

tended with more tools as they are developed. For instance, several authors have discussed the construction of solids by sweeping a curve along a complex path [Coq 87] or by joining two curves of different shape [WeF 88].

As better schemes for editing points without losing visual continuity are developed, they may also be incorporated, making the program much more versatile. Work is currently underway in this direction, and also to provide flexible tetrahedrization routines for the sake of post-processing.

3 Constructive Solid Geometry Models

The Constructive Solid Geometry model is specially apt for the definition of geometry by the user, since it seems to be closer to the way a person thinks about describing an object's shape. Classical CSG systems operate upon primitives from a certain predefined family of basic shapes, like cylinders, spheres, prisms wedges and cones. This family could nonetheless be extended arbitrarily, to include complex shapes, whether defined by implicit equations or by collections of parametric patches.

Because of the nature of the CSG model, one major problem needs to be solved in order for this to be feasible: the robust computation of boolean operations between solids of irregular shape. Of course this problem is not exclusive to the CSG approach. Boolean operations are basic to any solid modeler, and one of the most delicate aspects in the design of these programs. Consequently, there are many references in the bibliography to these problem. For example [BaF 87] and [PrG 86] address the problem of finding the intersection of two surfaces, an elementary operation needed in any approach to boolean operations evaluation for solids with free-form boundaries. In general this problem is delicate because the intersection of two relatively simple surfaces can be rather intricate. The actual solutions proposed so far rely on approximations of the solution and *repatching*, i.e. redefining the patches near the intersection curve.

The problem of evaluating boolean expressions involving solids with free-form boundaries has been studied by Thomas in [Tho 87]. His approach consists in first simplifying the expression, and then evaluating the intersections of the boundary pieces. This produces all the pieces of the boundary of the result, which is then assembled by classifying points, curves and surfaces according to a certain graph traversal algorithm.

The interesting feature of his approach is that it allows him to evaluate boolean operations involving solids which are not completely defined. This is a very valuable feature in a context where the adequate definition of portions of the intervening solids which are irrelevant to the final result may be fairly cumbersome. His proposal covers delicate aspects of the problem, like coincidences, in a formal manner, but its implementation in a modeler may be costly due to the delicate problem of classifying the portions of the boundary that determine the boundary of the result.

4 Octree Representations

An octree is a tree that represents solid objects through the recursive subdivision of a finite cubic universe. In this structure, each tree node is terminal or has eight descendants. The tree divides the space of the universe into cubes of different sizes. The root of the tree represents the universe, a cube with a 2^N edge length. This cube is divided into eight identical cubes, called octants, with an edge length of 2^{N-1}. Each octant is represented by one of the eight descendants of the root. If an octant contains a too-complex part of the solid (Grey node), it is divided into another eight identical cubes which are represented as descendants of the octant in question. This process is repeated recursively until valid terminal nodes are obtained, [Mea 82]; the process also stops when octants of a minimum edge length called resolution (usually $2^0 = 1$) appear. Consequently, a specific octree representation is characterized by the definition of the types of terminal nodes which are allowed. The more extensive the set of geometric cases considered in the definition of the allowed terminal nodes, the fewer the intermediate (Grey) nodes in the octree representation of a particular object. On the other hand, the size and location of a cubic octant are determined by the level and the position within the octree of its associated node.

In the simplest octree representation, which we will call classical octrees in the rest of the paper, only Black and White terminal nodes are allowed. Both types of nodes are homogeneous terminal nodes. Nodes with associated cubic octants which are completely inside the object are coded as Black nodes, whereas those completely outside it are called White nodes [Mea 82].

On the other hand, octree models that incorporate new terminal nodes containing parts of the object surface can also be derived. They can be called vector octrees [Sam 89]. They present some advantages over classical octrees, as it can be shown [Nav 86] that they can yield exact representations for polyhedra. On the other hand, they also reduce the degree of subdivision and therefore require less storage than classical octrees [Nav 86]. Different models can be considered, depending on the specific node types used in the representation. Restricting ourselves to the representation of polyhedral solids, Face, Edge and Vertex nodes can be used in adition to the classical White and Black terminal nodes, Fig. 7. Face nodes are crossed only by a single planar face of the solid, whereas Edge nodes contain two neighboring faces and a part of their common edge, and Vertex nodes contain one vertex of the polyhedron and part of the faces and edges converging to it, Fig. 7. In this context, several new octree models can be defined, [Nav 86]. Face octrees are octrees in which the set of terminal nodes includes White, Black and Face nodes. If edge nodes are added to the face octree representation, the face-and-edge octree representation follows. Finally, an extended octree is obtained if vertex nodes are in turn incorporated [ABJN 85], [BrN 85] (A similar representation scheme, called Polytrees, was independently proposed in [CCV 85]; see for instance [BrN 90] for a discussion on the relative performances of both schemes.) On the other hand, in face octrees minimum size nodes are generated along the edges of the polyhedron, whereas in the face-and-edge octrees they appear near the vertices; extended octrees have no minimal size nodes, in general. See [Nav 86] and [BrN 90] for a discussion concerning relative performances and storage requirements of the extended octrees.

Extended octrees can also be used for the representation of free-form surfaces. In this case, the following types of terminal nodes containing parts of the surface are allowed [BrA 87]:

- Face nodes, containing part of a single patch of the surface.
- Edge-a nodes, which contain a piece of the common boundary of two adjacent patches.
- Vertex-a nodes, with one vertex of the initial surface and part of the patches converging to it. All edges in this type of node are patch boundaries.
- Edge-b nodes, in which the edge is a piece of the intersection curve (in general not isoparametric) between two patches of different surfaces which interfere.
- Vertex-b nodes, similar to vertex-a nodes but with some of their edges representing intersection curves (as in edge-b nodes).

In [BrA 87], an extended octree representation for surfaces based on these node types is proposed. A surface octree is first generated by recursive clipping of the surface within the octree nodes during the subdivision process. In the surface octree, terminal nodes not intersecting the surface are undetermined: they can be considered either as white or black. However, when the surface closes a region inside the root node and defines two halfspaces, a volume octree that distinguishes between White and Black nodes can be obtained from the surface octree. This is the case of closed parametric surfaces, or of surfaces that completely traverse the initial node. The conversion algorithm is simple, and is based on the neighborhood relations implicit in the octree [BrA 87]. Volume octrees model the solid region enclosed by the surface, rather than the surface itself; as a consequence, volume

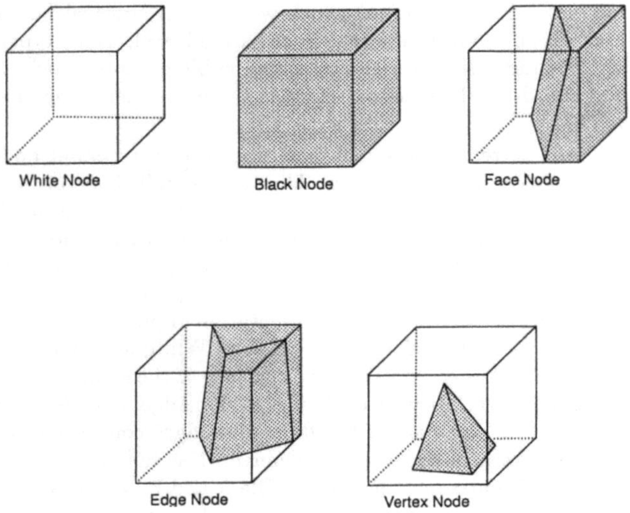

Fig. 7 Face, Edge and Vertex nodes in the extended octree representation

interrogations and point-in-solid classifications can be performed. They add orientation information to the initial parametric surfaces. Algorithms for boolean operations between volume octrees that represent surfaces and extended octrees have also been proposed [Aya 88].

The approach presented in [BrA 87] is restricted to parametric surfaces composed of biquadratic patches, as both the clipping and planarity tests can be simpler. In particular, the four boundary curves are planar parabolas, bounds can be obtained for the number of intersections between a straight line and the patch, and a geometric algorithm for this intersection based on the implicitation in a pencil of parametric planes can be derived. However, and in spite of the simplicity of the algorithms, this approach has two main drawbacks: the clipping process is computationally expensive, and the spatial subdivision is forced both in non-planar regions and in the boundaries of smoothly connected patches.

Face octrees provide an alternative representation for solids bounded by free-form surfaces. A face octree [Bru 90] can be defined as an octree with White, Black, Grey and Face nodes, together with a tolerance ε. White, Black and Grey nodes have the same meaning as in classical octrees. Face nodes contain part of the surface of the object, and they include in their encoding the geometric information about an associated oriented plane π_C, such that for every point P of the object surface S in the cube associated with the node, $dist(P, \pi_C) \leq \varepsilon$, Fig. 8.

Consequently, the part of the surface of the object within a face node must be sufficiently planar. The value ε associated with the face octree controls the degree of approximation of the representation together with the depth of the tree. The encoding of face nodes must include both the corresponding type of node and geometric information on the

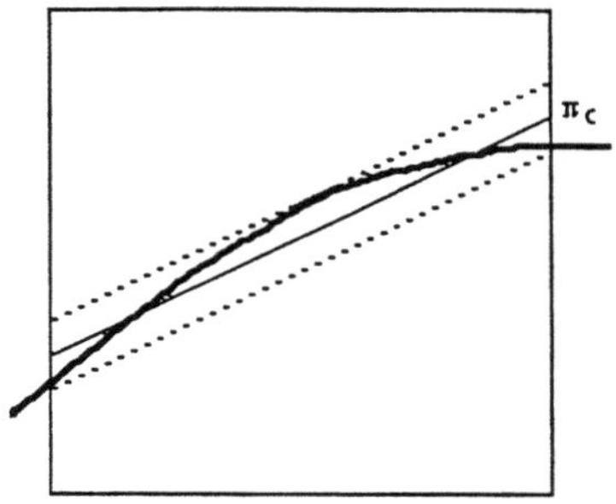

Fig. 8 The surface in a face node of a face octree must be closer than ε to the corresponding associated plane π_c

associated plane. This can be either the explicit plane equation or a pointer to a table of face plane equations.

Face octrees can be used for the representation of polyhedra. In this case, the representation is more compact than the corresponding classical octree and is exact, provided that every face of the polyhedron has at least one face node pointing to it. Algorithms can be derived [Nav 86] for the re-computation of the boundary representation of the object.

On the other hand, face octrees can be used for the representation of objects limited by complex (free form) surfaces. As only plane faces are allowed, the octree representation is obviously approximate. However, the piece of surface represented in a face node depends only on its flatness; the artificial bondaries between smoothly connected patches in the surface model disappear in the face octreė representation. Decreasing the tolerance leads to deeper trees since previous terminal face nodes have to be subdivided. In this case, as the number of different associated planes in all face nodes can be very high, it is better to include the explicit equation of the associated plane in every face node. And, as it is only required that the plane approximates the surface within the node, a restricted discrete set of candidate planes can be used by limiting both the set of feasible normal directions and the distance to the origin [Bru 90].

Every face node in the face octree representation of a closed free-form surface can be interpreted as a band which spans a distance equal to the tolerance at both sides of the plane associated with the node, and which is limited by two planes parallel to it. Because of the definition of a face node, the exact surface must lie in the band. Therefore, the whole boundary of the object is contained in the region defined by the set of bands of the face nodes in the octree, Fig. 9. Bands of neighbor nodes must overlap at the commom boundary, although not completely. The union of all face node bands defines a 'thick surface' with a width of twice the tolerance, which contains the true surface of the object. The whole surface can be analyzed by studying the properties of the coded thick surface, and no artificial geometric elements such as boundaries between smoothly connecting patches need to be considered.

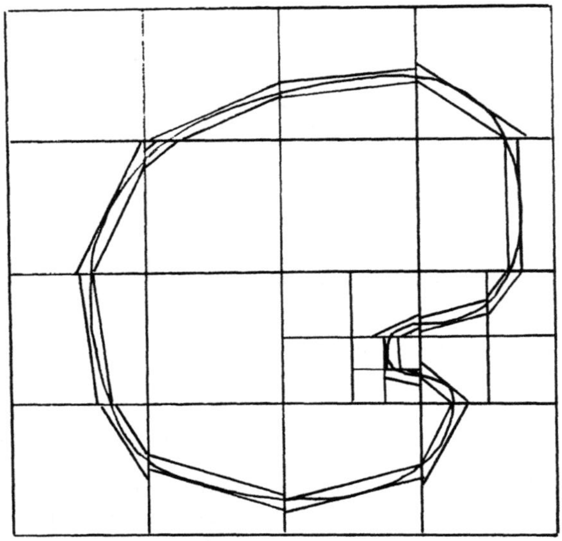

Fig. 9 The overall surface of the object being represented is contained in the union of a set of bands in the cubes associated with face nodes (2D example)

```
procedure build_FO (list_faces,x,y,z,scale)
          clipping (list_faces,x,y,z,scale,list_real_faces)
          if no faces and node_is_outside then
                write_node (White)
          elseif no faces and node_is_inside then
                write_node (Black)
          elseif flat_enough then
                Compute_discrete_plane; write_node (Face)
          else
             write_node (Grey)
             for each of the eight subnodes do
                 compute_node_coordinates (xi,yi,zi)
                 build_FO (list_real_faces,xi,yi,zi,scale/2)
             enddo
          endif
end_procedure
```

Fig. 10 The algorithm for generating a face octree

Face octrees can be generated from a set of patches enclosing a finite region of the space. The algorithm corresponding to the generation of a classical octree can be described by the recursive algorithm (see Fig. 10) that generates the octree in preorder.

The two main parts of the algorithm are the clipping procedure, and the function that must detect if the portion of surface within the node is flat enough. Both algorithms are based on the use of bounding boxes of the patches in the list of faces correponding to the node. On the other hand, the problem of determining if a node which does not contain any part of the surface is White (outside the object) or Black (inside), can be solved by a propagation algorithm in a similar way as in the generation of the volume octree from the surface octree, in the extended octree representation of surfaces [BrA 87].

The number of nodes in the face octree representation of a curved object depends on its geometry. Some bounds can be obtained [Bru 90] for the total number of nodes, as a function of the thickness of the solid and the distribution of the curvature on its surface.

For some purposes, a boundary model of the solid may be needed. It can be useful in order to exchange information with other systems, or for rendering purposes. One possibility is to keep it along all the way; in this case, the Face octree acts more as an auxiliary model that aids in certain computations, but the original boundary model is available and therefore need not be rebuilt. On the other hand, if the Face octree representation keeps no reference to the original surface, a canonic surface must be generated from the octree. The canonic surface of a Face octree can be interpreted as the representative of all the surfaces that generate the same octree because they lie within its set of bands.

One possible algorithm for the computation of the canonic surface [Bru 90] is based on the degree elevation of the planes in face nodes. These planes can be considered as algebraic surfaces, controlled by weights in the vertices of the cubes associated with the nodes. Enlargement of the cubic domain together with the degree elevation of these surfaces leads to coincident domains for neighbor planes. The canonic surface can then be obtained from weight averaging while preserving the tolerance of the bands.

Among the applications of the canonic surface of the Face octree we can quote the conversion to a boundary representation, the refinement of the Face octree in order to obtain a Face octree representation with a smaller tolerance, local refinement for boolean operation purposes, precise geometric interrogations and rendering. Rendering requires both the computation of precise profiles from the point of view (silhouettes) and the determination of normal vectors. These can be computed either by a 'Gouraud style' averaging of the normals at the significative points of the canonic surface, or by computing the canonic surface itself. On the other hand, face octrees can be used for geometric and volume interrogations; in most cases, an algorithm involving simple geometric computations can generate the correct answer, although critical cases require the computation of the canonic surface of the octree. A typical example is the intersection between a straight line and the solid: the question whether the line intersects the solid or not can be answered in most cases by using linear algebra, as the line can only intersect the solid if it goes through a face node. Moreover, if the line intersects both planes of the band, then it intersects the surface of the solid. Similar interrogations [Bru 90] are the point-solid classification (in order to know if the point is in, on or out the solid) which use the space subdivision generated by the octree, volume computations, and plane sections of the solid.

5 Conclusions

We have discussed aspects of the implementation of free-form solids in three well established solid models, reviewing the current results. Although much has been done, the dust has not yet definitely settled on the problem. The path to the most features in the best way (least cost) remains unclear, although the increase in performance of the hardware together with better theoretical approaches will produce very powerful solid modelers that support free-form boundaries in the near future.

References

[Aya 88] Ayala, D., *Boolean operation between solids and surfaces by octrees: models and algorithms*, Computer Aided Design **20** (1988), pp 452–465 and c1–c2.

[ABJN 85] Ayala, D., Brunet, P., Juan, R., Navazo, I., *Object representation by means of non-minimal division quadtrees and octrees*, ACM Transactions on Graphics, **4** (1985) pp 41–59.

[BaF 87] Barnhill, R. E., Farin, G. E., Jordan, M., Piper, B. R. *Surface/surface intersections*, Computer Aided Geometric Design, **4** (1987), pp 3–16.

[BBK 78] Barnhill, R. E., Brown, J. H. and Klucewicz, I. M., *A new twist in computer-aided geometric design*, Computer Graphics and Image Processing **8** (1978), pp 78–91.

[Bat 88] Batlle, F., *Sistema de disseny de superfícies amb splins racionals*, PFC (Oct 1988), School of Industrial Engineering, U.P.C., Barcelona.

[BFK 84] Boehm, W., Farin, G., Kahmann, J., *A survey of curve and surface methods in CAGD*, Computer Aided Geometric Design **1** (1984), pp 1–60.

[Bru 88] Brunet, P., *Including shape handles in recursive subdivision surfaces*, Computer Aided Geometric Design, **5** (1988), pp 41–50.

[Bru 90] Brunet, P., *Face octrees. Involved algorithms and applications*, Report LSI–90–14, Dept de Llenguatges i Sistemes Informatics, Univ. Politècnica de Catalunya (1990).

[BrA 87] Brunet, P., Ayala, D., *Extended octree representation of free form surfaces*, Computer Aided Geometric Design, **4** (1987) pp 141–154.

[BrN 85] Brunet, P., Navazo, I., *Geometric modeling using exact octree representation of polyhedral objects*, Proc. Eurographics'85, North-Holland, Nice 1985, pp 159–169.

[BrN 90] Brunet, P., Navazo, I., *Solid representation and operation using extended octrees*, ACM Transactions on Graphics, **9** April 1990.

[CCV 85] Carlbom, I., Chakravarty, I., Vandersschel, D. A., *A hierarchical data structure for representing the spatial decomposition of 3D objects*, IEEE Computer Graphics and Applications, **5** (1985) pp 24–31.

[CaC 78] Catmull, E. E., Clark, J. H., *Recursively generated B-spline surfaces on arbitrary topological meshes*, Computer Aided Design, **10** (1978), pp 350–355.

[Cha 74] Chaikin, G. M., *An algorithm for high speed curve generation*, Computer Graphics and Image Processing, **3** (1974) pp 346–349.

[ChK 84] Chiyokura, H., Kimura, F., *A new surface interpolation method for irregular curve models*, Computer Graphics Forum **3** (1984) pp 209–218.

[Coo 69] Coons, S. A., *Surfaces for the computer-aided design of space forms*, Report MAC–TR–41, MIT (1967).

[Coq 87] Coquillart, S., *A control-point based sweeping technique*, IEEE Computer Graphics and Applications **7** 11, pp. 36–45.

[Dah 89] Dahmen, W., *Smooth piecewise quadric surfaces*, in: T. Lyche and L. L. Schumaker (eds.), Mathematical methods in computer aided geometric design, Academic Press, 1989. pp 181–194.

[DoS 78] Doo, D. V. H., Sabin, M., *Behaviour of recursive subdivision surfaces near extraordinary points*, Computer Aided Design **10** (1978) p 356–360.

[Far 82] Farin, G. E., *Visually C^2 cubic splines*, Computer Aided Design **14**(3) pp 137–139.

[Gol 87] Goldman, R. N., *The role of surfaces in solid modeling*, in: G. Farin (ed.), Geometric modeling: algorithms and new trends, SIAM 1987.

[Gre 74] Gregory, J. A., *Smooth interpolation without twist constraints*, in: Barnhill and Riesenfeld (eds.), Computer Aided Geometric Design, Academic Press (1974), pp 71–87.

[Her 85] Herron, G., *Smooth closed surfaces with discrete triangular interpolants*, Computer Aided Geometric Design, **2** (1985), pp 297–306.

[Mea 82] Meagher, D., *Efficient synthetic image generation of arbitrary 3D objects*, Proc. IEEE Computer Society, Conf. on pattern recognition and image processing (1982) pp 473–478.

[MiP 87] Michelli, C. A., Prautzsch, H., *Computing surfaces invariant under subdivision*, Computer Aided Geometric Design, **4** (1987) pp 321–328.

[Nas 87] Nasri, A., *Polyhedral subdivision methods for free-form surfaces*, ACM Transactions on Graphics **6** (1987) pp 29–73.

[Nav 86] Navazo, I., *Geometric modeling of octree encoded polyhedral objects*, Doctoral thesis, Univ. Politècnica de Catalunya, Barcelona 1986.

[Nie 87] Nielson, G. M., *A Transfinite, Visually Continuous, Triangular Interpolant*, in: G. E. Farin (ed.), Geometric Modeling: algorithms and new trends, SIAM, Philadelphia, 1987, pp 235–246.

[Pet 90] Peters, J., *Smooth mesh interpolation with cubic patches*, Computer Aided Design, **22** (1990) pp 109–120.

[Pip 87] Piper, B. R., *Visually Smooth Interpolation with Triangular Bézier Patches*, in: G. E. Farin (ed.), Geometric Modeling: algorithms and new trends, SIAM, Philadelphia, 1987, pp 221–233.

[PrG 86] Pratt, M. J., Geisow, A. D., *Surface/surface intersection problems*, in: J. A. Gregory (ed.), The mathematics of surfaces, Oxford University Press, Oxford, 1986, pp 117–141.

[Req80] Requicha, A., *Representations for rigid solids: theory, methods and systems*, Computing Surveys of the ACM, **12** (1980) pp 437–464.

[Sam 89] Samet, H., *The design and analysis of spatial data structures*, Addison-Wesley, 1989.

[Sar 86] Sarraga, R., *G1 interpolation of generally unrestricted cubic Bézier curves*, Report GMR–5424, Computer Science Dept, General Motors Research Laboratories, Warren, Michigan (1986). Also in Computer Aided Geometric Design, **4** (1987), pp 23–39.

[Sar 89] Sarraga, R., *Errata: G1 interpolation of generally unrestricted cubic Bézier curves*, Computer Aided Geometric Design, **6** (1989) pp 167–172.

[Sed 85] Sederberg, T. W., *Piecewise algebraic surface patches*, Computer Aided Geometric Design, **2** (1985), pp 53–59.

[Tho 87] Thomas, S. W., *Boundary expressions for set operations on sculptured solids*,

[VaP84] Varady, T., Pratt, M. J., *Design techniques for the definition of solid objects with free-form geometry*, Computer Aided Geometric Design **1** (1984) pp 207–226.

[Wat 88] Watkins, M. A., *Problems in geometric continuity*, Computer Aided Design **20** (1988), pp 499–502.

[WeF 88] Weiss, G., Furtner, P., *Computer-aided treatment of developable surfaces*, Computers and Graphics **12**, pp 39–52.

Box Splines and Applications

M. Dæhlen and T. Lyche

T. Lyche
Institutt for informatikk
Box 1080, Blindern, 0316 Oslo 3
Norway

M. Dæhlen
Center for Industrial Research
Box 124, Blindern, 0314 Oslo 3
Norway

Abstract. We give an elementary introduction to box spline methods for the representation of surfaces. First, we derive basic properties of box splines starting with the univariate cardinal case. Proofs of most of the results are included. We proceed with a detailed presentation of refinement and evaluation methods for box splines. We discuss shape preserving properties, the construction of non-rectangular box spline surfaces, applications of box splines to surface modelling and problems related to an imbedding of box spline surfaces within a tensor product surface.

1. Introduction

B-splines in one variable and tensor product B-spline surfaces have found numerous applications in areas like curve and surface fitting, computer aided geometric design (CAGD), signal processing, mathematical modelling, etc.. For splines in more than one variable there are several candidates analogous to B-splines, one of which is the box spline.

Box splines are locally supported piecewise polynomials on uniform grids. In the class of splines defined from polyhedra, where to box splines belongs, we also find cone splines and simplex splines. Cone splines are dealt with in this paper, since the theory of polyhedral splines posesses a number of useful relations between cone splines and box splines. We will not discuss simplex splines. Although, a beautiful theory of simplex splines has been created, the development of practical applications has been more difficult.

In this paper we give an introduction to box splines via univariate cardinal splines, bivariate tensor product cardinal splines and bivariate box splines. Moreover, we give a detailed introduction to knotline insertion or subdivision algorithms for box splines and methods for evaluation of box splines.

In addition to a very careful introduction to box splines we also consider a number of methods and applications. In particular, we cover interpolation and approximation with linear combinations of box splines, modelling of box spline surfaces with special emphasis on non-rectangular surfaces, and the use of box splines in a tensor product environment. We illustrate the discussions with several examples.

The emphasis in this paper is on basic properties, approximation methods and applications of box splines to modelling. There has been a large amount of research on cardinal interpolation, approximation order, and exponential box splines. See [6,34,43] and references therein.

2. Splines

Bivariate box splines on a three direction mesh were first considered systematically in [44]. The general multivariate definition were first given in [7] and their basic properties discussed in [8] and [27]. A detailed analysis of bivariate smooth piecewise polynomials on a three direction mesh using box spline techniques were given in [9].

There is a close connection between box splines and univariate and tensor product B-splines. We start by reviewing some facts about B-splines on a uniform knot sequence (cardinal splines).

2.1 Univariate Cardinal Splines

Cardinal B-splines were introduced by Schoenberg in 1946 ([46]) and studied in his monograph ([48]). In this section we will give a detailed introduction to univariate cardinal splines and we shall be particularly interested in refinement problems as initiated in [36].

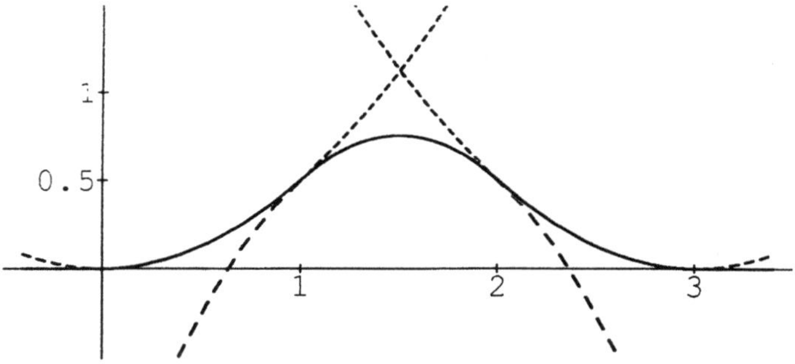

Figure 2.1. The Cardinal B-spline of order 3

Univariate B-splines are often defined by a recurrence relation expressing a B-spline as a linear combination of two B-splines of one degree lower (cf. Property (vii) below). In the uniform case we can instead use the following recurrence relation, also refered to as the convolution definition of univariate cardinal splines.

Definition 2.1 . *The function $M_k : \mathbb{R} \to \mathbb{R}$ defined recursively by*

$$M_k(u) = \int_0^1 M_{k-1}(u-t)\,dt, \tag{2.1}$$

where

$$M_1(u) = \begin{cases} 1, & \text{if } 0 \le u < 1, \\ 0, & \text{otherwise}, \end{cases}$$

is called the Cardinal B-spline of order k.

M_1 is the indicator function of the half open interval $[0,1)$, and it is not too hard to see that M_2 is a piecewise linear hat function:

$$M_2(u) = uM_1(u) + (2-u)M_1(u-1) = \begin{cases} u, & \text{if } 0 \le u < 1, \\ 2-u, & \text{if } 1 \le u < 2, \\ 0, & \text{otherwise}. \end{cases}$$

It takes a little more work to show that M_3 can be written in the form

$$M_3(u) = q_0(u)M_1(u) + q_1(u)M_1(u-1) + q_2(u)M_1(u-2),$$

where

$$q_0(u) = u^2/2, \qquad q_1(u) = -u^2 + 3u - 3/2, \qquad q_2(u) = (3-u)^2/2.$$

Figure 2.1 shows M_3 and the three parabolas q_0, q_1, and q_2.

We see that M_k is a polynomial of degree at most $k-1$ on each of the intervals $(i-1, i), i = 1, 2, \ldots, k$. The numbers $0, 1, \ldots, k$ where we change from one polynomial to another are called *knots*. Since the knots are uniformly spaced the term *uniform B-spline* is often used for M_k.

The following properties follow directly from (2.1) by induction on k.

Theorem 2.2 . *We have*

(i) $M_k(u) \equiv 0$ *for* $u \notin [0,k)$. *(Local support)*

(ii) $M_k(u) > 0$ *for* $u \in (0,k)$. *(Positivity)*

(iii) $\int_0^k M_k(u)\,du = 1$. *(Normalization)*

(iv) $\sum_i M_k(u-i) \equiv 1$. *(Partition of unity)*

(v) $\sum_i iM_k(u-i) = u - k/2$, $\sum_i i^2 M_k(u-i) = (u-k/2)^2 + k/12$.

(vi) $DM_k(u) = \frac{d}{du}M_k(u) = M_{k-1}(u) - M_{k-1}(u-1)$. *(Differentiation formula).*

It is clear that M_2 is a continuous function, and that the higher order functions must be continuous since they are obtained by integration. From Figure 2.1 it is appearant that the parabolas making up M_3 join with continuous tangent. This means that M_3 and M_k for $k \geq 3$ is of smoothness class C^1, i.e. they are continuous and have a continuous first derivative everywhere. In general, the polynomial pieces of M_k join with continuity C^{k-2}. This follows by induction on k using the differentiation formula (vi).

The next result, the uniform case of the deBoor/Cox/Mansfield recurrence relation, is a little harder to show.

Theorem 2.3 . *We have*

(vii) $M_k(u) = (uM_{k-1}(u) + (k-u)M_{k-1}(u-1))/(k-1)$. *(Recurrence relation).*

Proof: (vii) follows by subtracting u times (vi) in Theorem 2.2 from the formula

$$uDM_k(u) = (k-1)M_k(u) - kM_{k-1}(u-1) \quad \text{for} \quad k = 2,3,\ldots \tag{2.2}$$

and rearranging. We have to prove (2.2). For $k = 2$ this is done by use of the explicit formulae for M_2 and M_1. Assume by induction on k that (2.2) holds for $k = n-1$. Differentiating under the integral sign in (2.1) we find

$$uDM_n(u) = uD\int_0^1 M_{n-1}(u-t)\,dt = \int_0^1 (u-t)DM_{n-1}(u-t)\,dt + \int_0^1 tDM_{n-1}(u-t)\,dt.$$

Using the induction hypothesis on the first term and an integration by parts on the second term we obtain

$$\begin{aligned} uDM_n(u) &= \int_0^1 [(n-2)M_{n-1}(u-t) - (n-1)M_{n-2}(u-1-t)]\,dt \\ &\quad - [tM_{n-1}(u-t)]_{t=0}^1 + \int_0^1 M_{n-1}(u-t)\,dt \\ &= (n-2)M_n(u) - (n-1)M_{n-1}(u-1) - M_{n-1}(u-1) + M_n(u) \\ &= (n-1)M_n(u) - nM_{n-1}(u-1). \quad \blacksquare \end{aligned}$$

Using Theorem 2.3 we can derive algorithms for computing cardinal B-splines. We refer to [49] for details.

Cardinal B-spline curves are constructed from one cardinal B-spline by translations $M_k(u) \to M_k(u+i)$ and dilations $M_k(u) \to M_k(\nu u)$. By taking linear combinations of such functions we have an efficient tool for approximating and modelling complex shapes.

Definition 2.4 . *Given a positive integer ν, and $c_i \in \mathbb{R}$ for all integers i. A function of the form*

$$p(u) = \sum_{i \in \mathbb{Z}} c_i M_k(\nu u - i), \quad u \in \mathbb{R} \tag{2.3}$$

is called a cardinal spline of order k. When the c_i's vary over $\mathbb{R}$ we obtain a linear space of functions, which we denote by $S_{k,\mathbb{Z}/\nu}$ or $S_{k,\nu}$ for short. We set $S_k = S_{k,1}$.

In practical applications only a finite number of the B-spline coefficients will be nonzero, and we sum over a finite index set J of consecutive integers.

Properties (iv) and (v) show that any straight line belongs to $S_{k,\nu}$. Indeed,

$$a + bu = \sum_i (a + b(i + k/2)/\nu) M_k(\nu u - i)$$

for any real a, b. It can be shown that any polynomial of degree $< k$ lies in $S_{k,\nu}$. In general $S_{k,\nu}$ consists of all piecewise polynomials of degree $< k$ and smoothness C^{k-2} on the grid

$$\mathbb{Z}/\nu = \{i/\nu \, : \, i \in \mathbb{Z}\}.$$

This follows from the Curry & Schoenberg Theorem [49].

If $p \in S_k$ then p is also a piecewise polynomial of degree $< k$ and smoothness C^{k-2} on the grid $\mathbb{Z}/\nu$ for any positive integer ν. In other words $S_k \subset S_{k,\nu}$. This means that any $p \in S_k$ can also be written in terms of the B-splines in $S_{k,\nu}$. Thus

$$p = \sum_j c_j M_k(u - j) = \sum_i b_i M_k(\nu u - i), \tag{2.4}$$

for suitable coefficients c_j and b_i. There is a simple averaging method ([36]) to find the coefficients b_i from the c_j's.

For derivation it is convenient to consider a family of cardinal splines with the same coefficients c_j. We define for $\ell = 1, 2, \ldots, k$

$$p_\ell(u) = \sum_j c_j M_\ell(u - j).$$

Using (2.1) we see that

$$p_{\ell+1}(u) = \int_0^1 p_\ell(u - t)\, dt. \tag{2.5}$$

Now let $b_{i,\ell}$ be such that

$$p_\ell(u) = \sum_i b_{i,\ell} M_\ell(\nu u - i).$$

Then

$$p_{\ell+1}(u) = \int_0^1 p_\ell(u - t)\, dt = \sum_i b_{i,\ell} [\int_0^1 M_\ell(\nu(u - t) - i)\, dt] \tag{2.6}$$

Now

$$\int_0^1 M_\ell(\nu u - \nu t - i)\,dt = \frac{1}{\nu}\int_0^\nu M_\ell(\nu u - i - t)\,dt$$

$$= \frac{1}{\nu}\sum_{j=0}^{\nu-1}\int_0^1 M_\ell(\nu u - i - j - t)\,dt = \frac{1}{\nu}\sum_{j=0}^{\nu-1} M_{\ell+1}(\nu u - i - j).$$

Inserting this in (2.6) we find

$$p_{\ell+1}(u) = \frac{1}{\nu}\sum_i b_{i,\ell}[\sum_{j=0}^{\nu-1} M_{\ell+1}(\nu u - i - j)] = \sum_i b_{i,\ell+1}M_{\ell+1}(\nu u - i),$$

where

$$b_{i,\ell+1} = \frac{1}{\nu}\sum_{j=0}^{\nu-1} b_{i-j,\ell}, \tag{2.7}$$

Given a cardinal spline $p \in S_k$ with coefficients c_j. The following algorithm finds the coefficients $b_i = b_{i,k}$ in (2.4).

Algorithm 2.5

$b_{\nu i+j,1} = c_i, \quad j = 0, 1, \ldots, \nu - 1, \quad$ all i

for $\ell = 1, \ldots, k-1$

$\quad b_{i,\ell+1} = \sum_{j=0}^{\nu-1} b_{i-j,\ell}/\nu, \qquad$ all i

If a large number of knots are going to be inserted, then for computational cost it is recommended to use the algorithm repeatedly with $\nu = 2$ rather than once with a large ν.

The following theorem shows that Algorithm 2.5 produces coefficients b_i which are close to the function if ν is large.

Theorem 2.6 . *Suppose $p \in S_k$ is given by (2.4). For any integer i and any $u \in [i/\nu, (i+k)/\nu]$ we have*

$$|p(u) - b_i| \leq \frac{k-1}{\nu}\max_j |c_{j+1} - c_j|. \tag{2.8}$$

Proof: Suppose $u \in [r/\nu, (r+1)/\nu)$ where $i \leq r \leq i + k - 1$. Using properties (i),(ii), and (iv) of M_k we obtain

$$|p(u) - b_i| = |\sum_{j=r-k+1}^{r}(b_j - b_i)M_k(\nu u - j)| \leq \max_j |b_j - b_i| \leq (k-1)\max|b_{i+1} - b_i|.$$

Figure 2.2. A cardinal spline with control polygon

The lemma will follow if we can show that

$$|b_{j+1,\ell} - b_{j,\ell}| \le \frac{1}{\nu} \max_j |c_{j+1} - c_j|, \quad \ell = 2, 3, \ldots, k. \tag{2.9}$$

Now by Algorithm 2.5

$$b_{\nu i+j,2} - b_{\nu i+j-1,2} = \frac{1}{\nu}(c_i - c_{i-1}), \quad j = 0, 1, \ldots, \nu - 1.$$

Thus (2.9) holds for $\ell = 2$. But for $\ell \ge 2$

$$|b_{j+1,\ell+1} - b_{j,\ell+1}| \le \max_j |b_{j+1,\ell} - b_{j,\ell}|.$$

Hence (2.9) follows and the proof is complete. ■

With a more careful choice of u it can be shown that

$$|p((i + k/2)/\nu) - b_i| \le \frac{d_k}{\nu^2} \max_j |c_{j+1} - 2c_j + c_{j-1}|, \tag{2.10}$$

where d_k only depends on k. Moreover, the exponent 2 in ν^2 is highest possible.

Definition 2.7 . *Given $p \in S_{k,\nu}$ with coefficients c_i as in (2.3). The polygon obtained by drawing straight line segments between neighbouring points $\big((i + k/2)/\nu, c_i\big) \in \mathbb{R}^2$ is called the control polygon of p in $S_{k,\nu}$.*

The previous discussion shows that the control polygon will be close to the curve if ν is large. Also by the positivity and partition of unity properties we have that p(u) is in the convex hull of the those points of the control polygon corresponding to B-splines which are nonzero at u. In symbols

$$p(u) \in \operatorname{conv}\{\big((i + k/2)/\nu, c_i\big) : M_k(u - i) \ne 0\}.$$

The fact that we have a bounding polygon for the curve which becomes smaller for larger ν is quite useful. It is used for rendering of curves, computing intersections, and for theoretical work. A cardinal spline and its control polygon is shown in Figure 2.2. Note that the control polygon extends beyond the ends of the curve, but that part of the polygon is close to, or models, the curve. For the control polygon of a box spline surface the same phenomena appear.

In applications one is often dealing with parametric spline curves.

Figure 2.3. Refinement of a linear B-spline

Definition 2.8 . *Given a positive integer ν, an $N \in \{2,3\}$, and $c_i \in \mathbb{R}^N$ for all integers i. A function of the form*

$$p(u) = \sum_{i \in \mathbb{Z}} c_i M_k(\nu u - i), \quad u \in \mathbb{R}, \tag{2.11}$$

is called a cardinal spline curve of order k. The polygon in $\mathbb{R}^N$ obtained by connecting neighbouring c_i by straight line segments is called the control polygon of the curve.

We see that p is a plane parametric curve for $N = 2$, and a space curve for $N = 3$. In the explicit case where the spline coefficients c_i are reals we defined the control polygon to have corners $((i + k/2)/\nu, c_i) \in \mathbb{R}^2$. Using property (v) we see that this is the control polygon of the parametric curve $(u, p(u))$.

Let us now return to the explicit case and take a closer look on how the b_i's are related to the c_j's in (2.4). We shall derive

Theorem 2.9 . *We have*

$$M_k(u) = \sum_j \beta_k^\nu(j) M_k(\nu u - j) \tag{2.12}$$

where

$$q_k^\nu(z) = \nu^{1-k}(1 + z + z^2 + \cdots z^{\nu-1})^k = \sum_j \beta_k^\nu(j) z^j. \tag{2.13}$$

Thus, the coefficients $\beta_k^\nu(j)$ used to express M_k as a sum of B-splines on the grid $\mathbb{Z}/\nu$ are simply the power form coefficients of the generating function $q_k^\nu(z)$. For $\nu = 2$ we find $q_k^2(z) = 2^{1-k}(1+z)^k = 2^{1-k}\sum_{j=0}^k \binom{k}{j} z^j$, leading to

$$M_k(u) = 2^{1-k} \sum_{j=0}^{k} \binom{k}{j} M_k(2u - j). \tag{2.14}$$

For $k = 2$ we have the situation in Figure 2.3. The hat function $M_2(u)$ is the sum of three smaller sized hats

$$M_2(u) = \frac{1}{2} M_2(2u) + M_2(2u - 1) + \frac{1}{2} M_2(2u - 2).$$

Using (2.13) we find

$$p(u) = \sum_j c_j M_k(u-j) = \sum_i b_i M_k(\nu u - i),$$

where

$$b_i = \sum_j c_j \alpha_{j,k}(i), \quad \text{and} \quad \alpha_{j,k}(i) = \beta_k^\nu(i+\nu j). \tag{2.15}$$

This gives the transformation formula for the B-spline coefficients on the grid $\mathbb{Z}/\nu$ in terms of those on $\mathbb{Z}$. The numbers $\alpha_{j,k}(i)$ are called *discrete cardinal B-splines*. For material on discrete B-splines, see [38] and references therin.

A proof of Theorem 2.9 can be given using (2.1) and an inductive argument. Instead we shall derive (2.12) using more illuminating Fourier transform techniques.

We start by observing that (2.1) can be written

$$M_k(u) = \int_{-\infty}^{\infty} M_{k-1}(u-t) M_1(t)\, dt.$$

This means that cardinal B-splines can be expressed in terms of convolutions.

Definition 2.10 . *The convolution $f * g$ of two sufficiently smooth functions $f, g : \mathbb{R} \to \mathbb{R}$ is defined by*

$$(f * g)(u) = \int_{-\infty}^{\infty} f(u-t) g(t)\, dt.$$

Theorem 2.11 . *For any integer ℓ with $1 \le \ell < k$ we have*

$$M_k = M_{k-1} * M_1 = \ldots = M_1 * M_1 * \cdots * M_1 = M_\ell * M_{k-\ell}.$$

Proof: Using (2.1) repeatedly and the properties $f * g = g * f$ and $f * (g * h) = (f * g) * h = f * g * h$ of convolutions the result follows. ■

To find the Fourier transform of M_k we can compute the Fourier transform of M_1 and use the fact that the Fourier transform of a convolution is the product of the Fourier transform of the factors in the convolution.

Theorem 2.12 . *The Fourier Transform of M_k is*

$$\widehat{M_k}(\eta) = \int_{-\infty}^{\infty} e^{iu\eta} M_k(u)\, du = [\widehat{M_1}(\eta)]^k = [\frac{e^{i\eta}-1}{i\eta}]^k,$$

where i is the imaginary unit.

We want to take the Fourier transform on both sides of (2.12). For this we also need the Fourier transform of $M_k(\nu u - j)$. Using Theorem 2.12 we find

$$\int_{-\infty}^{\infty} e^{iu\eta} M_k(\nu u - j)\, du = \frac{1}{\nu} e^{ij\eta/\nu} \int_{-\infty}^{\infty} M_k(t) e^{it\eta/\nu}\, dt = \frac{1}{\nu} e^{ij\eta/\nu} \Big(\frac{e^{i\eta/\nu}-1}{i\eta/\nu}\Big)^k.$$

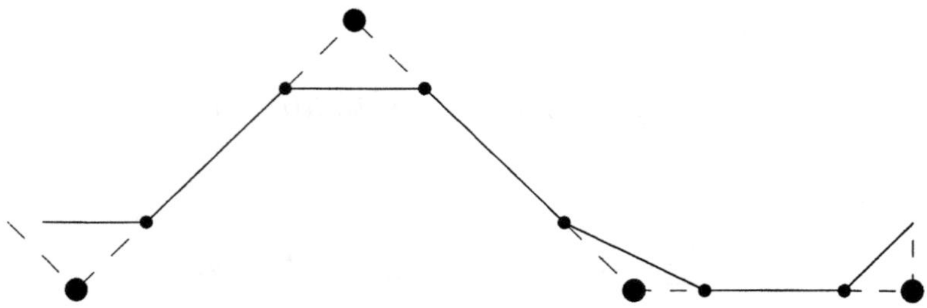

Figure 2.4. Corner cutting

Letting $z = e^{i\eta/\nu}$, the Fourier transform of (2.12) can be written

$$(\frac{z^\nu - 1}{i\eta})^k = \sum_j \beta_k^\nu(j)\nu^{k-1}z^j(\frac{z-1}{i\eta})^k.$$

Multiplying both sides by $(i\eta)^k/(\nu^{k-1}(z-1)^k)$ we obtain

$$\nu^{1-k}(\frac{z^\nu - 1}{z-1})^k = \sum_j \beta_k^\nu(j)z^j.$$

Since $(z^\nu - 1)/(z-1) = 1 + z + z^2 + \cdots z^{\nu-1}$ for $z \neq 1$ we obtain Theorem 2.9 .

Example 2.13 . *Taking $k = 3$ in Theorem 2.9 we obtain*

$$p(u) = \sum_j c_j M_3(u-j) = \frac{1}{4}\sum_j [M_3(2u-2j) + 3M_3(2u-2j-1)$$
$$+ 3M_3(2u-2j-2) + M_3(2u-2j-3)] = \sum_i b_i M_3(2u-i),$$

where

$$b_{2j} = \frac{1}{4}(3c_{j-1} + c_j), \qquad b_{2j+1} = \frac{1}{4}(c_{j-1} + 3c_j), \tag{2.16}$$

If we apply (2.16) iteratively we obtain a sequence $(b^n)_{n\geq 1}$ of control polygons given by $b_j^0 = c_j$ and for $n \geq 1$

$$b_{2j}^n = \frac{1}{4}(3b_{j-1}^{n-1} + d_j^{n-1}), \qquad b_{2j+1} = \frac{1}{4}(d_{j-1}^{n-1} + 3d_j^{n-1}).$$

This is known as Chaikin's algorithm [11]. By (2.8) the b^n's converge uniformly to p. One application of Chakin's algorithm is illustrated in Figure 2.4. The polygon $(j + 3/2, c_j)$'s is indicated by circles, while the polygon $((j+3/2)/2, b_j)$ with b_j given by (2.16) is indicated by smaller circles. This process is known as corner cutting.

2.2 Tensor Product Cardinal Splines

In this section we will show that B-splines on a uniform quadrilateral grid, or tensor product cardinal B-splines, can be defined by a recurrence relation of a form similar to the definition of univariate cardinal splines given in (2.1). In addition to providing an important example of box splines in two variables this will motivate the general definition of box splines.

Definition 2.14 . *A bivariate tensor product cardinal B-spline of order* $\mathbf{k} = (k_1, k_2)$ *is a function of two variables* (u, v) *given by*

$$M_{\mathbf{k}}(u,v) = M_{k_1}(u)M_{k_2}(v).$$

Here the functions on the right are univariate cardinal B-splines of order k_1 *and* k_2 *given by (2.1).*

The functions $M_{\mathbf{k}}$ can be defined recursively by successive integration using (2.1) either on the u or the v variable. Thus

$$(2.17) \qquad M_{(k_1,k_2)}(u,v) = \int_0^1 M_{k_1-1,k_2}(u-t,v)\,dt = \int_0^1 M_{k_1,k_2-1}(u,v-t)\,dt,$$

starting with

$$(2.18) \qquad M_{(1,1)}(u,v) = \begin{cases} 1, & \text{if } (u,v) \in [0,1)^2, \\ 0, & \text{otherwise.} \end{cases}$$

We see that $M_{(k_1,k_2)}$ is defined by two directions $\mathbf{e}^1 = (1,0)$ and $\mathbf{e}^2 = (0,1)$. We can compute $M_{(k_1,k_2)}$ either from $M_{(k_1-1,k_2)}$ by integrating in the direction $\mathbf{e}^1$ or from $M_{(k_1,k_2-1)}$ by integrating in the direction $\mathbf{e}^2$. Starting with $M_{1,1}$ we obtain $M_{(k_1,k_2)}$ after $k_1 - 1$ integrations in the $\mathbf{e}^1$ direction and $k_2 - 1$ integrations in the $\mathbf{e}^2$ direction. Thus, $M_{(k_1,k_2)}$ is defined by k_1 repetitions of $\mathbf{e}^1$ and k_2 repetitions of $\mathbf{e}^2$. Let for $|\mathbf{k}| = k_1 + k_2$, $\mathbf{X} = (\mathbf{x}^1, \mathbf{x}^2, \ldots, \mathbf{x}^{|\mathbf{k}|})$ denote the collection of these vectors. We require $\mathbf{x}^1 = \mathbf{e}^1$ and $\mathbf{x}^2 = \mathbf{e}^2$, but otherwise the ordering is arbitrary. In this paper we will use the alternative notations

$$M_{\mathbf{k}}(u,v) = M(u,v\,|\,\mathbf{X})$$

for a tensor product cardinal B-spline. Corresponding to (2.17) we have for any integer μ with $2 \le \mu \le |\mathbf{k}|$

$$(2.19) \qquad M(u,v\,|\,\mathbf{X}_\mu) = \int_0^1 M(u - tx_\mu, v - ty_\mu\,|\,\mathbf{X}_{\mu-1})\,dt,$$

where $\mathbf{X}_\mu = (\mathbf{x}^1, \mathbf{x}^2, \ldots, \mathbf{x}^\mu)$, and $\mathbf{x}^\mu = (x_\mu, y_\mu)$. This defines tensor product B-splines recursively. The recurrence starts with (2.18).

Figure 2.5. *The support of the Courant finite element*

2.3 Bivariate Box Splines

The tensor product cardinal splines defined in the previous section is an important special case box spline defined on a two directional grid. By allowing allowing arbitrary vectors x^i in (2.19) we can define general bivariate box splines as follows:

Definition 2.15 . *Suppose $\mathbf{X}_\mu = (\mathbf{x}^1, \mathbf{x}^2, \ldots, \mathbf{x}^\mu)$ with $\mathbf{x}^i = (x_i, y_i)$ are $\mu \geq 2$ vectors in $\mathbb{R}^2$ with $\mathbf{x}^1$ and $\mathbf{x}^2$ linearly independent. A bivariate box spline with direction vectors $\mathbf{X}_\mu$, is a function $M(u, v \,|\, \mathbf{X}_\mu) : \mathbb{R}^2 \to \mathbb{R}$ defined recursively by*

$$M(u, v \,|\, \mathbf{X}_\mu) = \int_0^1 M(u - tx_\mu, v - ty_\mu \,|\, \mathbf{X}_{\mu-1})\, dt, \tag{2.20}$$

with

$$M(u, v \,|\, \mathbf{X}_2) = \begin{cases} 1/|\det(\mathbf{X}_2)|, & \text{if } (u, v) \in [\mathbf{X}_2[, \\ 0, & \text{otherwise,} \end{cases} \tag{2.21}$$

and where

$$[\mathbf{X}_\mu[= \{t_1\mathbf{x}^1 + \ldots + t_\mu\mathbf{x}^\mu \;:\; 0 \leq t_j < 1, \quad 1 \leq j \leq \mu\}. \tag{2.22}$$

Example 2.16 . *The function $M(u, v \,|\, \mathbf{X}_2)$ is a constant times the indicator function of the half open parallelogram $[\mathbf{X}_2[$ with sides parallel to $\mathbf{x}^1$ and $\mathbf{x}^2$. The constant is one over the area of the parallelogram. If for positive integers k_1 and k_2 we set $\mu = k_1 + k_2$ and $\mathbf{X}_\mu$ consists of the unit vectors $\mathbf{e}^1$ and $\mathbf{e}^2$ repeated k_1 and k_2 times, we obtain the tensor product cardinal B-splines of the previous section. In this case $[\mathbf{X}_\mu[$ is a rectangle with sides k_1 and k_2.*

In Theorem 2.21 below we show that $M(u, v \,|\, \mathbf{X}_\mu) = 0$ if (u, v) does not belong to $[\mathbf{X}_\mu[$.

Definition 2.17 . *The set*

$$[\mathbf{X}_\mu] = \{t_1\mathbf{x}^1 + \ldots + t_\mu\mathbf{x}^\mu \;:\; 0 \leq t_j \leq 1, \quad 1 \leq j \leq \mu\}$$

is called the support of $M(u, v \,|\, \mathbf{X}_\mu)$.

Example 2.18 . *Let $M(u, v) = M(u, v \,|\, \mathbf{X}_3)$ with $\mathbf{X}_3 = (\binom{1}{0}, \binom{0}{1}, \binom{1}{1})$. From (2.20) we have*

$$M(u, v) = \int_0^1 M(u - t, v - t \,|\, \begin{pmatrix}1\\0\end{pmatrix}, \begin{pmatrix}0\\1\end{pmatrix})\, dt.$$

By (2.21) the integrand is the indicator function of the half open unit square. Therefore, $M(u, v)$ can only be nonzero if (u, v) belongs to the interior of the hexagonal support set $[\mathbf{X}_3]$ looking as in Figure 2.5.

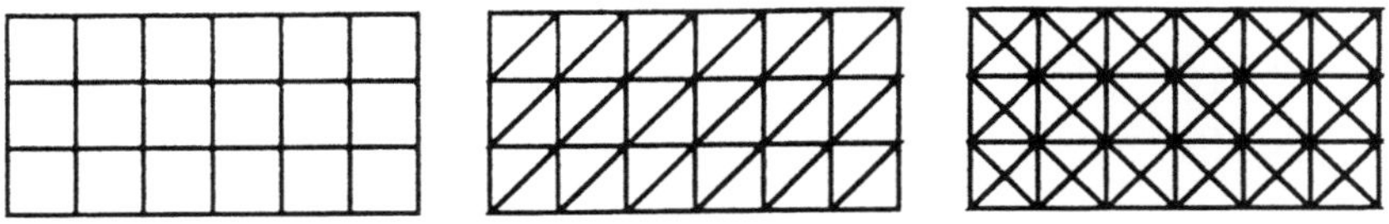

Figure 2.6. Two, three, and four directional grids

After some calculation we find

$$M = u\chi_A + v\chi_B + (u+1-v)\chi_C + (v+1-u)\chi_D + (2-v)\chi_E + (2-u)\chi_F. \tag{2.23}$$

where for any region $G \subset \mathbb{R}^2$ we have the indicator function

$$\chi_G(u,v) = \begin{cases} 1, & (u,v) \in G; \\ 0, & \text{otherwise.} \end{cases}$$

The graph of M is shaped like a hexagonal pyramid and is known as the Courant finite element.

We will now take a closer look at the situation where the direction vectors of a box spline consists of repetitions of a few nonparallel directions.

Definition 2.19 . *Suppose for an integer $r \geq 2$, that*

$$\mathbf{X}_\mu = (\overbrace{\mathbf{d}^1, \cdots, \mathbf{d}^1}^{k_1 \text{ times}}, \overbrace{\mathbf{d}^2, \cdots, \mathbf{d}^2}^{k_2 \text{ times}}, \ldots, \overbrace{\mathbf{d}^r, \cdots, \mathbf{d}^r}^{k_r \text{ times}}),$$

where $\mathbf{E} = (\mathbf{d}^1, \mathbf{d}^2, \ldots, \mathbf{d}^r)$ are pairwise nonparallel vectors in $\mathbb{R}^2$, and $k_1, k_2, \ldots, k_r$ are positive integers with $\mu = \sum_i k_i$. We write

$$\mathbf{X}_\mu = \mathbf{E}^{\mathbf{k}} = \mathbf{E}^{(k_1,\ldots,k_r)}$$

and call $M(u,v \mid \mathbf{E}^{\mathbf{k}})$ an r directional box spline. If for $r \leq 4$ we have the standard directions

$$\mathbf{d}^1 = \begin{pmatrix}1\\0\end{pmatrix} = \rightarrow, \quad \mathbf{d}^2 = \begin{pmatrix}0\\1\end{pmatrix} = \uparrow, \quad \mathbf{d}^3 = \begin{pmatrix}1\\1\end{pmatrix} = \nearrow, \quad \mathbf{d}^4 = \begin{pmatrix}1\\-1\end{pmatrix} = \searrow, \tag{2.24}$$

then we denote $= M(u,v \mid \mathbf{E}^{\mathbf{k}})$ by $M_{\mathbf{k}}(u,v)$. We define the value of $M_{\mathbf{k}}$ or one of its derivatives on a grid line by taking limits from the right and/or above.

We see that the Courant finite element $M_{(1,1,1)}$ is a three directional box spline. The standard directions generate regular r directional grids G_r, $r = 2,3,4$. To obtain G_r we draw straight lines through each integer grid point in the r first standard directions. These grids are shown in Figure 2.6. In the literature the regular 3 and 4 directional grids are sometimes called type-1 and type-2 triangulations. The terms unidiagonal- and crisscross partitions are also used ([13]).

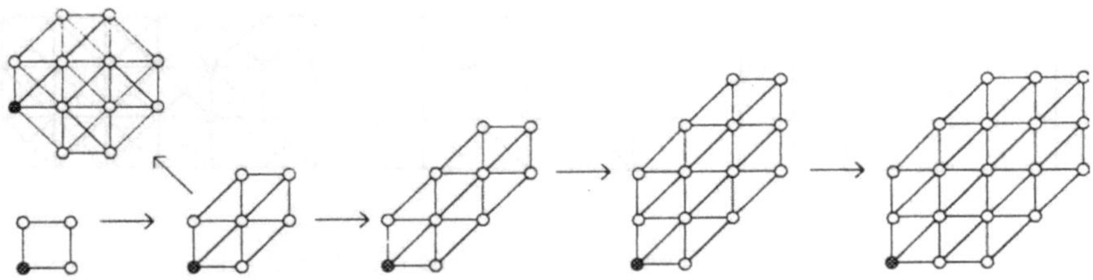

Figure 2.7. The support of some box splines

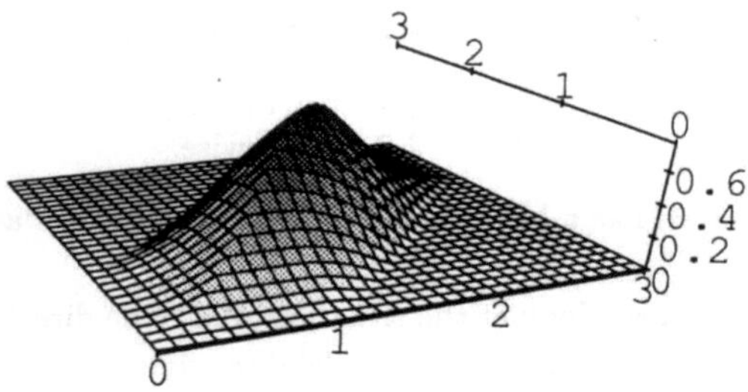

Figure 2.8. The C^0 quadratic box spline

When $r = 3$ the support of the box spline $M_{\mathbf{k}} = M_{(k_1,k_2,k_3)}$ is

$$[\mathbf{E}^{k_1,k_2,k_3)}] = \{(t_1 + t_3, t_2 + t_3) \,:\, 0 \le t_i \le k_i, \quad 1 \le i \le 3\}.$$

A sequence of such hexagonal sets and corresponding grids are shown from left to right in the bottom row of Figure 2.7.

At the bottom left in Figure 2.7 we find the unit square which is the support of $M_{(1,1)}$. (The darker grid point indicates the position of the origin.) To the right of the unit square we see the support of $M_{(1,1,1)}(u,v)$. This box spline is given explicitly by (2.23). Continuing towards the right we obtain the C^0 *quadratic box spline*

$$M_{(1,1,2)}(u,v) = \int_0^1 M_{(1,1,1)}(u-t, v-t)\,dt.$$

The graph of this surface is shown in Figure 2.8.

It appears that the surface has continuous first derivatives across the horizontal and vertical grid lines, but only continuity in position across diagonals. This is in agreement with the statement of the following theorem. For $r = 3$ the proof can be found in [9].

Theorem 2.20 . *$M_{\mathbf{k}}$ is a piecewise polynomial of degree at most $\mu - 2 = \sum_{i=1}^{r} k_i - 2$ on $G_r, r = 2,3,4$. Moreover, $M_{\mathbf{k}} \in C^{\mu-2-k_i}$ across a direction d_i.*

For the C^0 quadratic box spline we have $\mu - 2 = 2$, $k_1 = k_2 = 1, k_3 = 2$. Therefore, according to Theorem 2.20 we have C^1 continuity across the u and v axes, and C^0 continuity across diagonals as asserted.

Other three directional box splines of practical interest are the three C^1 cubics with direction multiplicities $(2,2,1)$, $(2,1,2)$, $(1,2,2)$, and the C^2 quartic $(2,2,2)$. The support of the functions $M_{(1,2,2)}$ and $M_{(2,2,2)}$ are shown to the right in Figure 2.7.

By using four directions we can obtain more smoothness with lower degree. The function $M_{(1,1,1,1)}(u,v)$ is called the Zwart Element. It is a piecewise quadratic of smoothness C^1. The support of this function is shown in the top left position in Figure 2.7.

The Definition 2.15 gives rise to the construction of box splines defined on a variaty of different grids. So far in this section we have considered the important standard grids given by the standard directions $\mathbf{d}^1, \mathbf{d}^2, \mathbf{d}^3$ and $\mathbf{d}^4$, see equation (2.24). The grids G_r, $r = 2,3,4$ shown in Figure 2.6 are the most well known because of their regularity and nice behaviour. However, introducing the four extra directions

$$\mathbf{d}^5 = \begin{pmatrix} 1/2 \\ 1 \end{pmatrix}, \quad \mathbf{d}^6 = \begin{pmatrix} 1 \\ 1/2 \end{pmatrix}, \quad \mathbf{d}^7 = \begin{pmatrix} 1 \\ -1/2 \end{pmatrix}, \quad \mathbf{d}^8 = \begin{pmatrix} 1/2 \\ -1 \end{pmatrix}$$

we can build a group of fairly well behaved grids. Figure 2.9(a)–(d) show four grids which are constructed by selecting vectors among $(\mathbf{d}_1, \ldots, \mathbf{d}^8)$,

$$\begin{array}{ll} \mathbf{d}^1, \mathbf{d}^2, \mathbf{d}^3, \mathbf{d}^5, \mathbf{d}^6, & \text{Figure 2.9(a)}, \\ \mathbf{d}^1, \mathbf{d}^2, \mathbf{d}^4, \mathbf{d}^5, \mathbf{d}^6, & \text{Figure 2.9(b)}, \\ \mathbf{d}^1, \mathbf{d}^2, \mathbf{d}^5, \mathbf{d}^6, \mathbf{d}^7, \mathbf{d}^8 & \text{Figure 2.9(c)}, \\ \mathbf{d}^1, \mathbf{d}^2, \mathbf{d}^3, \mathbf{d}^4, \mathbf{d}^5, \mathbf{d}^6, \mathbf{d}^7, \mathbf{d}^8 & \text{Figure 2.9(d)}. \end{array}$$

We observe the particularly nice behaviour of the grid shown in Figure 2.9(b). On this grid we can define the C^2-cubic box spline $M(u,v\,|\,2\mathbf{X}_\mu)$ with $\mathbf{X}_\mu = \{\mathbf{d}^1, \mathbf{d}^2, \mathbf{d}^4, \mathbf{d}^5, \mathbf{d}^6\}$.

We now show some properties of box splines, cf. Theorem 2.2 in the univariate case.

Theorem 2.21 . *For $M(u,v) = M(u,v\,|\,\mathbf{X}_\mu)$ we have*

(i) $M(u,v) \equiv 0$ for $(u,v) \notin [\mathbf{X}_\mu[$. (Local support)

(ii) $M(u,v) > 0$ for $(u,v) \in]\mathbf{X}_\mu[$. (Positivity)

(iii) $\int\int_{[\mathbf{X}_\mu]} M(u,v)\, du dv = 1$. (Normalization)

Here

$$]\mathbf{X}_\mu[= \{t_1\mathbf{x}^1 + \ldots + t_\mu \mathbf{x}^\mu \;:\; 0 < t_j < 1, \quad 1 \le j \le \mu\}.$$

Proof: From (2.21) it is not hard to see that (i),(ii), and (iii) all hold for $\mu = 2$. Suppose by induction that they all hold for $\mu = m-1$. We prove for each property in turn that they hold for $\mu = m$ using (2.20) and the induction hypothesis.

(i) Suppose $(u,v) \notin [\mathbf{X}_m[$. Then $(u - tx_m, v - ty_m) \notin [\mathbf{X}_{m-1}[$ for all $t \in [0,1)$. But then the integrand in (2.20) is zero.

(ii) If $(u,v) \in]\mathbf{X}_m[$ then $(u - sx_m, v - sy_m) = (x,y)$ for some $(x,y) \in]\mathbf{X}_{m-1}[$ and some $s \in (0,1)$. We can therefore find an interval $[a,b]$ with $0 \le a < b \le 1$ such

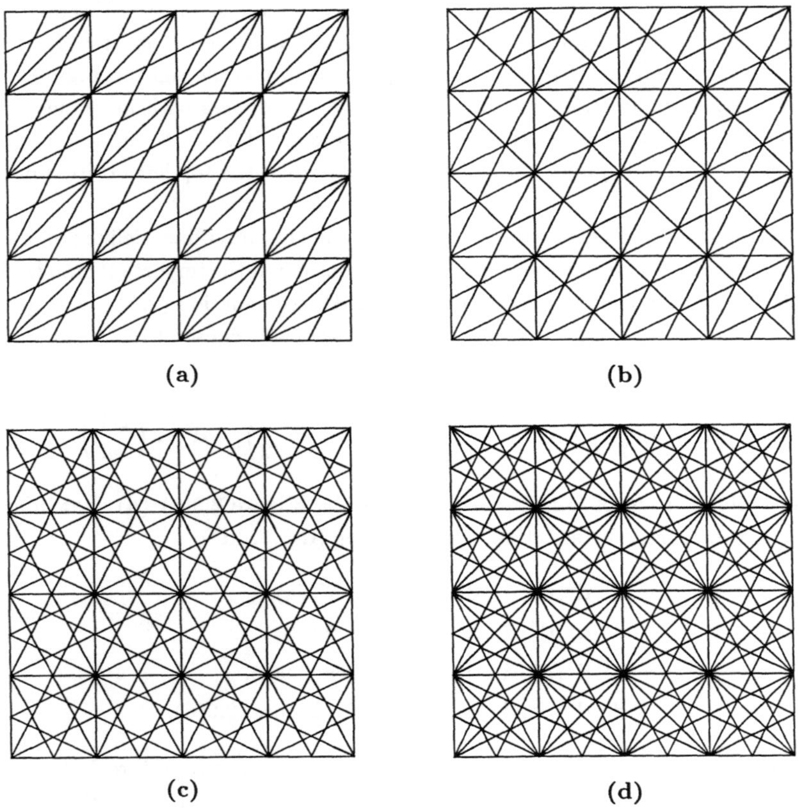

Figure 2.9. Various box spline grids

that $(u - tx_m, v - ty_m) \in]\mathbf{X}_{m-1}[$ for $t \in [a, b]$. But then the integrand in (2.20) is nonnegative and positive for $t \in [a, b]$.

(iii) We integrate both sides of (2.20) over all of $\mathbb{R}^2$. Changing the order of integration gives the result. ■

The next result shows that under mild restrictions on the direction vectors the translates of box splines form a partition of unity.

Theorem 2.22 . *Suppose that* $\mathbf{x}^1$ *and* $\mathbf{x}^2$ *in addition to being linearly independent have integer components. Then for each* $(u, v) \in \mathbb{R}^2$ *we have*

(iv) $\sum_{(i,j)\in\mathbb{Z}^2} M(u - i, v - j \,|\, \mathbf{X}_\mu) \equiv 1,$ *(Partition of unity)*

Proof: We first show that when we sum over the subgrid

$$\mathbf{X}_2\mathbb{Z}^2 = \{ix_1 + jx_2, iy_1 + jy_2 \,:\, (i,j) \in \mathbb{Z}^2\},$$

of $\mathbb{Z}^2$ then

$$\sum_{(\ell,m)\in \mathbf{X}_2\mathbb{Z}^2} M(u-\ell, v-m\,|\,\mathbf{X}_\mu) \equiv 1/|\det(\mathbf{X}_2)|. \tag{2.25}$$

For $\mu = 2$ this follows since the supports of $M(u-\ell, v-m\,|\,\mathbf{X}_2)$ for $(\ell, m) \in \mathbf{X}_2\mathbb{Z}^2$, form an essentially disjoint union, or a tiling, of the plane. For $\mu > 2$ equation (2.25) follows by summing both sides of (2.20) and using induction. To show that (iv) follows from (2.25) we observe that $\mathbb{Z}^2$ is the disjoint union of sets of the form $((g,h)+\mathbf{X}_2\mathbb{Z}^2)_{(g,h)\in G}$, with

$$G = \{(g,h) \in \mathbb{Z}^2 \,:\, (g,h) \in [\mathbf{X}_2[\}.$$

Since it can be shown that G contains $|\det(\mathbf{X}_2)|$ elements we obtain (iv) by summing both sides of (2.25) over G. ■

We next turn to differentiation of box splines.

Definition 2.23 . *For a sufficiently smooth function f we let*

$$D_{\mathbf{x}} f(u,v) = \lim_{\substack{h\to 0\\ h>0}} \frac{f(u+hx, v+hy) - f(u,v)}{h}$$

denote the one sided derivative in the direction $\mathbf{x} = (x,y)$.

If $f \in C^1$ then $D_{\mathbf{x}} f = x\partial f/\partial u + y\partial f/\partial v$ is the usual directional derivative of f.

To state a differentiation formula we need the reduced direction vectors

$$\mathbf{X}_\mu^i = \mathbf{X}_\mu \setminus \{\mathbf{x}^i\} = (\mathbf{x}^1, \ldots, \mathbf{x}^{i-1}, \mathbf{x}^{i+1}, \ldots, \mathbf{x}^\mu).$$

If $\mathbf{x}^1$ and $\mathbf{x}^2$ are linearly independent then $M(u,v\,|\,\mathbf{X}_\mu^i)$ is well defined for $i = 3, \ldots, \mu$. However, it is possible that $M(u,v\,|\,\mathbf{X}_\mu^1)$ and $M(u,v\,|\,\mathbf{X}_\mu^2)$ are not well defined. For example, if $\mathbf{X}_3 = (\binom{1}{0}, \binom{0}{1}, \binom{0}{1})$ then $\mathbf{X}_3^1 = (\binom{0}{1}, \binom{0}{1})$ and these vectors are not linearly independent.

Theorem 2.24 . *If $\mathbf{X}_\mu^i$ contains at least two linearly independent vectors then*

$$D_{\mathbf{x}^i} M(u,v\,|\,\mathbf{X}_\mu) = M(u,v\,|\,\mathbf{X}_\mu^i) - M(u-x_i, v-y_i\,|\,\mathbf{X}_\mu^i). \tag{2.26}$$

If $\mathbf{x} = \sum_{i=1}^\mu \omega_i \mathbf{x}^i$ then

$$D_{\mathbf{x}} M(u,v\,|\,\mathbf{X}_\mu) = \sum_{i=1}^{\mu} \omega_i [M(u,v\,|\,\mathbf{X}_\mu^i) - M(u-x_i, v-y_i\,|\,\mathbf{X}_\mu^i)], \tag{2.27}$$

provided $M(u,v\,|\,\mathbf{X}_\mu^i)$ is well defined for all i with $\omega_i \neq 0$.

Proof: Assume without loss of generality that $i = \mu$. By (2.20) we have

$$D_{\mathbf{x}_\mu} M(u,v\,|\,\mathbf{X}_\mu) = D_{\mathbf{x}_\mu} \int_0^1 M(u-tx_\mu, v-ty_\mu\,|\,\mathbf{X}_{\mu-1})\, dt.$$

Differentiating under the integral sign and observing that

$$D_{\mathbf{x}_\mu} M(u - tx_\mu, v - ty_\mu \,|\, \mathbf{X}_{\mu-1}) = -\frac{\partial}{\partial t} M(u - tx_\mu, v - ty_\mu \,|\, \mathbf{X}_{\mu-1})$$

(2.26) follows. To prove (2.27) we multiply both sides of (2.26) by ω_i and sum over i. ■

It is perhaps worth pointing out that one sided derivatives of box splines always exist. For example, $D_{\mathbf{x}} M_{(1,1)}(u,v) \equiv 0$ for all directions $\mathbf{x}$. The conditions in the previous Theorem are necessary because we have defined box splines as functions. It is possible to define box splines for any collection of direction vectors by treating them as *distributions* [8].

Further properties of bivariate box splines can be found in the section on multivariate box splines.

Box spline spaces can be formed from one or more box splines by using translations and dilations. We restrict our attention to box splines on the regular grids G_r, $r = 2, 3, 4$.

Definition 2.25 . *An element of the space*

$$S_{\mathbf{k},\nu} = \{p(u,v) = \sum_{(i,j)\in\mathbb{Z}^2} c_{i,j} M_{\mathbf{k}}(\nu u - i, \nu v - j) \,:\, c_{i,j} \in \mathbb{R}^N, u, v \in \mathbb{R}\}, \tag{2.28}$$

is called a box spline surface . The surface is called explicit if $N = 1$ and parametric if $N = 3$. We set $S_{\mathbf{k},1} = S_{\mathbf{k}}$.

We see that a box spline surface is a piecewise polynomial on the grid G_r/ν. Also from Theorem 2.21 it follows that a box spline surface lies in the convex hull of its coefficients

$$p(u,v) \in \operatorname{conv}\{c_{i,j} \,:\, M_{\mathbf{k}}(\nu u - i, \nu v - j) \neq 0\}.$$

We next define the control polygon for 2 and 3 directional box splines. We consider only the explicit surface case.

Definition 2.26 . *For the two directional box spline surface $p \in S_{(k_1,k_2)}$ the control polygon is the piecewise bilinear surface*

$$P_{(k_1,k_2),\nu}(u,v) = \sum_{i,j} c_{i,j} M_{(2,2)}(\nu u - i - k_1/2, \nu v - j - k_2/2). \tag{2.29}$$

For a three directional box spline surface the control polygon

$$P_{(k_1,k_2,k_3),\nu}(u,v) = \sum_{i,j} c_{i,j} M_{(1,1,1)}(\nu u - i - (k_1+k_3)/2, \nu v - j - (k_2+k_3)/2), \tag{2.30}$$

is a linear surface.

The control polygon $P_{\mathbf{k},\nu}$ of p on the refined grid will converge to p as ν tends to infinity. For a proofs, see [16,28,30]. Thus, we can use $P_{\mathbf{k},\nu}$ for a suitable ν to obtain a sufficiently good approximation to p.

Corresponding to a region $\Omega \subset \mathbb{R}^2$ we have the box spline space

$$(2.31)\quad S_{\mathbf{k},\nu}(\Omega) = \{p(u,v) = \sum_{(i,j)\in J_{\mathbf{k},\nu}(\Omega)} c_{i,j} M_{\mathbf{k}}(\nu u - i, \nu v - j) \quad c_{i,j} \in \mathbb{R}^N, u, v \in \mathbb{R}\},$$

where

$$(2.32)\qquad J_{\mathbf{k},\nu}(\Omega) = \{(i,j) : M_{\mathbf{k}}(\nu u - i, \nu v - j) \neq 0 \quad \text{for some} \quad (u,v) \in \Omega^0\}.$$

Here Ω^0 is the interior of the set Ω.

For approximation it is important to know when the box splines spanning $S_{\mathbf{k},\nu}(\Omega)$ are linearly independent. A statement for the general case can be found in the next section.

Theorem 2.27 . *For a region $\Omega \in \mathbb{R}^2$ consider the set $F = (M_{\mathbf{k}}(\nu u - i, \nu v - j))_{(i,j)} \in J_{\mathbf{k},\nu}(\Omega)$ of r directional box splines on the standard directions (2.24). Then F is a linearly independent set for $r = 2, 3$, but linearly dependent for $r = 4$.*

Similar to the one variable case it is possible to express box splines as convolutions. We refer to [31] for further details.

2.4 Multivariate Box Splines

We will now give a number of useful results on multivariate box splines, and we start by stating the general definition.

Definition 2.28 . *Suppose for positive integers μ, d that $\mathbf{X}_\mu = (\mathbf{x}^1, \mathbf{x}^2, \ldots, \mathbf{x}^\mu)$ with $\mathbf{x}^i = (x_1^i, x_2^i, \ldots, x_d^i)$ are $\mu \geq d$ vectors in $\mathbb{R}^d$ with $\det(\mathbf{X}_d) \neq 0$. A d-variate box spline with direction vectors $\mathbf{X}_\mu$, is a function $M(\mathbf{u}\,|\,\mathbf{X}_\mu) : \mathbb{R}^d \to \mathbb{R}$ defined recursively by*

$$(2.33)\qquad M(\mathbf{u}\,|\,\mathbf{X}_\mu) = \int_0^1 M(\mathbf{u} - t\mathbf{x}^\mu\,|\,\mathbf{X}_{\mu-1})\,dt,$$

with

$$(2.34)\qquad M(\mathbf{u}\,|\,\mathbf{X}_d) = \begin{cases} 1/|\det(\mathbf{X}_d)|, & \text{if } \mathbf{u} \in [\mathbf{X}_d[, \\ 0, & \text{otherwise,} \end{cases}$$

and where

$$(2.35)\qquad [\mathbf{X}_\mu[= \{t_1\mathbf{x}^1 + \ldots + t_\mu \mathbf{x}^\mu : 0 \leq t_j < 1, \quad 1 \leq j \leq \mu\}.$$

Example 2.29 . *Suppose $d = 1$ and $\mathbf{X}_\mu = (1, 1, \ldots, 1)$. Comparing Definitions 2.28 and 2.1 we see that the box spline in this case is identical to the univariate cardinal B-spline. For $d = 2$ we clearly have the bivariate box splines of Section 2.4. For $d = 3$ the functions in Definition 2.28 are called trivariate box splines.*

By induction it is not too hard to see that $M(\mathbf{u}\,|\,\mathbf{X}_\mu)$ is positive on

$$]\mathbf{X}_\mu[= \{t_1\mathbf{x}^1 + \ldots + t_\mu \mathbf{x}^\mu : 0 < t_j < 1, \quad 1 \leq j \leq \mu\}$$

and zero for any $\mathbf{u}$ not in the set

$$[\mathbf{X}_\mu] = \{t_1\mathbf{x}^1 + \ldots + t_\mu\mathbf{x}^\mu \,:\, 0 \le t_j \le 1, \quad 1 \le j \le \mu\}.$$

The set $[\mathbf{X}_\mu]$ is called the *support* of $M(\mathbf{u}\,|\,\mathbf{X}_\mu)$.

It can be shown ([8]) that box splines are piecewise polynomials of degree $\mu - d$. Moreover, $M(\mathbf{u}\,|\,\mathbf{X}_\mu) \in C^{\xi-1}(\mathbb{R}^d)$ where

$$\xi = \max\bigl\{r \,:\, \text{all selections } \mathbf{Y} \text{ of } r \text{ elements from } X_\mu \text{ are such that } \mathbf{X}_\mu \backslash \mathbf{Y} \text{ spans } \mathbb{R}^d\bigr\}.$$

The proof of these properties are quite complicated.

It is useful to have criteria for when translates of box splines are linearly independent. The following theorem states the necessary and sufficient conditions to ensure global linear independence. A simple proof can be found in [31].

Theorem 2.30 . *The set*

$$\Bigl\{M(\mathbf{u} - \mathbf{j}\,|\,\mathbf{X}_\mu), \quad \mathbf{j} \in \mathbb{Z}^d\Bigr\}$$

is linearly independent on $\mathbb{R}^d$ *if and only if all selections* $\mathbf{Y}$ *of* d *vectors from* $\mathbf{X}_\mu$ *are such that* $\det\mathbf{Y}$ *takes one of the three values* -1, 0 *or* 1.

It is also useful to have criteria for local linear independence of box splines ([29,35]).

Theorem 2.31 . *Suppose* Ω *is any region on which all translates of a box spline are polynomials. Then the set of box splines which are nonzero on* Ω *is linearly independent on* Ω *if and only if the condition in Theorem 2.30 holds.*

The formula given in the following theorem is basic and it is often used as the definition of box splines. It defines a box spline as a distribution, or generalized function.

Theorem 2.32 . *Suppose* $\det(\mathbf{X}_d) \neq 0$. *For any function* f *which is continuous on* $[\mathbf{X}_\mu]$

$$\int_{[\mathbf{X}_\mu]} M(\mathbf{u}\,|\,\mathbf{X}_\mu) f(\mathbf{u})\, d\mathbf{u} = \int_{[0,1]^\mu} f(\mathbf{X}_\mu \mathbf{t})\, d\mathbf{t} \tag{2.36}$$

where $\mathbf{t} = (t_1, \ldots t_\mu)$, $[0,1]^\mu$ *is the unit cube in* $\mathbb{R}^\mu$, *and*

$$\mathbf{X}_\mu \mathbf{t} = \sum_{i=1}^{\mu} t_i \mathbf{x}^i.$$

Proof: For $\mu = d$ we obtain (2.36) by the change of variable $\mathbf{u} = t_1\mathbf{x}^1 + \cdots t_d\mathbf{x}^d$ in (2.34). The factor $1/|\det(\mathbf{X}_d)|$ is cancelled by the Jacobian of the transformation.

Assume by induction that (2.36) holds for $\mu = m-1$. Defining $f(\mathbf{u}) = 0$ for $\mathbf{u} \notin [\mathbf{X}_m]$ we obtain from (2.33) and the induction hypothesis

$$\begin{aligned}
&\int_{\mathbb{R}^d} M(\mathbf{u}\,|\,\mathbf{X}_m) f(\mathbf{u})\, d\mathbf{u} \\
&= \int_{\mathbb{R}^d} [\int_0^1 M(\mathbf{u} - t\mathbf{x}^m\,|\,\mathbf{X}_{m-1})\, dt] f(\mathbf{u})\, d\mathbf{u} \\
&= \int_0^1 [\int_{\mathbb{R}^d} M(\mathbf{u} - t\mathbf{x}^m\,|\,\mathbf{X}_{m-1}) f(\mathbf{u})\, d\mathbf{u}] dt \\
&= \int_0^1 [\int_{\mathbb{R}^d} M(\mathbf{u}\,|\,\mathbf{X}_{m-1}) f(\mathbf{u} + t\mathbf{x}^m)\, d\mathbf{u}]\, dt \\
&= \int_0^1 [\int_{[0,1]^{m-1}} f(\sum_{i=1}^{m-1} t_i x^i + t x^m)\, dt_1 \ldots dt_{m-1}]\, dt \\
&= \int_{[0,1]^m} f(\mathbf{X}_m \mathbf{t})\, dt. \quad \blacksquare
\end{aligned}$$

In the definition (2.33) we assumed a particular ordering of the vectors in $\mathbf{X}_\mu$. Using (2.36) we obtain as a corollary that the value of M is independent of the ordering of the direction vectors.

Corollary 1. *We have*

$$M(\mathbf{u}\,|\,\mathbf{X}_\mu) = M(\mathbf{u}\,|\,\mathbf{Y}_\mu)$$

for any permutation $\mathbf{Y}_\mu = (\mathbf{y}^1, \mathbf{y}^2 \ldots, \mathbf{y}^\mu)$ *of* $\mathbf{X}_\mu$.

The Fourier transform of a box spline is given next.

Corollary 2. *Suppose* $\mathbf{X}_\mu$ *contains a basis for* $\mathbb{R}^d$. *Then*

$$\widehat{M}(\mathbf{z}\,|\,\mathbf{X}_\mu)) = \int_{\mathbb{R}^d} e^{i(\mathbf{u}\cdot\mathbf{z})} M_k(\mathbf{u}\,|\,\mathbf{X}_\mu)\, d\mathbf{u} = \prod_{j=1}^{\mu} \frac{e^{i\mathbf{z}\cdot\mathbf{x}^j} - 1}{i\mathbf{z}\cdot\mathbf{x}^j}.$$

where $\mathbf{x}\cdot\mathbf{y} = \sum_{j=1}^d x_j y_j$ *for any vectors* $\mathbf{x} = (x_1, \ldots, x_d)$ *and* $\mathbf{y} = (y_1, \ldots, y_d)$ *in* $\mathbb{R}^d$.

Theorem 2.32 can also be used to give a geometric interpretation of a box spline.

Corollary 3. *Let* $\mathbf{Y}_\mu = (\mathbf{y}^1, \mathbf{y}^2, \ldots, \mathbf{y}^\mu)$ *with* $\mathbf{y}^i = (\mathbf{x}^i, \mathbf{z}^i) \in \mathbb{R}^\mu$ *and* $\mathbf{z}^i \in \mathbb{R}^{\mu-d}$, *be a lifting of* $\mathbf{X}_\mu$ *from* $\mathbb{R}^d$ *to* $\mathbb{R}^\mu$ *such that* $\det(\mathbf{Y}_\mu) \neq 0$. *Then*

$$M(\mathbf{u}\,|\,\mathbf{X}_\mu) = \frac{1}{|\det(\mathbf{Y}_\mu)|} \int_{\mathbb{R}^{\mu-d}} \chi_{[\mathbf{Y}_\mu]}(\mathbf{u}, \mathbf{w})\, d\mathbf{w}, \qquad \text{for} \quad \mathbf{u} \in \mathbb{R}^d, \tag{2.37}$$

where $\chi_{[\mathbf{Y}_\mu]}$ *is the indicator function of the parallelepiped*

$$[\mathbf{Y}_\mu] = \{t_1\mathbf{y}^1 + t_2\mathbf{y}^2 + \ldots + t_\mu\mathbf{y}^\mu \,:\, 0 \le t_i \le 1, \quad i = 1, 2, \ldots, n\}.$$

Proof: Let f be an arbitrary function which is continuous on $[\mathbf{X}_\mu]$. We define $f(\mathbf{u}) = 0$ for $\mathbf{u} \notin [\mathbf{X}_\mu]$. We use (2.36) and the change of variable

$$\mathbf{y} = \sum_{i=1}^{\mu} t_i \mathbf{y}^i.$$

Partitioning y as $y = (\mathbf{u}, \mathbf{w}) = (\sum_{i=1}^{\mu} t_i \mathbf{x}^i, \sum_{i=1}^{\mu} t_i \mathbf{z}^i)$ we find

$$\begin{aligned}\int_{\mathbb{R}^d} M(\mathbf{u} \,|\, \mathbf{X}_\mu) f(\mathbf{u})\, d\mathbf{u} &= \int_{[0,1]^\mu} f(\mathbf{X}_\mu \mathbf{t})\, dt \\ &= \frac{1}{|\det([\mathbf{Y}_\mu])|} \int_{[\mathbf{Y}_\mu]} f(\mathbf{u})\, d\mathbf{u} d\mathbf{w} \\ &= \frac{1}{|\det([\mathbf{Y}_\mu])|} \int_{\mathbb{R}^\mu} \chi_{[\mathbf{Y}_\mu]}(\mathbf{u}, \mathbf{w}) f(\mathbf{u})\, d\mathbf{u} d\mathbf{w} \\ &= \frac{1}{|\det([\mathbf{Y}_\mu])|} \int_{\mathbb{R}^d} [\int_{\mathbb{R}^{\mu-d}} \chi_{[\mathbf{Y}_\mu]}(\mathbf{u}, \mathbf{w})\, d\mathbf{w}] f(\mathbf{u})\, d\mathbf{u}.\end{aligned}$$

Comparing the first and last expression and observing that f is arbitrary the Corollary follows. ■

Geometrically we can construct $M(\mathbf{u} \mid \mathbf{X}_\mu)$ as follows. We first lift the vectors $\mathbf{x}^1, \mathbf{x}^2, \ldots, \mathbf{x}^\mu$ in $\mathbb{R}^d$ to vectors $\mathbf{Y}_\mu = (\mathbf{y}^1, \mathbf{y}^2, \ldots, \mathbf{y}^\mu)$ in $\mathbb{R}^\mu$. A possible lifting is given by $\mathbf{y}^i = (\mathbf{x}^i, \mathbf{0}), i = 1, 2, \ldots, d$ and $\mathbf{y}^i = (\mathbf{x}^i, \mathbf{e}^{i-d}), i = d+1, \ldots, n$, where the $\mathbf{e}^{i-d}$ are the unit vectors in $\mathbb{R}^{\mu-d}$. Now for each $\mathbf{u} \in \mathbb{R}^d$ we obtain the value $M(\mathbf{u} \,|\, \mathbf{X}_\mu)$ as the $\mu - d$ dimensional volume of those points in $[\mathbf{Y}_\mu]$ which project to $\mathbf{u}$. In symbols

$$M(\mathbf{u} \,|\, \mathbf{X}_\mu) = \frac{\operatorname{vol}_{n-d}(\{\mathbf{w} \in \mathbb{R}^{\mu-d} : (\mathbf{u}, \mathbf{w}) \in [\mathbf{Y}_\mu]\})}{\operatorname{vol}_\mu([\mathbf{Y}_\mu])}, \quad \mathbf{u} \in \mathbb{R}^d, \tag{2.38}$$

where $\operatorname{vol}_\mu([\mathbf{Y}_\mu]) = \det([\mathbf{Y}_\mu]) \neq 0$. A similar geometric definition of univariate B-splines was given by I. J. Schoenberg in a letter to P. J. Davis in 1965. This was used by de Boor to give a definition of multivariate B-splines in 1975 [4].

Example 2.33 . *For the $M_{(1,1,1)}$ bivariate box spline and the lifting $\mathbf{Y}_3 = (1,0,0)^T, (0,1,0)^T, (1,1,1)^T$ the equation (2.37) takes the form*

$$M_{(1,1,1)}(u, v) = \frac{1}{|\det(\mathbf{Y}_3)|} \int_{\mathbb{R}} \chi_{[\mathbf{Y}_3]}(u, v, w)\, dw, \qquad \text{for} \quad (u, v) \in \mathbb{R}^2.$$

$[\mathbf{Y}_3]$ is a parallelepiped obtained from the unit cube in $\mathbb{R}^3$ by moving the top facet horizontally in the $(1,1)$ direction. At (u, v) the value of the box spline is given by the length of that part of the vertical line through $(u, v, 0)$ which lies in $[\mathbf{Y}_3]$.

Corollary 2 can be used to derive knot line refinement algorithms for box splines. The following result can be found in [10] and generalizes Theorem 2.9 .

Theorem 2.34 . *For $\nu \in \mathbb{N}$ and $\mathbf{x}^\ell \in \mathbb{Z}^d, \ell = 1, 2, \ldots \mu$*

$$M(\mathbf{u} \,|\, \mathbf{X}_\mu) = \sum_{\mathbf{j} \in \mathbb{Z}^d} \beta(\mathbf{j} \,|\, \mathbf{X}_\mu) M(\nu \mathbf{u} - \mathbf{j} \,|\, \mathbf{X}_\mu), \quad \mathbf{u} \in \mathbb{R}^d, \tag{2.39}$$

where the generating function for the $\beta(\mathbf{j} \,|\, \mathbf{X}_\mu) = \beta(\mathbf{j} \,|\, \nu, \mathbf{X}_\mu)$ is

$$q(\mathbf{z} \,|\, \mathbf{X}_\mu) = \sum_{\mathbf{j}} \beta(\mathbf{j} \,|\, \mathbf{X}_\mu) \mathbf{z}^j = \nu^{d-\mu} \prod_{\ell=1}^{\mu} (1 + \mathbf{z}^{\mathbf{x}^\ell} + \mathbf{z}^{2\mathbf{x}^\ell} + \ldots + \mathbf{z}^{(\nu-1)\mathbf{x}^\ell}), \tag{2.40}$$

$\mathbf{z} = (z_1, \ldots, z_d)$ and $\mathbf{z}^{\mathbf{y}} = z_1^{y_1} \cdot z_2^{y_2} \cdots z_d^{y_d}$ *for any* $\mathbf{y} = (y_1, \ldots, y_d)$.

Proof: Using Corollary 2 we find

$$\int_{\mathbb{R}^d} e^{i(\mathbf{u}\cdot\eta)} M(\nu\mathbf{u} - \mathbf{j} \,|\, \mathbf{X}_\mu)\, d\mathbf{u} = \nu^{-d} e^{i(\mathbf{j}\cdot\eta)/\nu} \prod_{\ell=1}^{\mu} \frac{e^{i(\eta\cdot\mathbf{x}^\ell)/\nu} - 1}{i(\eta\cdot\mathbf{x}^\ell)/\nu}.$$

Letting $\mathbf{z} = (z_1, \ldots, z_d)$ with $z_\ell = e^{\eta_\ell/\nu}$ and $\eta = (\eta_1, \ldots, \eta_d)$ the Fourier transform of (2.39) can be written

$$\prod_{\ell=1}^{\mu} \frac{\mathbf{z}^{\nu\mathbf{x}^\ell} - 1}{i(\eta\cdot\mathbf{x}^\ell)} = \nu^{\mu-d} \prod_{\ell=1}^{\mu} \frac{\mathbf{z}^{\mathbf{x}^\ell} - 1}{i(\eta\cdot\mathbf{x}^\ell)} \sum_{\mathbf{j}} \beta(\mathbf{j} \,|\, \mathbf{X}_\mu)\mathbf{z}^{\mathbf{j}}.$$

Solving for the sum term we obtain (2.40). ■

We can also give a recurrence relation for the β's. This recurrence relation was first proved in [16,28]. We define $\beta(\mathbf{j} \,|\, \mathbf{X}_\mu)$ by (2.40) also for $\mu = 1, 2, \ldots, d-1$.

Theorem 2.35 . *For* $\mu > 1$ *and* $\mathbf{x}^\ell \in \mathbb{Z}^d, \ell = 1, 2, \ldots \mu$ *we have*

$$\beta(\mathbf{j} \,|\, \nu, \mathbf{X}_\mu) = \frac{1}{\nu} \sum_{\ell=0}^{\nu-1} \beta(\mathbf{j} - \ell\mathbf{x}^\mu \,|\, \nu, \mathbf{X}_{\mu-1}). \tag{2.41}$$

Proof: From (2.40) we have

$$q(\mathbf{z} \,|\, \mathbf{X}_\mu) = q(\mathbf{z} \,|\, \mathbf{X}_{\mu-1})(1 + \mathbf{z}^{\mathbf{x}^\mu} + \mathbf{z}^{2\mathbf{x}^\mu} + \ldots + \mathbf{z}^{(\nu-1)\mathbf{x}^\mu})/\nu.$$

Inserting

$$q(\mathbf{z} \,|\, \mathbf{X}_\ell) = \sum_{\mathbf{j}} \beta(\mathbf{j} \,|\, \nu, \mathbf{X}_\ell)\mathbf{z}^j, \qquad \ell = \mu - 1, \mu,$$

and comparing coefficients of equal powers of $\mathbf{z}$ we obtain the theorem. ■

The following transformation formula for box splines is sometimes useful. For a $d \times d$ matrix $\mathbf{A}$ we let $\mathbf{AX}_\mu$ denote the direction vectors $\mathbf{Ax}^1, \mathbf{Ax}^2, \ldots, \mathbf{Ax}^\mu$.

Theorem 2.36 . *Suppose* $\mathbf{A}$ *is a nonsingular* $d \times d$ *matrix. Then*
(vii) $M(\mathbf{u} \,|\, \mathbf{X}_\mu) = |\det \mathbf{A}| M(\mathbf{Au} \,|\, \mathbf{AX}_\mu)$ *(Transformation Formula)*

Proof: Since $\det(\mathbf{AX}_d) = \det(\mathbf{A})\det(\mathbf{X}_d)$ the result follows for $\mu = d$. Suppose by induction that the result holds for $\mu = m - 1$. Then by (2.33) and the induction hypothesis

$$\begin{aligned} &|\det \mathbf{A}| M(\mathbf{Au} \,|\, \mathbf{AX}_m) \\ &\quad = |\det \mathbf{A}| \int_0^1 M(\mathbf{Au} - t\mathbf{Ax}^m \,|\, \mathbf{AX}_{m-1})\, dt \\ &\quad = \int_0^1 M(\mathbf{u} - t\mathbf{x}^m \,|\, \mathbf{X}_{m-1})\, dt = M(\mathbf{u} \,|\, \mathbf{X}_m). \quad ■ \end{aligned}$$

Next we give a symmetry property of box splines. We need left continuous versions $\tilde{M}$ of box splines. To define $\tilde{M}$ we use (2.33) and replace $[\mathbf{X}_d[$ by $]\mathbf{X}_d] = \{t_1\mathbf{x}^1 + \cdots t_d\mathbf{x}^d : 0 < t_i \le 1, \quad i = 1, 2, \ldots, d\}$ in (2.34).

Theorem 2.37 . *For any* $\mathbf{X}_\mu$ *we have*

$$M(\mathbf{u}\,|\,\mathbf{X}_\mu) = \tilde{M}(\sum_{i=1}^{\mu} \mathbf{x}^i - \mathbf{u}\,|\,\mathbf{X}_\mu). \tag{2.42}$$

Proof: Since $\mathbf{u} \in [\mathbf{X}_d[\iff (\mathbf{x}^1 + \cdots \mathbf{x}^d - \mathbf{u}) \in]\mathbf{X}_d]$ we see that (2.42) holds for $\mu = d$. Suppose by induction that (2.42) holds for $\mu = m-1$. Then by (2.33) and the induction hypothesis

$$\begin{aligned}
&\tilde{M}(\sum_{i=1}^{m} \mathbf{x}^i - \mathbf{u}\,|\,\mathbf{X}_m) \\
&= \int_0^1 \tilde{M}(\sum_{i=1}^{m-1} \mathbf{x}^i + (1-t)\mathbf{x}^m - \mathbf{u}\,|\,\mathbf{X}_{m-1})\,dt \\
&= \int_0^1 M(\mathbf{u} - (1-t)\mathbf{x}^m\,|\,\mathbf{X}_{m-1})\,dt = M(\mathbf{u}\,|\,\mathbf{X}_m). \quad \blacksquare
\end{aligned}$$

It is also of interest to see what happens if we change sign, or flip one of the direction vectors.

Theorem 2.38 . *For* $\mathbf{u} \in \mathbb{R}^d$ *and* $i = 1, \ldots, n$ *we have*

$$M(\mathbf{u}\,|\,\mathbf{x}^1, \ldots, \mathbf{x}^\mu) = M(\mathbf{u} - \mathbf{x}^i\,|\,\mathbf{x}^1, \ldots, \mathbf{x}^{i-1}, -\mathbf{x}^i, \mathbf{x}^{i+1}, \ldots, \mathbf{x}^\mu).$$

Proof: Because M is independent of the ordering of the direction vectors it is enough to prove the result for $i = \mu$. Since $|\det(\mathbf{x}^1, \ldots, \mathbf{x}^{d-1}, -\mathbf{x}^d)| = |\det(\mathbf{x}^1, \ldots, \mathbf{x}^d)|$ and $\mathbf{u} \in [\mathbf{x}^1, \ldots, \mathbf{x}^d[\iff \mathbf{u} - \mathbf{x}^d \in [\mathbf{x}^1, \ldots, \mathbf{x}^{d-1}, -\mathbf{x}^d[$ the result follows for $\mu = d$. We use (2.33) and the substitution $s = 1 - t$ for the general case. $\blacksquare$

2.5 Cone Splines

Our reason for including cone splines in this paper is that there exists a number of interesting relation between box splines and cone splines. Later we will use one of these results to deduce a simple and fast algorithm for the evaluation of box splines. We start with some preliminaries.

Definition 2.39 . *Given* $\mathbf{y} \in \mathbb{R}^d$ *the set* $\{\mathbf{u} \in \mathbb{R}^d : \mathbf{y} \cdot \mathbf{u} > 0\}$ *is called an open halfspace. We say that* $\mu \geq 1$ *vectors* $\mathbf{X}_\mu = (\mathbf{x}^1, \mathbf{x}^2, \ldots, \mathbf{x}^\mu)$ *in* $\mathbb{R}^d$ *satisfy condition H if they lie in an open halfspace. The vectors* $\mathbf{X}_\mu$ *satisfy condition HI if in addition* $\mu \geq d$ *and* $\det(\mathbf{x}^1, \ldots, \mathbf{x}^d) \neq 0$.

In $\mathbb{R}$ condition H means that all the direction vectors should be nonzero and have the same sign. In $\mathbb{R}^2$ the vectors $\binom{1}{0}, \binom{0}{1}, \binom{1}{1}$ satisfy both condition H and HI. The three vectors belong to the halfspace $\{(u,v) : u + v > 0\}$. The three vectors $\binom{1}{0}, \binom{0}{1}, \binom{-1}{-1}$ do not satisfy condition H or HI. Condition H is sometimes stated in the form $0 \notin \operatorname{conv}(\mathbf{X}_\mu)$.

Definition 2.40 . *Suppose* $\mathbf{X}_\mu = (\mathbf{x}^1, \mathbf{x}^2, \ldots, \mathbf{x}^\mu)$ *are* $\mu \geq d$ *vectors in* $\mathbb{R}^d$ *satisfying condition HI. A d variate cone spline with direction vectors* $\mathbf{X}_\mu$ *is a function* $C(\mathbf{u}) = C(\mathbf{u} \,|\, \mathbf{X}_\mu) : \mathbb{R}^d \to \mathbb{R}$ *defined recursively by*

$$C(\mathbf{u} \,|\, \mathbf{X}_\mu) = \int_0^\infty C(\mathbf{u} - t\mathbf{x}^\mu \,|\, \mathbf{X}_{\mu-1})\, dt, \tag{2.43}$$

with

$$C(\mathbf{u} \,|\, \mathbf{X}_d) = \begin{cases} 1/|\det(\mathbf{X}_d)|, & \text{if } \mathbf{u} \in < \mathbf{X}_d >_+, \\ 0, & \text{otherwise,} \end{cases} \tag{2.44}$$

and where for $\ell \geq 1$

$$< \mathbf{X}_\ell >_+ = \{t_1\mathbf{x}^1 + \ldots + t_\ell \mathbf{x}^\ell \,:\, 0 \leq t_j, \quad 1 \leq j \leq \ell\}. \tag{2.45}$$

The graph of the bivariate function $C(u, v \,|\, \mathbf{X}_2)$ looks like an infinite piece of cake. The height of the piece is the same as the height of the box spline $M(u, v \mid \mathbf{X}_2)$ (cf. (2.34)).

Theorem 2.41 . *Suppose for* $d = 1$ *that* $x^i > 0$ *for* $i = 1, 2, \ldots, \mu$*. A univariate cone spline is given by*

$$C(u \,|\, x^1, \cdots, x^\mu) = \frac{1}{(\mu-1)!} \frac{u_+^{\mu-1}}{x^1 \cdots x^\mu}, \qquad u \in \mathbb{R} \tag{2.46}$$

where

$$u_+^{\mu-1} = \begin{cases} u^{\mu-1}, & \text{if } u \geq 0, \\ 0, & \text{otherwise.} \end{cases} \tag{2.47}$$

Proof: From (2.44) we have $M(u \,|\, \mathbf{X}_1) = 1/x^1$ for $u \geq 0$. Thus (2.46) holds for $\mu = 1$. The general case follows by induction from (2.43). ∎

The function $u_+^{\mu-1}$ is called a truncated power. For this reason Cone splines are also called truncated powers. These functions were introduced in [26].

Corresponding to the box splines $M_\mathbf{k}$ we have the bivariate r directional cone splines $C_\mathbf{k}$. The explicit form of a bivariate 2 directional cone spline ($r = 2$) is given next.

Theorem 2.42 . *The bivariate tensor product cone spline is given by*

$$C_{(k_1,k_2)}(u, v) = \frac{u_+^{k_1-1}}{(k_1-1)!} \frac{v_+^{k_2-1}}{(k_2-1)!}, \quad (u, v) \in \mathbb{R}^2.$$

Proof: By induction it is easily seen that

$$C_{(k_1,k_2)}(u, v) = C_{k_1}(u) C_{k_2}(v),$$

where

$$C_k(u) = C(u \,|\, \overbrace{1, 1 \ldots, 1}^{k \text{ times}})$$

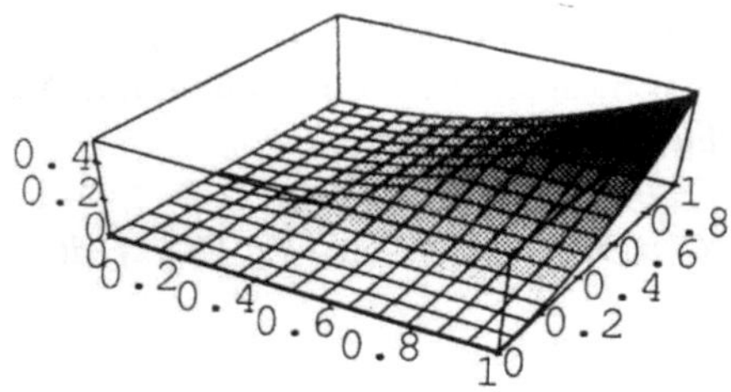

Figure 2.10. The Cone spline $C_{(1,1,2)}$

is a univariate cardinal cone spline. The theorem now follows from Theorem 2.41 . ■

Example 2.43 . *Corresponding to 3 directional bivariate box splines we have the cone splines $C_{(k_1,k_2,k_3)}$. These functions are nonzero over the two sectors $v \geq u > 0$ and $u \geq v > 0$. On the upper sector $v \geq u \geq 0$ we find from (2.43)*

$$C_{(1,1,1)}(u,v) = u, \quad C_{(1,1,2)}(u,v) = \frac{1}{2}u^2, \quad C_{(2,2,1)}(u,v) = \frac{1}{6}u^2(3v-u), \tag{2.48}$$
$$C_{(2,2,2)}(u,v) = \frac{1}{12}u^3(2v-u).$$

For the lower sector we can use the symmetry $C_{(k_1,k_2,k_3)}(u,v) = C_{(k_1,k_2,k_3)}(v,u)$, valid if $k_1 = k_2$ for all $(u,v) \in \mathbb{R}^2$. The graph of the C^0 cone spline $C_{(1,1,2)}$ is shown in Figure 2.10.

Example 2.44 . *The cone spline $C_{\mathbf{k}}$, $\mathbf{k} = (2,2,1,1)$ on the 4 direction mesh (cf. Figure 2.5) is nonzero over the three sectors $v \geq u \geq 0$, $u \geq v \geq 0$ and $-u \leq v \leq 0$. We have*

$$C_{(2,2,1,1)} = \begin{cases} (\sqrt{2}/3)u^3v + ((\sqrt{2}-2)/3)u^4, & v \geq u \geq 0 \\ ((\sqrt{2}-1)/12)(u+v)^4 - (\sqrt{2}/3)v^3u - ((\sqrt{2}-2)/3)v^4, & u \geq v \geq 0 \\ ((\sqrt{2}-1)/12)(u+v)^4, & -u \leq v \leq 0. \end{cases}$$

Cone splines and box splines are related. The following results of this section was first stated in [27]. In the following we let $\mathbb{Z}_+^\mu$ denote the set of all μ tuplets with nonnegative integer valued components.

Theorem 2.45 . *Suppose $\mathbf{x}^i \in \mathbb{Z}^d, i = 1, \ldots, \mu$ satisfy condition HI. Then for $\mathbf{u} \in \mathbb{R}^d$*

$$C(\mathbf{u}\,|\,\mathbf{X}_\mu) = \sum_{\mathbf{j}\in\mathbb{Z}_+^\mu} M(\mathbf{u} - \mathbf{X}_\mu\mathbf{j}\,|\,\mathbf{X}_\mu). \tag{2.49}$$

Proof: For $\mu = d$ we observe that the cone $< \mathbf{X}_d >_+$ is the union of the half open translated boxes $B_{\mathbf{j}} = \{\mathbf{X}_d\mathbf{j} + \sum_{i=1}^d t_i\mathbf{x}^i \,:\, 0 \leq t_i < 1\}$, $\mathbf{j} \in \mathbb{Z}_+^d$. Therefore,

$$\begin{aligned} C(\mathbf{u}\,|\,\mathbf{X}_d) &= \chi_{<\mathbf{X}_d>_+}(\mathbf{u})/|\det(\mathbf{X}_d)| \\ &= \sum_{\mathbf{j}\in\mathbb{Z}_+^d} \chi_{B_{\mathbf{j}}}(\mathbf{u})/|\det(\mathbf{X}_d)| = \sum_{\mathbf{j}\in\mathbb{Z}_+^d} M(\mathbf{u} - \mathbf{X}_d\mathbf{j}\,|\,\mathbf{X}_d). \end{aligned}$$

This proves (2.49) for $\mu = d$. Assume by induction that (2.49) holds for $\mu = m-1$. Then

$$\sum_{\mathbf{s}\in\mathbb{Z}_+^m} M(\mathbf{u}-\mathbf{X}_m\mathbf{s}\,|\,\mathbf{X}_m) \overset{\text{def\_M}}{=} \sum_{\mathbf{s}\in\mathbb{Z}_+^m} \int_0^1 M(\mathbf{u}-t\mathbf{x}^m-\mathbf{X}_m\mathbf{s}\,|\,\mathbf{X}_{m-1})\,dt$$

$$= \sum_{\sigma=0}^{\infty} \int_0^1 \sum_{\mathbf{s}\in\mathbb{Z}_+^{m-1}} M(\mathbf{u}-t\mathbf{x}^m-\sigma\mathbf{x}^m-\mathbf{X}_{m-1}\mathbf{s}\,|\,\mathbf{X}_{m-1})\,dt$$

$$\overset{\text{ind\_hyp}}{=} \sum_{\sigma=0}^{\infty} \int_0^1 C(\mathbf{u}-(t+\sigma)\mathbf{x}^m\,|\,\mathbf{X}_{m-1})\,dt$$

$$= \int_0^{\infty} C(\mathbf{u}-t\mathbf{x}^m\,|\,\mathbf{X}_{m-1})\,dt \overset{\text{def\_C}}{=} C(\mathbf{u}\,|\,\mathbf{X}_m). \quad \blacksquare$$

We next introduce a discrete version of cone splines. We denote by $\mathcal{C}_0$ the space of all sequences with at most a finite number of nonzero terms.

Definition 2.46 . *Suppose $\mathbf{X}_\mu = (\mathbf{x}^1, \mathbf{x}^2, \ldots, \mathbf{x}^\mu)$ are $\mu \geq 1$ vectors in $\mathbb{Z}^d$ satisfying condition H. A d variate discrete cone spline with direction vectors $\mathbf{X}_\mu$ is a sequence $\gamma(\mathbf{j}) = \gamma(\mathbf{j}\,|\,\mathbf{X}_\mu) : \mathbb{Z}^d \to \mathbb{R}$ defined by*

$$\sum_{\mathbf{j}\in\mathbb{Z}^d} \gamma(\mathbf{j}\,|\,\mathbf{X}_\mu) f(\mathbf{j}) = \sum_{\mathbf{t}\in\mathbb{Z}_+^\mu} f(\mathbf{X}_\mu\mathbf{t}), \tag{2.50}$$

for all $f \in \mathcal{C}_0$.

Taking for fixed $\mathbf{i} \in \mathbb{Z}^d$ the sequence $f(\mathbf{j}) = \delta_{\mathbf{i},\mathbf{j}}$ we find

$$\gamma(\mathbf{i}\,|\,\mathbf{X}_\mu) = \#\{\mathbf{t} \in \mathbb{Z}_+^\mu \,:\, \mathbf{X}_\mu\mathbf{t} = \mathbf{i}.\} \tag{2.51}$$

Thus $\gamma(\mathbf{i})$ is uniquely defined as the number of nonnegative integer solutions $\mathbf{t}$ of the $\mu \times \mu$ linear system $\mathbf{X}_\mu\mathbf{t} = \mathbf{i}$.

Combining for fixed $\mathbf{u}$ (2.50) with $f(\mathbf{j}) = M(\mathbf{u}-\mathbf{j}\,|\,\mathbf{X}_\mu)$ in Definition and (2.49) we obtain

Theorem 2.47 . *Suppose $\mathbf{x}^i \in \mathbb{Z}^d, i = 1, \ldots, \mu$ satisfy condition HI. Then for $\mathbf{u} \in \mathbb{R}^d$*

$$C(\mathbf{u}\,|\,\mathbf{X}_\mu) = \sum_{\mathbf{j}\in\mathbb{Z}^d} \gamma(\mathbf{j}\,|\,\mathbf{X}_\mu) M(\mathbf{u}-\mathbf{j}\,|\,\mathbf{X}_\mu). \tag{2.52}$$

Let us take a closer look at discrete cone splines. First we show that they can be given by a discrete analog of the recurrence relation (2.43) for cone splines.

Theorem 2.48 . *If $\det(\mathbf{X}_\ell) \neq 0$ then*

$$\gamma(\mathbf{j}\,|\,\mathbf{X}_\ell) = \begin{cases} 1, & \text{if } \mathbf{j} = \mathbf{X}_\ell\mathbf{t} \text{ for some } \mathbf{t} \in \mathbb{Z}_+^\ell, \\ 0, & \text{otherwise.} \end{cases} \qquad \ell = 1, \ldots, d \tag{2.53}$$

For $\mu > 1$ we have

$$\gamma(\mathbf{j}\,|\,\mathbf{X}_\mu) = \sum_{\ell=0}^{\infty} \gamma(\mathbf{j} - \ell\mathbf{x}^\mu\,|\,\mathbf{X}_{\mu-1}). \tag{2.54}$$

Proof: (2.53) follows from (2.51). For any $f \in \mathcal{C}_0$

$$\sum_{\mathbf{j}\in\mathbb{Z}^d} [\sum_{\ell=0}^{\infty} \gamma(\mathbf{j} - \ell\mathbf{x}^\mu\,|\,\mathbf{X}_{\mu-1})] f(\mathbf{j}) = \sum_{\ell=0}^{\infty} [\sum_{\mathbf{t}\in\mathbb{Z}_+^{\mu-1}} f(\mathbf{X}_{\mu-1}\mathbf{t} + \ell\mathbf{x}^\mu)] = \sum_{\mathbf{t}\in\mathbb{Z}_+^\mu} f(\mathbf{X}_\mu \mathbf{t}).$$

Hence (2.54) follows from (2.50). ■

We consider next differences of discrete cone splines.

Definition 2.49 . *For any sequence $f : \mathbb{Z}^d \to \mathbb{R}$ and $\mathbf{x} \in \mathbb{Z}^d$ we let*

$$\nabla_{\mathbf{x}} f(\mathbf{j}) = f(\mathbf{j}) - f(\mathbf{j} - \mathbf{x})$$

denote the backward difference of f with spacing x. We set

$$\nabla_{\mathbf{X}_\mu} = \nabla_{\mathbf{x}^1} \nabla_{\mathbf{x}^2} \cdots \nabla_{\mathbf{x}^\mu}.$$

Note that $\nabla_{\mathbf{X}_\mu} = \nabla_{\mathbf{Y}_\mu}$ for any permutation $\mathbf{Y}_\mu$ of $\mathbf{X}_\mu$.

Theorem 2.50 . *For $\mathbf{j} \in \mathbb{Z}^d$*

$$\nabla_{\mathbf{X}_\mu} \gamma(\mathbf{j}\,|\,\mathbf{X}_\mu) = \delta_{\mathbf{0},\mathbf{j}} = \begin{cases} 1, & \textit{if } \mathbf{j} = \mathbf{0}, \\ 0, & \textit{otherwise}. \end{cases} \tag{2.55}$$

Proof: Taking differences on both sides of (2.54) we obtain

$$\nabla_{\mathbf{x}^\mu} \gamma(\mathbf{j}\,|\,\mathbf{X}_\mu) = \gamma(\mathbf{j}\,|\,\mathbf{X}_{\mu-1}). \tag{2.56}$$

Using (2.53) we see that (2.55) holds for $\mu = 1$. Since by (2.56) $\nabla_{\mathbf{X}_\mu} \gamma(\mathbf{j} \mid \mathbf{X}_\mu) = \nabla_{\mathbf{X}_{\mu-1}} \gamma(\mathbf{j}\,|\,\mathbf{X}_{\mu-1})$ we obtain (2.55) for any $\mu \geq 1$. ■

We are now in position to invert the relationship in Theorem 2.48 .

Theorem 2.51 . *Suppose $\mathbf{x}^i \in \mathbb{Z}^d, i = 1, \ldots, \mu$ satisfy condition HI. Then for $\mathbf{u} \in \mathbb{R}^d$*

$$M(\mathbf{u}\,|\,\mathbf{X}_\mu) = \nabla_{\mathbf{X}_\mu} C(\mathbf{u}\,|\,\mathbf{X}_\mu). \tag{2.57}$$

Proof: Let $\mathbf{x}$ be one of the direction vectors. We apply $\nabla_{\mathbf{x}}$ to both sides of (2.52). Then

$$\nabla_{\mathbf{x}} C(\mathbf{u}) = C(\mathbf{u}) - C(\mathbf{u} - \mathbf{x}) = \sum \gamma(\mathbf{j})[M(\mathbf{u} - \mathbf{j}) - M(\mathbf{u} - \mathbf{j} - \mathbf{x})]$$
$$= \sum [\gamma(\mathbf{j}) - \gamma(\mathbf{j} - \mathbf{x})] M(\mathbf{u} - \mathbf{j}) = \sum \nabla_{\mathbf{x}} \gamma(\mathbf{j}) M(\mathbf{u} - \mathbf{j}).$$

This implies that

$$\nabla_{\mathbf{X}_\mu} C(\mathbf{u}) = \sum \nabla_{\mathbf{X}_\mu} \gamma(\mathbf{j}) M(\mathbf{u} - \mathbf{j}).$$

Using (2.55) we see that this is the same as (2.57). ■

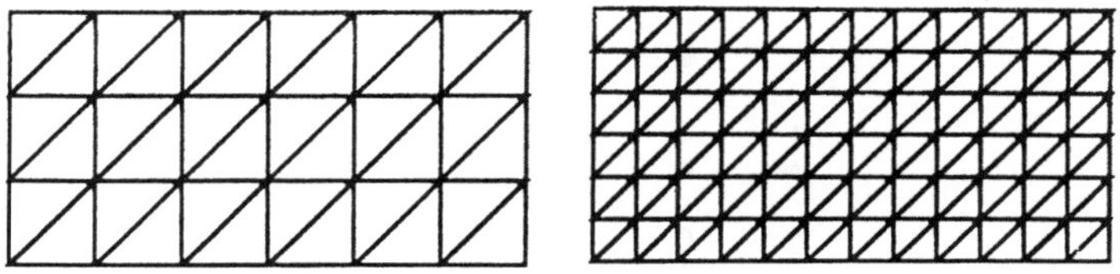

Figure 3.1. Refinement of a three directional grid

3. Refinement and Evaluation

In the previous section we gave the basic concepts of univariate cardinal splines and multivariate box splines with special emphasis on box splines in two varaiables. We will here return to box splines in two variables. In the first subsection we discuss knotline refinement or subdivision, which is the basis for efficient rendering of spline surfaces, computations of intersection, etc.. These algorithms play a key role in CAGD, and were developed in [2,41], and in full generality in [16,28]. All surface examples in this text are accomplished by the use of knotline insertion and the efficient method for box spline evaluation given in Section 3.2 ([20]).

3.1 Knotline Refinement

Consider again the space $S_{\mathbf{k},\nu}$ given by (2.28). For $\nu > 1$ the grid G_r/ν corresponding to $S_{\mathbf{k},\nu}$ will be a refinement of the grid G_r. The situation for a three directional grid is shown for $\nu = 2$ in Figure 3.1.

In Theorem 2.34 we showed how in general one could write a box spline in S_k as a sum of box splines in $S_{k,\nu}$. For the standard grids $G_r, r \leq 4$ and multiplicities $\mathbf{k} = (k_1, k_2, k_3, k_4)$ with $k_i \geq 0$ this relation can be written

$$(3.1) \qquad M_{\mathbf{k}}(u,v) = \sum_{(i,j)\in \mathbb{Z}^2} \beta_{\mathbf{k}}^{\nu}(i,j) M_{\mathbf{k}}(\nu u - i, \nu v - j), \quad (u,v) \in \mathbb{R}^2,$$

where the generating function q in (2.40) takes the simple form

$$(3.2) \qquad q_{\mathbf{k}}^{\nu}(w,z) = \sum_{i,j} \beta_{\mathbf{k}}^{\nu}(i,j) w^i z^j = \nu^{2-\mu}(1+w)^{k_1}(1+z)^{k_2}(1+wz)^{k_3}(1+w/z)^{k_4}.$$

We call $\{(i,j), \beta_{\mathbf{k}}^{\nu}(i,j)\}$ a mask for $M_{\mathbf{k}}$.

Example 3.1 . *For the Courant element $M = M_{(1,1,1,0)}$ we find*

$$\begin{aligned} q(w,z) &= (1+w)(1+z)(1+wz) \\ &= 1 + w + z + 2wz + w^2 z + wz^2 + w^2 z^2 \end{aligned}$$

This means that

$$\begin{aligned} M(u,v) =& M(2u,2v) + M(2u-1,2v) + M(2u,2v-1) + 2M(2u-1,2v-1) \\ &+ M(2u-2,2v-1) + M(2u-1,2v-2) + M(2u-2,2v-2). \end{aligned}$$

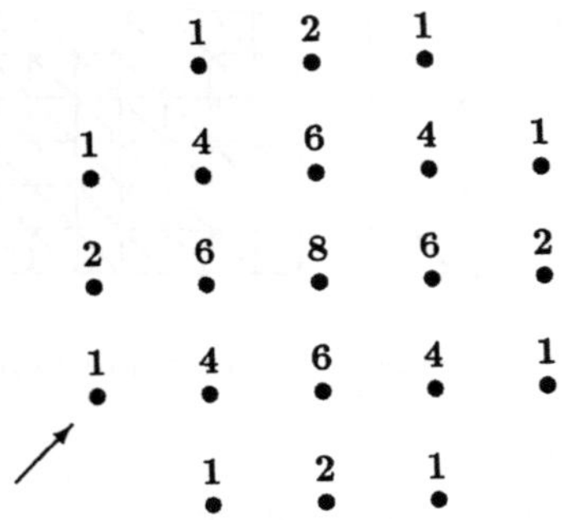

Figure 3.2. *The mask of the C^3-quartic box spline $M_{(2,2,1,1)}$*

The mask of the quartic C^3 box spline $M_{(2,2,1,1)}$ is shown in Figure 3.2. The mask is computed by expanding the function $(1+w)^2(1+z)^2(1+wz)(1+w/z)$. The arrow indicates the position of the origin.

Suppose $p \in S_{\mathbf{k}}$. From (3.1) it follows that we also have $p \in S_{\mathbf{k},\nu}$ for all $\nu \in \mathbb{N}$. Therefore,

$$p(u,v) = \sum_{m,n} c_{m,n} M_{\mathbf{k}}(u-m, v-n) = \sum_{i,j} b_{i,j} M_{\mathbf{k}}(\nu u - i, \nu v - j), \tag{3.3}$$

for certain coefficients $c_{m,n}$ and $b_{i,j}$. Using (3.1) we obtain a relation between the two sets of coefficients. Indeed, since

$$p(u,v) = \sum_{m,n} c_{m,n} \sum_{i,j} \beta^{\nu}_{\mathbf{k}}(i,j) M_{\mathbf{k}}(\nu(u-m)-i, \nu(v-n)-j)$$

we find after rearranging sums

$$b_{i,j} = \sum_{m,n} \alpha^{\nu}_{m,n,\mathbf{k}}(i,j) c_{m,n}, \tag{3.4}$$

where

$$\alpha^{\nu}_{m,n,\mathbf{k}}(i,j) = \beta^{\nu}_{\mathbf{k}}(i-\nu m, j-\nu n) \tag{3.5}$$

is called a *discrete box spline.*

Discrete box splines have many properties similar to box splines.

Theorem 3.2 . *We have*

$$(i)\quad \alpha^{\nu}_{m,n,\mathbf{k}}(i,j) \geq 0, \qquad m,n,i,j \in \mathbb{Z}, \quad \nu \in \mathbb{N},$$

$$(ii)\quad \sum_{m,n \in \mathbb{Z}^2} \alpha^{\nu}_{m,n,\mathbf{k}}(i,j) \equiv 1, \qquad (i,j) \in \mathbb{Z}^2.$$

Proof: Since all coefficients in the q polynomial are nonnegative we have $\beta_{\mathbf{k}}^{\nu}(i,j) \geq 0$, for all i, j, and (i) follows. From Theorem 2.22 we have

$$\sum_{m,n} M_{\mathbf{k}}(u-m, v-n) \equiv 1 \equiv \sum_{i,j} M_{\mathbf{k}}(\nu u - i, \nu v - j), \quad (u,v) \in \mathbb{R}^2.$$

Taking $c_{m,n} = b_{i,j} \equiv 1$ in (3.3) we obtain (ii). ■

The α's also satisfy a simple recurrence relation which is a discrete analog of (2.10). For the following theorem we arrange the standard directions in $\mathbf{X}_\mu$ such that $\mathbf{x}^1 = (1,0)^T$ and $\mathbf{x}^2 = (0,1)$. We then set

$$\alpha_{m,n}(i,j \,|\, \mathbf{X}_\mu) = \alpha_{m,n,\mathbf{k}}^{\nu}(i,j).$$

Theorem 3.3 . *The recurrence relation for discrete box splines is given by*

$$\alpha_{m,n}(i,j \,|\, \mathbf{X}_\mu) = \frac{1}{\nu} \sum_{\ell=0}^{\nu-1} \alpha_{m,n}(i - \ell x_\mu, j - \ell y_\mu \,|\, \mathbf{X}_{\mu-1}), \tag{3.6}$$

with

$$\alpha_{m,n}(i,j \,|\, \mathbf{X}_2) = \begin{cases} 1, & \text{if } (i,j) \in [\nu m, (\nu+1)m) \times [\nu n, (\nu+1)n) \cap \mathbb{Z}^2, \\ 0, & \text{otherwise.} \end{cases} \tag{3.7}$$

Proof: Combining (2.41) and (3.5) we obtain (3.6). To show (3.7) we first observe that $q(w,z \,|\, \mathbf{X}_2) = \sum_{m=0}^{\nu-1} \sum_{n=0}^{\nu-1} w^m z^n$. Hence,

$$\beta(i,j \,|\, \mathbf{X}_2) = \begin{cases} 1, & \text{if } i,j = 0,1,\ldots,\nu-1, \\ 0, & \text{otherwise,} \end{cases}$$

and (3.7) follows. ■

Based on the previous discussion we obtain a simple, fast, and stable *line averaging algorithm* for computing the $b_{i,j}$'s in (3.4) from the $c_{m,n}$'s. In this algorithm we first (cf. 3.7)) make ν^2 copies of each $c_{i,j}$ to obtain coefficients $b_{i,j}^2$. For $\ell = 3, \ldots, \mu$ we average successively in the directions $\mathbf{x}^3, \ldots, \mathbf{x}^\mu$ using the formula

$$b_{i,j}^{\ell} = \frac{1}{\nu} \sum_{r=0}^{\nu-1} b_{i-rx_\ell, j-ry_\ell}^{\ell-1}.$$

Usually the averaging is repeated several times with $\nu = 2$ rather than one application with a large ν.

Given

$$p(u,v) = \sum_{i,j} c_{i,j} M_{\mathbf{k}}(u-i, v-j),$$

the following algorithm computes $b_{i,j} = b_{i,j}^{\mu}$ such that

$$p(u,v) = \sum_{i,j} b_{i,j} M_{\mathbf{k}}(2u-i, 2v-j).$$

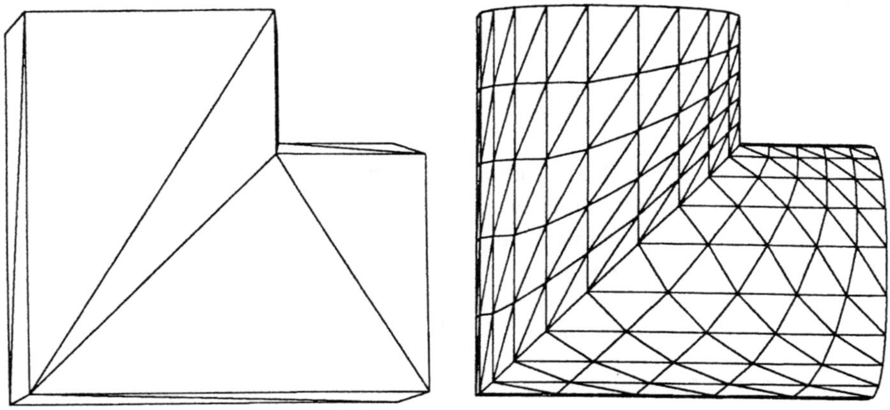

(a) (b)

Figure 3.3. **(a)** Coefficients in $\mathbb{R}^3$ of a C^0–quadratic box spline surface **(b)** The coefficients of the same surface after knotline refinement with $\nu = 5$

Algorithm 3.4

For $r, s = 0, 1$

$$b^2_{2i+r,2j+s} = c_i, \qquad \text{all } i, j$$

For $\ell = 3, \ldots, \mu$

$$b^\ell_{i,j} = (b^{\ell-1}_{i-x_\ell,j-y_\ell} + b^{\ell-1}_{i,j})/2, \qquad \text{all } i, j$$

This algorithm is fast, numerically stable and easy to program. If it is repeated several times we obtain a sequence of coefficients $b_{i,j}$ which converges to the exact surface.

Figure 3.3(a) and 3.3(b) show coefficients of a box spline surface before and after knotline refinement.

3.2 The Box Spline Recurrence Relation

Evaluation of box splines can also be based on a recurrence relation. We recall the notation used in the differentiation formula, where

$$\mathbf{X}^i_\mu = \{\mathbf{x}^1, \ldots, \mathbf{x}^{i-1}, \mathbf{x}^{i+1}, \ldots, \mathbf{x}^\mu\}.$$

We also recall that $M(u, v \,|\, \mathbf{X}^i_\mu)$ is not always well defined.

Theorem 3.5 . *Suppose all the direction vectors* $\mathbf{X}_\mu^i$ *contain at least two linearly independent direction vectors. Then for each* $(u,v) \in \mathbb{R}^2$

(vi) $M(u,v\,|\,\mathbf{X}_\mu) = \frac{1}{\mu-2}\sum_{i=1}^{\mu}[\omega_i M(u,v\,|\,\mathbf{X}_\mu^i) + (1-\omega_i)M(u-x_i, v-y_i\,|\,\mathbf{X}_\mu^i)]$,

where the ω_i, are any numbers satisfying the equations

$$u = \sum_{i=1}^{\mu} \omega_i x_i, \qquad v = \sum_{i=1}^{\mu} \omega_i y_i. \tag{3.8}$$

A proof of (vi) based on the distributional definition of box splines can be found in [33].

Example 3.6 . *Consider again* $M(u,v) = M(u,v\,|\,\mathbf{X}_3)$ *with* $\mathbf{X}_3 = (\binom{1}{0}, \binom{0}{1}, \binom{1}{1})$. *Given* (u,v) *the equations to determine* $\omega_1, \omega_2, \omega_3$ are

$$u = \omega_1 + \omega_3, \qquad v = \omega_2 + \omega_3.$$

A simple solution is $\omega_1 = u, \omega_2 = v, \omega_3 = 0$. *Let* $M^i(u,v) = M(u,v\,|\,\mathbf{X}_\mu^i)$. *Then the recurrence relation takes the form*

$$\begin{aligned} M(u,v) = uM^1(u,v) + (1-u)M^1(u-1,v) + vM^2(u,v) \\ + (1-v)M^2(u,v-1) + M^3(u-1,v-1), \end{aligned}$$

where

$$\begin{aligned} M^1(u,v) &= M(u,v\,|\begin{pmatrix}0\\1\end{pmatrix}, \begin{pmatrix}1\\1\end{pmatrix}), \\ M^2(u,v) &= M(u,v\,|\begin{pmatrix}1\\0\end{pmatrix}, \begin{pmatrix}1\\1\end{pmatrix}), \\ M^3(u,v) &= M(u,v\,|\begin{pmatrix}0\\1\end{pmatrix}, \begin{pmatrix}1\\0\end{pmatrix}). \end{aligned}$$

Using (2.21) and the notation from Figure 2.5 we find

$$\begin{aligned} M(u,v) &= u\chi_{A\cup C} + (1-u)\chi_{D\cup F} + v\chi_{B\cup D} + (1-v)\chi_{C\cup E} + \chi_{E\cup F} \\ &= u\chi_A + v\chi_B + (u+1-v)\chi_C + (v+1-u)\chi_D + (2-v)\chi_E + (2-u)\chi_F. \end{aligned}$$

This is the same as what we found in (2.23).

An algorithm for the evaluation of box splines based on the recurrence relation (vi) is numerically stable, since the sum consists of only convex combinations of box splines.

The right hand side of (vi) contains at least $\mu+1$ terms, and thus a straightforward implementation gives rise to a heavy bulk of work, at least for large μ. Moreover, box splines are defined as distributions, and the recurrence relation treats them as such. This makes the evaluation of box splines on joints between polynomial pieces very difficult. With the method presented in the next section this problem does not occur.

3.3 Evaluation via Cone Splines

Since cone splines are fairly easy to determine explicitly (cf. Theorems and Examples 2.41 ,2.42 , 2.43 and [20]), Theorem 2.51 provides a simple method for computing box splines. We present here an algorithm (cf. [20]) which to given multiplicities $\mathbf{k} = (k_1, \ldots, k_r), r \leq 4$, computes all box splines $M_{i,j} = M_{\mathbf{k}}(u-i, v-j)$ which can be nonzero at the point (u, v). We let $\mathbf{x}^i = (x_i, y_i), i = 1, 2, \ldots, \mu$ be the $\mu = \sum_i k_i$ direction vectors of $M_{\mathbf{k}}$.

Algorithm 3.7

$$\mu_1 = \mu - k_2 - 1$$
$$\mu_2 = \mu - k_1 - 1$$
$$w_1 = u - \lfloor u \rfloor + \mu_1$$
$$w_2 = v - \lfloor v \rfloor + \mu_2 - k_4$$

for $j = 0, \ldots, \mu_2$
  for $i = 0, \ldots, \mu_1$
    $M_{i,j} = C_{\mathbf{k}}(w_1 - i, w_2 - j)$

for $\ell = 1, \ldots, \mu$
  for $i = 0, \ldots, \mu_1$
    for $j = 0, \ldots, \mu_2$
      $M_{i,j} = M_{i,j} - M_{i+x_\ell, j+y_\ell}$

The numbers μ_1 and μ_2 are the number of cone splines which have to be computed to evaluate all non-zero box splines at a fixed point (u, v). To initialize the procedure we also translate the point (u, v) to (w_1, w_2), which becomes a point in the half open square $[\theta_1 - 1, \theta_1) \times [\theta_2 - 1, \theta_2)$, where

$$\theta = \begin{pmatrix} \theta_1 \\ \theta_2 \end{pmatrix} = \sum_{i=1}^{4} k_i \mathbf{d}^i.$$

The following comments are useful:

- The algorithm is numerically unstable for large μ, since differences have to be taken to get from cone spline values to box spline values.
- For the most common choices of multiplicities k_1, k_2, k_3 and k_4 explicit expressions for the cone splines are easily obtained and programmed, This makes the algorithm easy to program.
- The bulk of the work is the evaluation of the cone spline values. However, the computational time is no more that of $O(\mu_1 \mu_2)$.
- The algorithm can easily be extended to evaluate box splines in more that two variables (cf. [20]).

Box spline values can also be computed by the use of an explicit representation in terms of the Bézier net of each polynomial piece of the box spline. This is done by several authors, see [3,13,44,45].

4. Approximation methods

In this section we are going to present some local approximation schemes using box splines. In particular we will concentrate on the construction of bivariate box spline surfaces. An approach based on least squares is described in Section 6.

4.1 Shape Preserving Properties

We first recall from [13] the shape preserving properties for univariate cardinal splines. Given the grid $G = \mathbb{Z}/\nu$. It is convenient to defined centered cardinal splines N_k by

$$N_k(u) = M_k(u + k/2).$$

Translating the cardinal spline in Definition 2.1 we find

$$N_k(u) = \int_{-1/2}^{1/2} N_{k-1}(u-t)dt \tag{4.1}$$

and

$$N_1(u) = \begin{cases} 1, & \text{if } -1/2 \leq u < 1/2, \\ 0, & \text{otherwise.} \end{cases}$$

Since $M_k(u) = M_k(k-u)$ (cf. Theorem 2.37) it can be shown that $N_k(u)$ takes it maximum at the origin.

Let $\lambda = \{\lambda_i, i \in \mathbb{Z}\}$ be a set of local linear functionals, and let f denote some function on $\mathbb{R}$. By local we shall mean that $\lambda_i f$ uses information about the function f only in a neighbourhood of the point i/ν where $N_k(\nu u - i)$ takes it maximum. We define a spline approximation to f by

$$Vf = \sum_{i \in \mathbb{Z}} \lambda_i(f) N_k(\nu u - i). \tag{4.2}$$

Such approximation schemes are called quasi-interpolation schemes and we are going to study these in some more detail in Section 4.2, see also [13] and [15].

Now, let

$$\lambda_i(f) = f(i/\nu). \tag{4.3}$$

In this case the formula in (4.2) becomes the so-called Schoenberg variation diminishing spline approximation to the function $f(x)$. In the following we denote this approximation by $V_k f(x)$. By using property (v) in Theorem 2.2, see [5], it can be shown that an approximation order of $O(1/\nu^2)$ is obtained. Note that $1/\nu$ is the relative spacing between the knots in the uniform knots sequence over which the shifted B-spline $N_k(\nu u - i)$ are defined. The following theorem shows that the approximation $V_k f$ in (4.2) preserves certain shape properties of f.

Theorem 4.1 .
(i) If $f \geq 0$ then $V_k f \geq 0$.
(ii) If f is monotone then $V_k f$ is monotone.
(iii) If f is convex then $V_k f$ is convex.

Proof: The piecewise linear approximation

$$V_2 f(x) = \sum_{i \in \mathbb{Z}} f(i/\nu) N_2(\nu u - i)$$

certainly preserves positivity, monotonicity and convexity of the function f itself. By the definition of the cardinal B-splines given in (4.1) we have that

$$V_{k+1} f(u) = \int_{-1/2}^{1/2} V_k f(u-t) dt,$$

which implies, that positivity, monotonicity and convexity are preserved for all k. ■

Now, let us study these shape properties for bivariate box spline surfaces As for the univariate case, let us first define shifted box splines $N_{\mathbf{k}}(\mathbf{u})$. Analogous to Definition 2.15 we have the following definition.

Definition 4.2 . *Suppose $\mathbf{X}_\mu = (\mathbf{x}^1, \mathbf{x}^2, \ldots, \mathbf{x}^\mu)$ with $\mathbf{x}^i = (x_i, y_i)$ are $\mu \geq 2$ vectors in $\mathbb{R}^2$ with $\mathbf{x}^1$ and $\mathbf{x}^2$ linearly independent. The function $N(\mathbf{u} \,|\, \mathbf{X}_\mu) : \mathbb{R}^2 \to \mathbb{R}$ is defined recursively by*

$$N(\mathbf{u} \,|\, \mathbf{X}_\mu) = \int_{-1/2}^{1/2} N(\mathbf{u} - t\mathbf{x}^\mu \,|\, \mathbf{X}_{\mu-1})\, dt, \tag{4.4}$$

with

$$N(\mathbf{u} \,|\, \mathbf{X}_2) = \begin{cases} 1/|\det(\mathbf{X}_2)|, & \text{if } \mathbf{u} \in [\mathbf{X}_2[, \\ 0, & \text{otherwise,} \end{cases} \tag{4.5}$$

and where

$$[\mathbf{X}_\mu[= \{t_1 \mathbf{x}^1 + \ldots + t_\mu \mathbf{x}^\mu \ : \ -1/2 \leq t_j < 1/2, \quad 1 \leq j \leq \mu\}. \tag{4.6}$$

We observe that

$$N_{\mathbf{k}}(\mathbf{u}) = M_{\mathbf{k}}(\mathbf{u} + \theta_{\mathbf{k}}/2)$$

where

$$\theta_{\mathbf{k}} = \sum_{i=1}^{4} k_i \mathbf{d}^i$$

is the upper right corner of the support $[\mathbf{E}^{\mathbf{k}}]$ of $M_{\mathbf{k}}(\mathbf{u})$. As in Definition 2.19 we use the notation $N_{\mathbf{k}}(\nu\mathbf{u} - \mathbf{i}) = N(\nu\mathbf{u} - \mathbf{i} \,|\, \mathbf{X}_\mu)$ for box splines, where $\mathbf{k} = (k_1, k_2, k_3, k_4)$ are the multiplicities of the standard directions $\mathbf{d}^1$, $\mathbf{d}^2$, $\mathbf{d}^3$ and $\mathbf{d}^4$, respectively.

For the approximation of function or data in the univariate case we saw that certain shape properties of the function or the data was preserved by the approximation $V_k f$. The following simple example shows that we have to be more careful in the bivariate case.

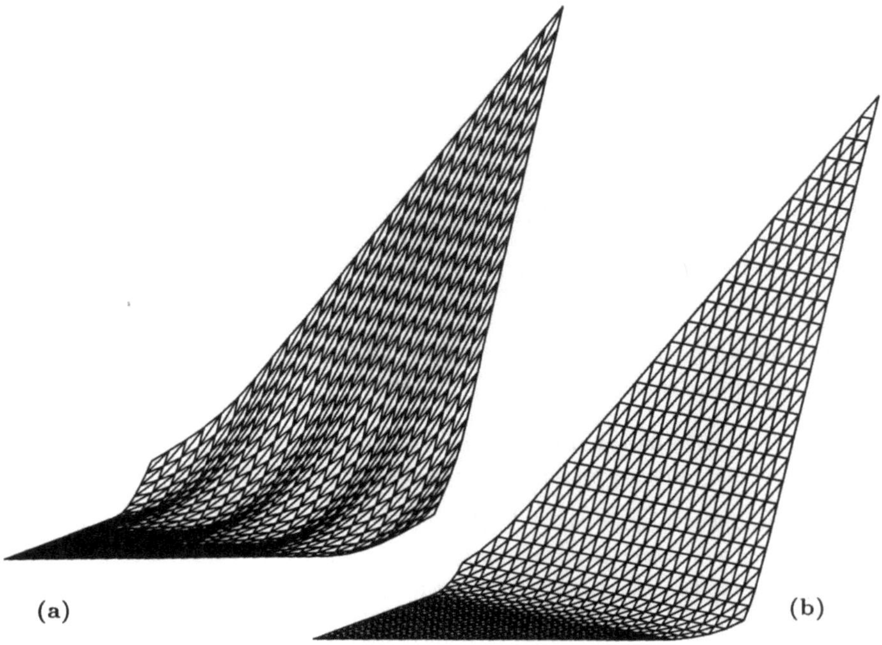

Figure 4.1. **(a)** The approximation Vf to f on the standard three direction mesh (cf. Figure 2.6) **(b)** The approximation Vf to f on the three direction mesh, where $\mathbf{d}^4$ is the diagonal direction instead of $\mathbf{d}^3$

Example 4.3 . *For* $\mathbf{k} = (2,2,1,0)$, *let us consider the box spline space* $S_{\mathbf{k},\nu}$, *see Definition 2.25 . This is a space of piecewise cubic polynomials on the standard three-direction grid shown in Figure 2.6. Moreover, let* f *be a function on* $\mathbb{R}^2$. *The bivariate analogue of the Schoenberg variation diminishing spline is given by*

$$Vf(\mathbf{u}) = \sum_{\mathbf{i} \in \mathbb{Z}^2} f(\mathbf{i}/\nu) N_{\mathbf{k}}(\nu\mathbf{u} - \mathbf{i}). \tag{4.7}$$

Now, let

$$f(\mathbf{u}) = f(u,v) = (u + v - 1)_+.$$

This function is monotone and convex in all directions. However, the approximation $Vf(\mathbf{u})$ *is not monotone and convex as is appearant in Figure 4.1(a). The three-direction grid in this case comes from the directions* $(\mathbf{d}^1, \mathbf{d}^2, \mathbf{d}^3)$, *while the diagonal crease in* $f(\mathbf{u})$ *is along the direction* $\mathbf{d}^4$. *So, if we instead consider the approximation (4.7) with* $\mathbf{k} = (2,2,0,1)$, *we obtain the surface shown in Figure 4.1(b), which is both monotone and convex in all directions.* ■

This example shows, that shape preserving properties of the surface Vf, depends on the grid over which the space of piecewise polynomials is defined. From now on we consider

Figure 4.2. A cubic box spline surface with 3484 coefficients approximating an octant of a sphere

$f_{\mathbf{i}} = f(\mathbf{i}/\nu)$ as given data and not necessarily samples on a function f. First, we give a definition of monotonicity and convexity in the bivariate case.

Definition 4.4 . *A set of data $\{z^{\mathbf{i}}, f_{\mathbf{i}}\}$, $z^{\mathbf{i}} \in \mathbb{R}^2$, is said to be monotone and/or convex in a direction $\mathbf{z}$, if there exists a triangulation of the set $\{z^{\mathbf{i}}\}$ such that the piecewise linear polynomial on this partition interpolating $f_{\mathbf{i}}$ at $z^{\mathbf{i}}$ is monotone and/or convex in the direction $\mathbf{z}$.*

Let $\mathbf{X}_\mu$ be as in Definition 2.15 with $\mu \geq 3$ and $\mathbf{x}^1, \mathbf{x}^2, \mathbf{x}^3$ beeing three pairwise non-parallel directions. Given data $(\mathbf{i}/\nu, f_{\mathbf{i}})_{\mathbf{i}\in\mathbb{Z}^2}$ we consider the box spline surfaces

$$V_\ell f(\mathbf{u}) = \sum_{\mathbf{i}\in\mathbb{Z}^2} f_{\mathbf{i}} N(\nu\mathbf{u} - \mathbf{i}|\mathbf{X}_\ell), \quad \ell = 3, \ldots \mu.$$

The piecewise linear surface $V_3 f(\mathbf{u})$ is called the control polygon of $V_\ell f(\mathbf{u})$, $\ell = 3, \ldots, \mu$ (cf. Definition 2.26). We can now state the shape properties of $V_\mu f(\mathbf{u})$ in terms of the control polygon $V_3 f(\mathbf{u})$.

Theorem 4.5 . *We have*

(i) if $V_3 f \geq 0$ then $V_\mu f \geq 0$,

(ii) if $V_3 f$ is monotone in a direction $\mathbf{z}$ then $V_\mu f$ is monotone in that direction.

(iii) if $V_3 f$ is convex in a direction $\mathbf{z}$ then $V_\mu f$ is convex in that direction.

Proof: We have

$$V_\ell f(\mathbf{u}) = \int_{-1/2}^{1/2} V_{\ell-1} f(\mathbf{u} - t\mathbf{x}^\ell)dt,$$

which, for $\ell = 4, \ldots, \mu$, preserves positivity, monotonicity and convexity of the control polygon $V_3 f(\mathbf{u})$. ■

We conclude this section by two examples where the operator V_ℓ is used for the approximation of an octant of a sphere and a set of discrete data, respectively. The first

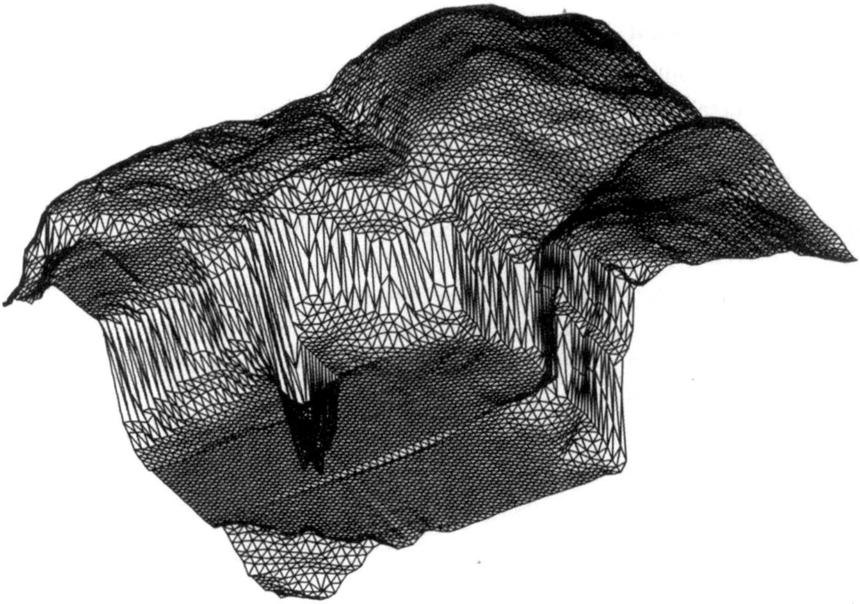

Figure 4.3. A cubic box spline surface approximating a terrain

example is a non-rectangular parametric surfaces in $\mathbb{R}^3$. This implies, that the coefficients $c_{\mathbf{j}}$, $\mathbf{j} \in J_{\mathbf{k},\nu}(\Omega)$ (cf. (2.32)) of the box spline in $S_{\mathbf{k},\nu}(\Omega)$ (cf. (2.31)) are points in $\mathbb{R}^3$.

Example 4.6 . *This example shows an approximation to an octant of sphere. The initial approximation is obtained by sampling the sphere at a huge number of points and taking these points as coefficients to a box spline surface defined on a triangular domain. Let Ω be equal to the triangle with corner $(0,0)$, $(1,0)$ and $(1,1)$. Using the C^1-cubic box splines $N_{\mathbf{k}}(80\mathbf{u}-\mathbf{i})$, with $\mathbf{k}=(2,2,1,0)$ we have by equation (5.2) in Section 5 that the dimension of $S_{\mathbf{k},\nu}(\Omega)$ is equal to 3484. (Take $\nu=80$, $n_1=n_2=\ell_2=1$ and $\ell_1=0$ in (5.2).) A uniform sample of 3484 points are taken from the octant. These points are ordered lexicographically as box spline coefficients starting in the corner $(0,0)$ of the triangular domain Ω. The approximation is shown in Figure 4.2.* ■

Example 4.7 . *The surface shown in Figure 4.3 is a box spline surface approximating a terrain given by 130×130 points. Let $\mathbf{k}=(2,2,1,0)$ and let the domain Ω be the unit square $[0,1]\times[0,1]$. Taking $\nu=128$, $n_1=n_2=1$ and $\ell_1=\ell_2=0$ in (5.2) we see that there is exactly 130×130 shifted C^1-cubic box splines $N_{\mathbf{k}}(128\mathbf{u}-\mathbf{i})$, defined on $\mathbb{Z}^2/128$, which is nonzero somewhere on Ω. The given points are taken as coefficients to the box spline surface. The distance between the given points and the approximation Vf is much less than the error caused by the digitizing process.* ■

4.2 A Quasi-interpolation Scheme

In this section we will give a brief discussion on quasi-interpolation schemes for box splines. More details can be found in [13,14], and references therein. Suppose $\mathbf{X}$ is a fixed set of directions vectors in $\mathbb{R}^d$. Let $\lambda_\mathbf{i}, \mathbf{i} \in \mathbb{Z}^d$ be local linear functionals and let f denote some function on $\mathbb{R}^d$. The quasi-interpolant to the function f is given by

$$Qf(\mathbf{u}) = \sum_{\mathbf{i}\in\mathbb{Z}^d} \lambda_\mathbf{i}(f) M(\mathbf{u} - \mathbf{i}|\mathbf{X}), \tag{4.8}$$

where the box spline $M(\mathbf{u} - \mathbf{i}|\mathbf{X})$ are shifted box splines defined on $\mathbb{Z}^d$, see Definition 2.28 . Given a discrete data set $F = \{f(\mathbf{i}), \mathbf{i} \in \mathbb{Z}^d\}$, we wish to find $\lambda_\mathbf{i}$ such that the distance between Qf and f is small at the grid points. To find a representative for $\lambda_\mathbf{i}$, we assume that the interpolant

$$Q_I f(\mathbf{u}) = \sum_{\mathbf{i}\in\mathbb{Z}^d} b_\mathbf{i} M(\mathbf{u} - \mathbf{i}|\mathbf{X}) \tag{4.9}$$

exist, that is, there exist coefficients $b_\mathbf{i}$ satisfying the conditions

$$\sum_{\mathbf{i}\in\mathbb{Z}^d} b_\mathbf{i} M(\mathbf{j} - \mathbf{i}|\mathbf{X}) = f(\mathbf{j}), \qquad \forall \mathbf{j} \in \mathbb{Z}^d. \tag{4.10}$$

We need the following definition.

Definition 4.8 . *With $A = (a(\mathbf{i}))_{\mathbf{i}\in\mathbb{Z}^d}$ and $B = (b(\mathbf{i}))_{\mathbf{i}\in\mathbb{Z}^d}$, the discrete convolution AB is given by*

$$(AB)(\mathbf{j}) = \sum_{\mathbf{i}\in\mathbb{Z}^d} b(\mathbf{i}) a(\mathbf{j} - \mathbf{i}), \quad \mathbf{j} \in \mathbb{Z}^d.$$

Substituting $\mathbf{q} = \mathbf{j} - \mathbf{i}$ we see that the discrete convolution comute, i.e., $AB = BA$.

Clearly (4.10) can be written as a discrete convolution $A_\mathbf{j} B = f(\mathbf{j})$ where $B = \{b_\mathbf{i}\}_\mathbf{i}$ and $A_\mathbf{j} = \{M(\mathbf{j} - \mathbf{i}|X)\}_\mathbf{i}$. Letting $A = \{A_\mathbf{j}\}_\mathbf{j}$ we can write (4.10) as

$$(I - H)B = F$$

where $H = I - A$, I is the identity in the convolution, i.e. $IB = B$ and $F = \{f(\mathbf{j})\}_\mathbf{j}$. Hence, we have the symbolic expression for the unknown coefficients B of the interpolant in (4.9), namely,

$$B = (I - H)^{-1} F. \tag{4.11}$$

If $||H|| < 1$ in some operator norm, we have the Neuman series expansion of $(I - H)^{-1}$ given by

$$(I - H)^{-1} = I + H + H^2 + H^3 + \dots.$$

By letting $P_\ell = I + H + H^2 + \ldots + H^\ell$ we obtain a well defined linear operator

$$Q_\ell f(\mathbf{u}) = \sum_{\mathbf{i}\in\mathbb{Z}^d} \lambda_\mathbf{i}(f) M(\mathbf{u} - \mathbf{i}|\mathbf{X}),$$

where $\lambda_\mathbf{i}(f) = (P_\ell F)(\mathbf{i})$. Assuming that an interpolant $Q_I f$ exist we have the following simple result.

Theorem 4.9 . *With* $A = \{\{M(\mathbf{j}-\mathbf{i}|X)\}_\mathbf{i}\}_\mathbf{j}$ *and* $F = \{f(\mathbf{j})\}_\mathbf{j}$ *we have that*

$$||Q_I f - Q_\ell f|| \leq ||H||^{\ell+1}||F||,$$

where

$$||Q_I f - Q_\ell f|| = ||\{(Q_I f - Q_\ell f)(\mathbf{j})\}_{\mathbf{j}\in\mathbb{Z}^d}||.$$

Proof: In terms of the convolution notation we have that

$$\{(Q_I f - Q_\ell f)(\mathbf{j})\}_{\mathbf{j}\in\mathbb{Z}^d} = (B - P_\ell F)A.$$

By (4.11), and since $P_\ell = I + H + H^2 + \ldots + H^\ell$ and $A = I - H$ we have that

$$\begin{aligned}\{(Q_I f - Q_\ell f)(\mathbf{j})\}_{\mathbf{j}\in\mathbb{Z}^d} &= \left[(I-H)^{-1} - (I + H + H^2 + \ldots + H^\ell)\right]F(I-H)\\ &= (I-H)^{-1}\left[I - (I-H)(I + H + H^2 + \ldots + H^\ell)\right]F(I-H)\\ &= (I-H)^{-1}H^{\ell+1}F(I-H) = H^{\ell+1}F,\end{aligned}$$

since the discrete convolution comute. It then follows that

$$\begin{aligned}||Q_I f - Q_\ell f|| &= ||H^{\ell+1}F||\\ &\leq ||H||^{\ell+1}||F||,\end{aligned}$$

which completes the proof. ■

Example 4.10 . *Evaluating the box spline* $M_\mathbf{k}(\mathbf{u}-\mathbf{i})$, $\mathbf{k} = (2,2,2)$ *at grid points in* $\mathbb{Z}^2$ *gives the following result:*

$$M_\mathbf{k}(\mathbf{u}-\mathbf{i}) = \begin{cases} 1/2, & \text{if } \mathbf{u} = \mathbf{i} + \theta_\mathbf{k}/2,\\ 1/12 \text{ or } 0, & \text{otherwise,}\end{cases}$$

where $\theta_\mathbf{k} = (4,4)^T$ *is the upper left corner of the support of* $M_\mathbf{k}(\mathbf{u})$. *Using the L-infinity norm and the fact that* $H = I - A$ *with* $A = \{\{M_\mathbf{k}(\mathbf{j}-\mathbf{i})\}_\mathbf{i}\}_\mathbf{j}$ *we obtain* $||H|| = 1/2$. *This implies, that*

$$||Q_I f - Q_\ell f|| \leq 2^{-(\ell+1)} \max_\mathbf{j} |f(\mathbf{j})|. \quad ■$$

5. Interpolation on Finite Domains

Existence and uniqueness questions for multivariate interpolation have received considerable attention, see [6,12,40] and references therein. In general, we are given an n dimensional space of functions

$$S = \text{span}\{\phi_1, \ldots, \phi_n\}$$

on a region Ω in $\mathbb{R}^d$. The Lagrange interpolation problem is to determine a subset P of Ω such that the n by n collocation matrix with elements $\phi_j(\mathbf{x}^i)$ is nonsingular for any choice of distinct $\mathbf{x}^1, \ldots, \mathbf{x}^n$ in P. In general, it is difficult to determine P. In [24], sufficient conditions on the distinct point set $\mathbf{x}^1, \ldots, \mathbf{x}^n$ are found for the interpolation problem using translates of bivariate C^0–quadratic box splines on a bounded convex set Ω in $\mathbb{R}^2$. In Section 5.1 we consider a related special case.

One way to obtain smooth box spline interpolants is to use cardinal interpolation schemes. Cardinal interpolation, i.e., interpolation with box splines at all grid points in $\mathbb{Z}^d$, is considered by several authors, see [34] and references therein. We will not consider it any further here.

5.1 Interpolation with $M_{(1,1,k_3)}$

In this and the next section we are going to look at interpolation with box splines on finite convex regions in the plane bounded by lines in the three direction mesh. The three direction mesh is the standard one given by the directions $\mathbf{d}^1$, $\mathbf{d}^2$ and $\mathbf{d}^1$, see Figure 2.5.

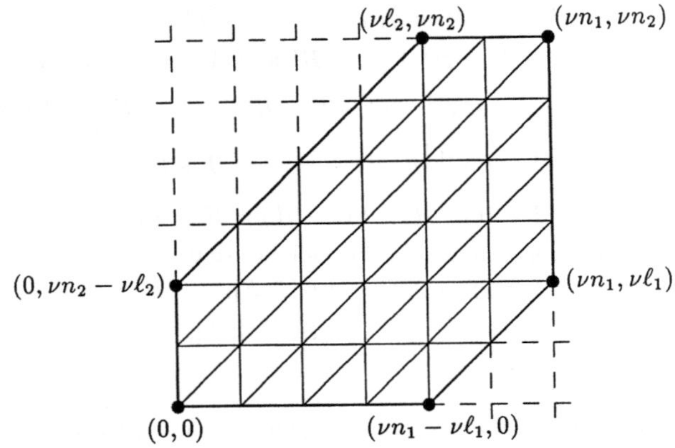

Figure 5.1. The domain Ω

The bounded convex regions of interest we denote by Ω, see Figure 5.1. Triangular-, trapeziodal-, pentagonal- and hexagonal domains are obtained by various values of the integers n_1, n_2, ℓ_1, ℓ_2 indicated in the figure. In terms of the integers n_1, n_2, ℓ_1, ℓ_2 and the multiplicities $\mathbf{k} = (k_1, k_2, k_3)$ we have from [24] that the dimension of the box spline space

$$S_{\mathbf{k}}, \nu(\Omega) = \Big\{ \sum_{\mathbf{i} \in J_{\mathbf{k},\nu}(\Omega)} c_{\mathbf{i}} M_{\mathbf{k}}(\nu\mathbf{u} - \mathbf{i}) \Big\} \tag{5.1}$$

is equal to

$$\begin{aligned} N =&(\nu n_1 + k_1 + k_3 - 1)(\nu n_2 + k_2 + k_3 - 1) \\ &- (\nu\ell_1 + k_3 - 1)(\nu\ell_1 + k_3)/2 - (\nu\ell_2 + k_3 - 1)(\nu\ell_2 + k_3)/2. \end{aligned} \tag{5.2}$$

The index set in (5.1) is

$$J_{\mathbf{k},\nu}(\Omega) = \{ \mathbf{i} \in \mathbb{Z}^2 \ : \ M_{\mathbf{k}}(\nu\mathbf{u} - \mathbf{i}) \neq 0 \ \text{ for some } \ \mathbf{u} \text{ in the interior of } \Omega \}.$$

The dimension N is equal to the number of elements in $J_{\mathbf{k},\nu}(\Omega)$.

In this section we will cover the result from [24] where interpolation points are given on grid lines in the mesh. It is also possible obtain non-singular collocation matrices if points are chosen outside the grid lines, but such choices are in most cases pathological, and we will therefore concentrate on the problem when interpolation points are on gridlines in spesified directions.

Suppose Ω is as in Figure 5.1 ℓ diagonals $\delta_1, \delta_2, \ldots, \delta_\ell$. Moreover, we assume that all interpolation points in Ω are given on the diagonals $\delta_1, \delta_2, \ldots, \delta_\ell$. If $\mathbf{p}^1$ and $\mathbf{p}^2$ are two points on δ_i, we say that $\mathbf{p}^1 < \mathbf{p}^2$ if $\mathbf{p}^1$ is below $\mathbf{p}^2$.

Assuming $\nu = 1$, we now consider a Hermite interpolation problem for the degree k_3 box spline $M_{\mathbf{k}}$, $\mathbf{k} = (1,1,k_3)$. It vanishes on all diagonals in the three direction mesh except one. Using the Definitions 2.1 and 2.15 it can be shown that

$$M_{(1,1,k_3)}(u,u) = M_{k_3+1}(u). \tag{5.3}$$

Thus, $M_{(1,1,k_3)}$ is equal to a univariate B-spline of order k_3+1 along the interior diagonal of its support.

Now for $i = 1, 2, \ldots, \ell$ let

$$\mathbf{x}^{i,1} \le \mathbf{x}^{i,2} \le \ldots \le \mathbf{x}^{i,\nu_i}$$

be given points on δ_i so that ν_i is the number of elements in $J_{\mathbf{k}}(\Omega)$ which lies on δ_i. Let, for suitable constants α_i and β_i

$$\phi_{i,j}(\mathbf{u}) = \phi_{i,j}(u,v) = M_{\mathbf{k}}(u - j - \alpha_i, v - j - \beta_i), \quad j = 1, 2, \ldots, \nu_i$$

be the nonzero box splines on δ_i ordered from left to right. We define linear functionals $\lambda_{i,j}$, $j = 1, 2, \ldots, \nu_i$ by

$$\lambda_{i,j} f = \max_p \{(D_u + D_v)^p f(\mathbf{x}^{i,j}) : \mathbf{x}^{i,j-p} = \mathbf{x}^{i,j}\},$$

where D_u and D_v denote partial derivatives in the u and v direction, respectively. We then have.

Theorem 5.1 . *For $i = 1, 2, \ldots, \ell$, suppose that at most k_3 of the points $\mathbf{x}^{i,1}, \ldots, \mathbf{x}^{i,\nu_i}$ are equal to one value. Then the above problem is unisolvent if and only if $\phi_{i,j}(\mathbf{x}^{i,j}) \neq 0$ for all i, j.*

Proof: Since $M_{\mathbf{k}}(\mathbf{u})$ is nonzero only over one diagonal we order the basis functions and the points $\mathbf{x}^{i,j}$ along diagonals of Ω moving from bottom left to top right. We continue this for all diagonals moving from bottom right to top left of Ω. Again, since each of the $M_{\mathbf{k}}$'s are nonzero only over one diagonal the collocation matrix for the interpolation problem is block diagonal

$$A = \operatorname{diag}(A_1, A_2, \ldots, A_\ell)$$

where each diagonal block A_i is a matrix of order ν_i with elements

$$\lambda_{i,p}\phi_{i,q}(\mathbf{x}^{i,p}), \quad p, q = 1, 2, \ldots \nu_i.$$

Now on each diagonal we have by (5.3) that $\phi_{i,1}, \ldots, \phi_{i,\nu_i}$ are consecutive shifted univariate B-splines of order $k_3 + 1$. By Theorem 4.67 in [49], A_i is nonsingular if and only if $\phi_{i,j}(\mathbf{x}^{i,j}) \neq 0$, $j = 1, 2, \ldots, \nu_i$. ∎

In [19,39], box splines of the above type have been used for modelling objects with slope discontinuities along diagonals. Similar results can be obtained for the C^0, degree k_1 and k_2 box splines $M_{(k_1,1,1)}$ and $M_{(1,k_2,1)}$, respectively. The bend shown in Figure 3.2(a) is constructed with this method.

5.2 Grid Point Interpolation with Smoothing

In the previous section we discussed interpolation on bounded regions, where we had the number of interpolation conditions equal to the dimension of the box splines space. If interpolation points are given on lines in one spesified direction in the three-direction grid, we found that interpolation using degree k_3 box splines $M_{(1,1,k_3)}$ reduces to a sequence of univariate interpolation problem.

One obvious restriction is that the box splines $M_{(1,1,k_3)}$ are C^0 across diagonals in the mesh, so a natural generalization is to try to formulate the interpolation problem using box splines which is C^1 or smoother. We know that box splines on a three-direction grid which is C^1 or smoother are nonzero somewhere along at least two lines in the mesh. This implies, that the problem can not be reduced to a sequence of univariate interpolation problems. The interpolation problem becomes global and it seems to be quite difficult to find necessary conditions to ensure unisolvence. One alternative choice is to interpolate only at grid points within a given bounded domain Ω.

The material in the rest of this section is taken from [1]. Let the domain $\Omega \in \mathbb{R}^2$ be as defined in the previous section, see Figure 5.1. We observe that the dimension of the box spline space $S_{\mathbf{k}}(\Omega)$ given in (5.2) exceed the number of grid points in Ω. Thus, we have extra degrees of freedom for the manipulation of an interpolant. One possibility is to introduce more interpolation points. However, this give us the problem of where to put these extra points, and it does not seem to be one natural way to introduce the necessary number of extra interpolation points. Alternatively we can define a smoothing functional and solve an optimization problem.

For a functional ψ on the box spline space $S_{\mathbf{k}}(\Omega)$ the idea is to select a grid point interpolant $f \in S_{\mathbf{k}}(\Omega)$ which minimize $\psi(f)$. Let $\mathbf{z}^1, \ldots, \mathbf{z}^\ell$ be an arbitrary subset of grid points in Ω. Given interpolation values ξ_i at $\mathbf{z}^i$, $i = 1, \ldots, \ell$, let

$$S_{\mathbf{k}}^I(\Omega) = \{f \in S_{\mathbf{k}}(\Omega) \ : \ f(\mathbf{z}^i) = \xi_i, \quad i = 1, \ldots, \ell\}.$$

$S_{\mathbf{k}}^I(\Omega)$ is the set of box spline functions

$$f(\mathbf{u}) = \sum_{\mathbf{j} \in J_{\mathbf{k}}(\Omega)} c_{\mathbf{j}} M_{\mathbf{k}}(\mathbf{u} - \mathbf{j}),$$

which interpolate ξ_i at grid points $\mathbf{z}^i$, $i = 1, \ldots, \ell$.

In order to define ψ, let

$$D_{\mathbf{d}^i}^2 f(\mathbf{u}) = \sum_{\mathbf{j} \in J_{\mathbf{k}}^{ii}(\Omega)} c_{\mathbf{j}}^{ii} M_{\mathbf{k}^{ii}}(\mathbf{u} - \mathbf{j}), \quad i = 1, 2, 3$$

denote the second order directional derivative in the standard directions $\mathbf{d}^1$, $\mathbf{d}^2$ and $\mathbf{d}^3$,respectively. The multiplicities for the derivative box splines $M_{\mathbf{k}^{ii}}(\mathbf{u}-\mathbf{j})$ are

$$\begin{aligned}\mathbf{k}^{11} &= (k_1-2, k_2, k_3),\\ \mathbf{k}^{22} &= (k_1, k_2-2, k_3),\\ \mathbf{k}^{33} &= (k_1, k_2, k_3-2).\end{aligned}$$

Using (2.26) its coefficients $c_{\mathbf{j}}^{ii}$ are

$$c_{\mathbf{j}}^{ii} = c_{\mathbf{j}-2\mathbf{d}^i} - 2c_{\mathbf{j}-\mathbf{d}^i} + c_{\mathbf{j}}. \tag{5.4}$$

The double indexing denote that second order differences is taken on the coefficients and that the multiplicity in direction $\mathbf{d}^i$ is reduced by 2.

This gives rise to the definition of a smoothing functional

$$\Psi(f) = \sum_{i=1}^{3} \int_{\Omega} \left(D^2_{\mathbf{d}^i} f(\mathbf{u})\right)^2 d\mathbf{u}.$$

A minimization of Ψ involves the computation of Gram matrices, so instead of minimizing Ψ we can define the more simple functional

$$\psi(f) = \sum_{i=1}^{3} \sum_{\mathbf{j}\in J_{\mathbf{k}^{ii}}(\Omega)} w_{\mathbf{j}}^{i} \left(c_{\mathbf{j}}^{ii}\right)^2, \tag{5.5}$$

where

$$w_{\mathbf{j}}^{i} = \int_{\Omega} M_{\mathbf{k}^{ii}}(\mathbf{u}-\mathbf{j})d\mathbf{u}.$$

Note that, by Theorem 2.21 (iii), $w_{\mathbf{j}}^{i} = 1$ for all i and $\mathbf{j}$ where the support of $M_{\mathbf{k}^{ii}}(\mathbf{u}-\mathbf{j})$ is contained in Ω. It can be shown that ψ is a good approximation to Ψ whenever f is a sufficiently smooth. This implies, that we can study the problem:

$$\begin{aligned}&\text{Find } f \in S^I_{\mathbf{k}}(\Omega) \text{ such that}\\ &\psi(f) \le \psi(g), \quad \forall g \in S^I_{\mathbf{k}}(\Omega).\end{aligned} \tag{5.6}$$

This quadratic optimization problem has a unique solution if and only if there exist at least three interpolation points $\mathbf{z}^1, \mathbf{z}^2, \mathbf{z}^3 \in \Omega$ which is not colinear. This is natural since the functional ψ vanish whenever f is planar.

To give a matrix formulation of (5.6) we have to order the basis $M_{\mathbf{k}}(\mathbf{u}-\mathbf{j}), \mathbf{j} \in J_{\mathbf{k}}(\Omega)$ and corresponding coefficients $c_{\mathbf{j}}$. Let $\mathbf{c} = (c_1, \ldots, c_n)$ be a one dimensional ordering of the coefficients (It is natural to order the basis along one of the three direction in the mesh). The matrix which performs second order differences on the coefficients in $\mathbf{c}$ can be written as

$$U = \begin{pmatrix} U^1 \\ U^2 \\ U^3 \end{pmatrix},$$

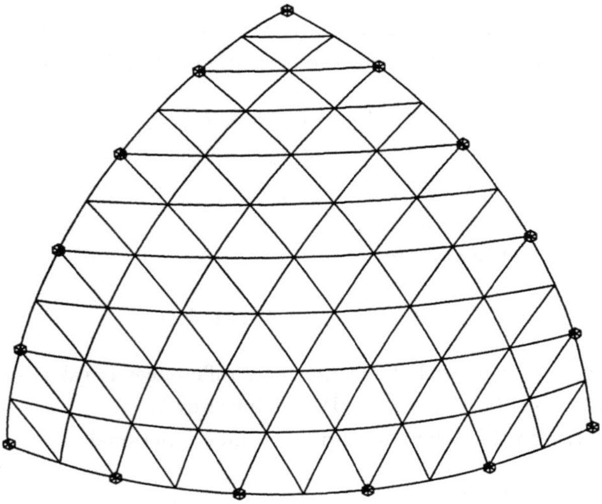

Figure 5.2. A triangular box spline surface interpolating at a set of grid points in Ω

where U^i is a $m_i \times n$ matrix with elements $1, -2, 1$ at positions according to equation (5.4). Here m_i is the number of elements in $J_{\mathbf{k}^{ii}}(\Omega)$. Moreover, let

$$W = \operatorname{diag}(w_1^1, \ldots, w_{m_1}^1, w_1^2, \ldots, w_{m_2}^2, w_1^3, \ldots, w_{m_3}^3).$$

We can now formulate (5.6) as the quadratic optimization problem:

$$\begin{aligned} &\text{Minimize } \mathbf{c}^T U^T W U \mathbf{c} \\ &\text{subject to } \mathbf{A}\mathbf{c} = \mathbf{b}, \end{aligned} \tag{5.7}$$

where $\mathbf{A}$ is the $\ell \times n$ collocation matrix corresponding to the ℓ interpolation conditions and where $\mathbf{b} = (\xi_1, \ldots, \xi_\ell)$ are the values to interpolate at the grid points $\mathbf{z}^1, \ldots, \mathbf{z}^\ell$, respectively. The Lagrange necessary condition for $\mathbf{c}$ to be a minimum for (5.7) is that we can find $\mathbf{y} \in \mathbb{R}^\ell$ such that

$$\begin{pmatrix} U^T W U & \mathbf{A}^T \\ \mathbf{A} & O \end{pmatrix} \begin{pmatrix} \mathbf{c} \\ \mathbf{y} \end{pmatrix} = \begin{pmatrix} \mathbf{0} \\ \mathbf{b} \end{pmatrix}. \tag{5.8}$$

The vector $\mathbf{y} = (y_1, \ldots, y_\ell)^T$ are the Lagrange multipliers. It is well known that the system (5.7) has a unique solution whenever $\mathbf{A}$ has full rank and when $U^T W U$ is positive definite on

$$L = \{\mathbf{a} \in \mathbb{R}^n \, : \, \mathbf{A}\mathbf{a} = 0\},$$

the null space of the matrix $\mathbf{A}$. This means that the problem has a unique solution if and only if there exist at least three interpolation points $\mathbf{z}^1, \mathbf{z}^2, \mathbf{z}^3 \in \Omega$, which is not colinear.

The following two examples show how this method can be used for the construction of non-rectangular surfaces. Both examples are parametric surfaces in $\mathbb{R}^3$. This implies,

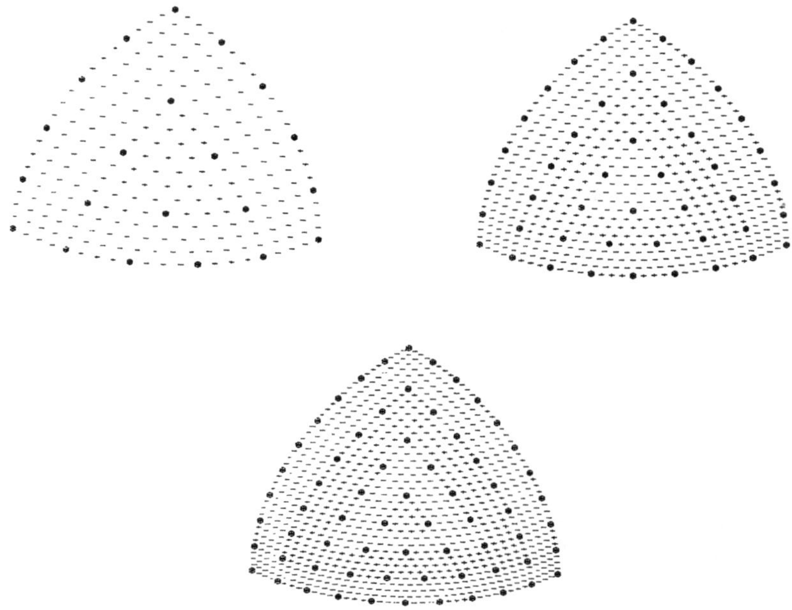

Figure 5.3. Three box spline surfaces interpolating points on an octant of a sphere. The interpolation points are given at all grid points of the respective domains.

that the coefficients $c_{\mathbf{j}}$, $\mathbf{j} \in J_{\mathbf{k},\nu}(\Omega)$ of the box spline in $S_{\mathbf{k},\nu}(\Omega)$ are points in $\mathbb{R}^3$. The interpolation is done in the $x-$, $y-$ and z–component, separately.

Example 5.2 . *We will interpolate at grid points given on the boundary of a triangular domain Ω with corners $(0,0)$, $(1,0)$ and $(1,1)$. The 15 interpolation points shown in Figure 5.2 are given along big-circles, such that the surface between the big-circles becomes an approximation to an octant of a sphere. Using translates of the C^2-quartic ($\mathbf{k} = (2,2,2)$) box spline $M_{\mathbf{k}}(5\mathbf{u} - \mathbf{i})$ defined on $\mathbb{Z}^2/5$, we have that the dimension of the box spline space $S_{\mathbf{k},5}(\Omega)$ given by (5.2) is equal to 42. By solving (5.7) we obtain the surface shown in Figure 5.2.* ■

Example 5.3 . *In addition to interpolation conditions corresponding to grid points on the boundary of Ω, we have given interpolation points corresponding to all interior grid points of Ω. Let Ω be a triangular domain with corners $(0,0)$, $(1,0)$ and $(1,1)$. Using the grid partitions corresponding to $\mathbb{Z}^2/5$, $\mathbb{Z}^2/8$ and $\mathbb{Z}^2/10$ we can define the shifted C^2-quartic ($\mathbf{k} = (2,2,2)$) box spline $M_{\mathbf{k}}(\nu\mathbf{u} - \mathbf{i})$, $\nu = 5, 8, 10$. The dimension of the box spline spaces $S_{\mathbf{k},\nu}(\Omega)$, for $\nu = 5, 8, 10$ are equal to 42, 75 and 102, respectively. Figure 5.3 shows the three C^2-quartic interpolants, where plus and minus signs superimposed* *on the surfaces indicate where each surface are outside and inside the unit sphere, respectively. We observe the nice oscillating behaviour of the interpolants. A numerical test indicates that the error is proportional to* ν^{-4}. ■

6. Approximation by Knotline Removal

In this section we describe a method, where a given box spline surface defined on a fine mesh is approximated by a box spline surface defined on a coarse mesh. The coarse mesh is contained in the fine mesh. The method can be regarded as an "inverse" refinement, where the approximation on the coarse mesh is found by a discrete least square method. With special emphasis on box splines defined on a three direction mesh we provide examples on smoothing, data reduction and approximation by knotline removal.

The basis for knotline removal are knotline insertion (refinement or subdivision), see Section 3.1. We will here give the basic ideas of the knotline removal approach on box spline surfaces. A detailed description is given in [22]. We restrict our treatment to bivariate surfaces. The technique can be extended to higher dimensions.

Given a domain $\Omega \in \mathbb{R}^2$ and integers $\nu \geq 1$ we consider the box spline spaces

$$S_{\mathbf{k},\nu}(\Omega) = \Big\{ \sum_{\mathbf{i}\in J_{\mathbf{k},\nu}(\Omega)} b_{\mathbf{i}} M_{\mathbf{k}}(\nu\mathbf{u} - \mathbf{i}) \; : \; b_{\mathbf{i}} \in \mathbb{R} \Big\}.$$

We assume that Ω is such that $S_{\mathbf{k}}(\Omega) = S_{\mathbf{k},1}(\Omega) \subset S_{\mathbf{k},\nu}(\Omega)$. This implies, that

$$f = \sum_{\mathbf{j}\in J_{\mathbf{k}}(\Omega)} c_{\mathbf{j}} M_{\mathbf{k}}(\mathbf{u} - \mathbf{j}), \; c_{\mathbf{j}} \in \mathbb{R}, \tag{6.1}$$

can be written as

$$f = \sum_{\mathbf{i}\in J_{\mathbf{k},\nu}(\Omega)} b_{\mathbf{i}} M_{\mathbf{k}}(\nu\mathbf{u} - \mathbf{i}), \; b_{\mathbf{i}} \in \mathbb{R}. \tag{6.2}$$

Here

$$b_{\mathbf{i}} = \sum_{\mathbf{j}} \alpha_{\mathbf{j}}(\mathbf{i}) c_{\mathbf{j}},$$

where $\alpha_{\mathbf{j}}(\mathbf{i}) = \alpha^{\nu}_{\mathbf{j},\mathbf{k}}(\mathbf{i})$ are the discrete box splines given in equation (3.5).

We can now state the following approach for knotline removal:
Given a box spline surface $f \in S_{\mathbf{k},\nu}(\Omega)$ defined on the grid $\mathbb{Z}^2/\nu$, say

$$f = \sum_{\mathbf{i}\in J_{\mathbf{k},\nu}(\Omega)} b_{\mathbf{i}} M_{\mathbf{k}}(\nu\mathbf{u} - \mathbf{i}), \; b_{\mathbf{i}} \in \mathbb{R}, \tag{6.3}$$

we wish to find new coefficients $\{c_{\mathbf{j}}\}$ to a function $g \in S_{\mathbf{k}}(\Omega)$ defined on the grid $\mathbb{Z}^2$, say

$$g = \sum_{\mathbf{j}\in J_{\mathbf{k}}(\Omega)} c_{\mathbf{j}} M_{\mathbf{k}}(\mathbf{u} - \mathbf{j}), \; c_{\mathbf{j}} \in \mathbb{R},$$

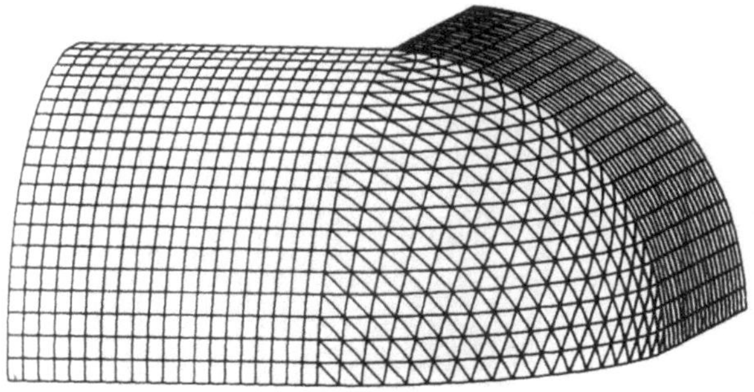

Figure 6.1. A cubic box spline surface with 89 coefficients approximating an octant of a sphere. The box spline surface is attached to two tensor product B–spline surfaces. A discussion on this is given in Section 7.

such that a certain distance between f and g is as small as possible. That is, we wish to approximate a function $f \in S_{\mathbf{k},\nu}$ with $g \in S_{\mathbf{k}}(\Omega)$. Using knotline insertion we can write $g \in S_{\mathbf{k}}(\Omega)$ as a function in $S_{\mathbf{k},\nu}(\Omega)$. Hence, we can minimize the distance between f and g in a convenient norm on $S_{\mathbf{k},\nu}(\Omega)$.

We define the ℓ_2-norm of $f \in S_{\mathbf{k},\nu}(\Omega)$ given by (6.1) as

$$||f||_{\ell^2} = \Big(\sum_{\mathbf{i} \in J_{\mathbf{k},\nu}(\Omega)} |b_{\mathbf{i}}|^2 \Big)^{1/2}.$$

Now, let $\mathbf{c} = (c_1, \ldots, c_n)^T$ be a convenient ordering of the box spline coefficients of $f \in S_{\mathbf{k}}(\Omega)$ given by (6.2) and let $\mathbf{b} = (b_1, \ldots, b_m)^T$ be a similar ordering of the box spline coefficients of $f \in S_{\mathbf{k},\nu}(\Omega)$. Then $\mathbf{b} = \mathbf{A}_\nu \mathbf{c}$, where $\mathbf{A}_\nu$ is the $m \times n$ knotline insertion matrix from $S_{\mathbf{k}}(\Omega)$ to $S_{\mathbf{k},\nu}(\Omega)$ given by

$$\mathbf{A}_\nu = \begin{pmatrix} \alpha_1(1) & \ldots & \alpha_n(1) \\ \vdots & & \vdots \\ \alpha_1(m) & \ldots & \alpha_n(m) \end{pmatrix}$$

Now, given $f \in S_{\mathbf{k},\nu}(\Omega)$, let $g^* \in S_{\mathbf{k}}(\Omega) \subset S_{\mathbf{k},\nu}(\Omega)$ denote the best approximation to f in the discrete least square norm ($||\cdot||_{\ell^2}$) on $S_{\mathbf{k},\nu}(\Omega)$. With $||v||_2 = \left(\sum |v|^2\right)^{1/2}$ (the ℓ^2 vector norm), the best approximation

$$g^* = \sum_{\mathbf{i} \in J_{\mathbf{k}}(\Omega)} c_{\mathbf{i}} M_{\mathbf{k}}(\mathbf{u} - \mathbf{i})$$

is given by the solution to least square problem

$$\min_{g \in S_{\mathbf{k}}(\Omega)} ||f - g||_{\ell^2} = \min_{\mathbf{c} \in \mathbb{R}^n} ||\mathbf{A}_\nu \mathbf{c} - \mathbf{b}||_2.$$

Figure 6.2. The cubic box spline face represented by $29 \times 47 = 1363$ coefficients. The plot is refined twice according to the method given in Section 3.1 with $\nu = 2$.

Example 6.1 . *Let us apply the above technique to the surface in Example 4.6. Let Ω be a triangle with corner $(0,0)$, $(1,0)$ and $(1,1)$. Using the shifted C^1-cubic box splines $M_{\mathbf{k}}(80\mathbf{u} - \mathbf{i})$, where $\mathbf{k} = (2,2,1)$, we recall that the dimension of $S_{\mathbf{k},\nu}(\Omega)$ is equal to 3484. A uniform sample of 3484 points are taken from an octant of the sphere. These points are ordered lexicographically as box spline coefficients starting in the corner $(0,0)$ of the three-sided domain Ω. From a subspace of $S_{\mathbf{k},80}(\Omega)$ we are going to find an approximation to the initial box spline surface. Keeping every tenth knotline in Ω, we obtain the subspace $S_{\mathbf{k},10}(\Omega)$ of $S_{\mathbf{k},80}(\Omega)$. By the formula (5.2) again, we have that the dimension of $S_{\mathbf{k},10}(\Omega)$ is equal to 89. Figure 6.1 shows the box spline approximation defined on Ω. The relative distance between the sphere and the box spline surface is less than 0.02. The non-reduced three-sided parametric surface is shown in Figure 4.2.*

From the shape preserving properties of splines discussed in Section 4.1 and basic properties of box splines it is easily seen that the initial box spline surface is a one-sided approximation to the sphere. Indeed, since the box spline surface is contained in the convex hull of its coefficients, we obtain a surface which is closer to the origin than the sphere itself. By estimating the maximum and minimum radial error between the initial surface and the sphere it is possible to find a better approximation by a sampling on a sphere with radius equal to the original radius plus the arithmetic mean of the maximum and minimum error.

To illustrate a possible use of this method we have constructed two tensor product B-spline surfaces which match with tangent plane continuity across two edges which is common with the three-sided box spline surface. A description of how to imbed box splines in a tensor product framework is given in [23,25,39]. Some more details on this is also given in Section 7. ■

In the above example we described a method for the construction of a box spline surface, which approximates a part of a function. This approximation method can be summarized as follows:

(i) We sample the function at a huge number of points and take these points as coefficients to a box spline surface defined on a suitable domain Ω. By the discussion in Section 4.1 this will give an approximation which in many cases has good shape preserving properties.

(ii) Step (i) will in general introduce redundant information on a surface with a huge number of coefficients. Using the knotline removal method we obtain a box spline approximation with "few" coefficients.

Example 6.2 . *The original data are points in* $\mathbb{R}^3$ *sampled by a sylindrical scanner from a bust. Originally,* $218 \times 362 = 78916$ *data points in* $\mathbb{R}^3$ *was given. These points was taken as coefficients to a* C^1*-cubic (*$\mathbf{k} = (2,2,1)$*) box spline surface in* $S_{\mathbf{k},8}(\Omega)$*, where* Ω *is the rectangle* $[0,27] \times [0,45]$*. The box spline approximation in* $S_{\mathbf{k}}(\Omega)$ *is shown in Figure 6.2. The number of coefficients is reduced to* $29 \times 47 = 1363$*. The relative maximum error between the box spline surface in* $S_{\mathbf{k},8}(\Omega)$ *and the approximation* $S_{\mathbf{k}}(\Omega)$ *is less that* 0.01. ■

Additional examples are given in [22].

7. Surface Modelling

An important problem in CAGD is the construction of non-rectangular surfaces. Tensor product B-spline surfaces are regarded as a powerful representation form in CAGD, but suffer restrictions because of its rectangular nature. In the previous sections we have discussed some methods for constructing box spline surfaces. In this section we will focus on problems which occur when box spline surfaces are imbedded within a tensor product framework. Moreover, we will use box splines for the construction of non-rectangular surfaces, which naturally arise when fillets and blends are constructed between tensor product surfaces. A more detailed discussion on this can be found in [23,25].

7.1 Matching Edges

The use of box spline surfaces together with tensor product B-spline surfaces requires a discussion on the problem of the construction of joints between the two surface types. We will consider the problem of extending a given box spline surface with tensor product surfaces. The converse is in general not possible since box splines are defined on uniform partitions and tensor products in general are defined on non-uniform partitions. We will therefore consider the following problems:

(i) The construction of a tensor product surface that exactly matches a partial derivative along an edge of a box spline surface.

(ii) The construction of a tensor product surface that mathces a linear combination of partial derivatives along an edge of a box spline surface, see Figure 5.2. In the literature this is refered to as geometric continuity or tangent plane continuity.

(iii) The construction of an overall tangent plane continuous surface consisting of tensor product surfaces, which completely surround a given box spline surface.

The differentiation formula for box spline is given in Theorem 2.24 . Given a box spline surface

$$f(\mathbf{u}) = \sum_{\mathbf{i}\in J_{\mathbf{k}}(\Omega)} c_{\mathbf{i}} M_{\mathbf{k}}(\mathbf{u}-\mathbf{i}), \quad \mathbf{u}\in \mathbb{R}^2,$$

we have the directional derivatives

$$(7.1) \qquad D_{\mathbf{d}^j} f(\mathbf{u}) = \sum_{\mathbf{i}\in J_{\mathbf{k}^j}(\Omega)} (c_{\mathbf{i}} - c_{\mathbf{i}-\mathbf{d}^j}) M_{\mathbf{k}^j}(\mathbf{u}-\mathbf{i}), \quad \mathbf{u}\in \mathbb{R}^2,\ j=1,2,3\ ,$$

where $\mathbf{k}^1 = (k_1-1, k_2, k_3)$, $\mathbf{k}^2 = (k_1, k_2-1, k_3)$ or $\mathbf{k}^3 = (k_1, k_2, k_3-1)$.

We note that cross-boundary partial derivatives of a tensor product surface are of the same degree and continuity as the boundary curve itself. Thus, to produce a tangent continuous match between a box spline surface and a tensor product B-spline surface we have to represent the boundary curve and the cross-boundary derivatives from the box spline surface on the same knotvector.

To explain this, let a C^2-quartic box spline surface be given. According to the derivative formula (7.1) the boundary curve is C^2-quartic and the cross-boundary partial derivatives are C^1-cubics. The C^2-quartic boundary curve can be written as a degree four B-spline curve with double knots. By inserting one knot at each double knot we obtain a degree four B-spline curve with triple knots. The C^1-cubic boundary curves of the cross-boundary derivative box spline surfaces can be written as degree three B-spline curves with double knots. By raising the degree ([18,42]) of the derivative curves we obtain degree four B-spline curves with triple knots. The boundary curve and the cross-boundary curves are now of the same degree, represented on the same knot vector, and we can construct a continuous match between the box spline surface and the tensor product B-spline surface.

The construction of tensor product B-spline surfaces that exactly matches only one partial derivative of box spline surfaces are in many cases not very useful. We will therefore discuss the construction of composite cross-boundary derivatives to obtain tangent plane continuity between a box spline surface and a tensor product surface.

Let Ω be a domain in $\mathbb{R}^2$ such that the boundary of Ω consists of lines in the three-direction grid. It follows that cross-boundary derivatives at a particular edge of the surface can be constructed as a linear combination of the partial derivatives in the two directions not parallel to the gridline along the edge. The edge of interest are parallel to one of the three standard vectors $\mathbf{d}^1$, $\mathbf{d}^2$ or $\mathbf{d}^3$. We want to construct a composite cross boundary derivative given by the two directions not parallel to the boundary. We denote these two directions by $\mathbf{x}$ and $\mathbf{y}$, where $\mathbf{x}, \mathbf{y} \in \{\mathbf{d}^1, \mathbf{d}^2, \mathbf{d}^3\}$. Let $\lambda_{\mathbf{x}}(t)$, $t \in [0,1]$ be the B-spline boundary curve of the derivative box spline surface $D_{\mathbf{x}}f$, and $\lambda_{\mathbf{y}}(t)$, $t \in [0,1]$ the B-spline boundary curve of the derivative box spline surface $D_{\mathbf{y}}f$. Without loss of generality we let $\lambda_{\mathbf{x}}$ and $\lambda_{\mathbf{y}}$ be B-spline curves defined on the same partition of the interval $[0,1]$. Now, let

$$C(t) = w_{\mathbf{x}}(t)\lambda_{\mathbf{x}}(t) + w_{\mathbf{y}}(t)\lambda_{\mathbf{y}}(t),$$

where $w_{\mathbf{x}}(t)$ and $w_{\mathbf{y}}(t)$ are suitable weight functions on the interval $[0,1]$. If $w_{\mathbf{x}}(t)$ and $w_{\mathbf{y}}(t)$ are polynomials of degree $\leq n$, it follows that $C(t)$ is a B-spline curve of degree $|\mathbf{k}|-3+n$, since $\lambda_{\mathbf{x}}$ and $\lambda_{\mathbf{y}}$ are degree $|\mathbf{k}|-3$ B-spline curves. With $w_{\mathbf{x}}(t) = (1-t)$ and $w_{\mathbf{y}}(t) = 1 - w_{\mathbf{x}}(t)$ we have that $C(t)$ is a degree $|\mathbf{k}|-2$ B-spline curve, where $C(0) = \lambda_{\mathbf{x}}$ and $C(1) = \lambda_{\mathbf{y}}$.

Since $C(t)$ is no longer a strictly partial derivative, but lies in the tangent plane of the surface f along the actual edge we obtain tangent plane continuity between the surfaces.

To completely surround a box spline surface with tensor product B-spline surfaces, such that the overall surface is G^1-continuous, we have to take care of the mixed partial derivatives, known as the twist vectors. At one corner of a box spline surface we have to attach three tensor product B-spline surfaces, two surfaces matching G^1 along each edge attached to the corner, and one surface which has the corner point in common with the box spline surface. To be able to construct tangent plane continuity at a corner between the three tensor product surfaces and the box spline surfaces, it is sufficient to ensure equal twist at the corner point of the four surfaces.

We have the derivative along the boundary of the box spline surface given by

$$\frac{\partial C(t)}{\partial t} = w'_{\mathbf{x}}(t)\lambda_{\mathbf{x}}(t) + w_{\mathbf{x}}(t)\lambda'_{\mathbf{x}}(t) + w'_{\mathbf{y}}(t)\lambda_{\mathbf{y}}(t) + w_{\mathbf{y}}(t)\lambda'_{\mathbf{y}}(t).$$

Evaluation at the endpoints $t = 0$ and $t = 1$ should give $\lambda'_{\mathbf{x}}(t)$ and $\lambda'_{\mathbf{y}}(t)$, which is the mixed partial derivatives (twist) of the box spline surface at the respective corners. To obtain this construction it is necessary that the weight functions $w_{\mathbf{x}}(t)$ and $w_{\mathbf{y}}(t)$ have zero derivatives at $t = 0$ and $t = 1$.

Let $w_{\mathbf{x}}(t) = 1 - 3t^3 + 2t^2$ and $w_{\mathbf{y}}(t) = 1 - w_{\mathbf{x}}(t)$. Then the twist of the surrounding surfaces at each corner of the box spline surface is equal to the twist of the box spline surface at the corner. To accomplish this we have to apply tensor product surfaces of degree $|\mathbf{k}|$ to surround a degree $|\mathbf{k}| - 2$ box spline surface to obtain an overall tangent plane continuous surface. Thus, to surround a C^2-quartic box spline surface we must use degree 6 tensor product B–spline surfaces.

7.2 Vertex Regions

In geometric modelling a frequently arising situation is the problem of constructing smooth blends in regions where a number of surfaces meet in one single point. Neighbourhoods of such regions are often called vertex regions. Figure 7.1(a) shows a vertex region where three tensor product surfaces meet in one single point.

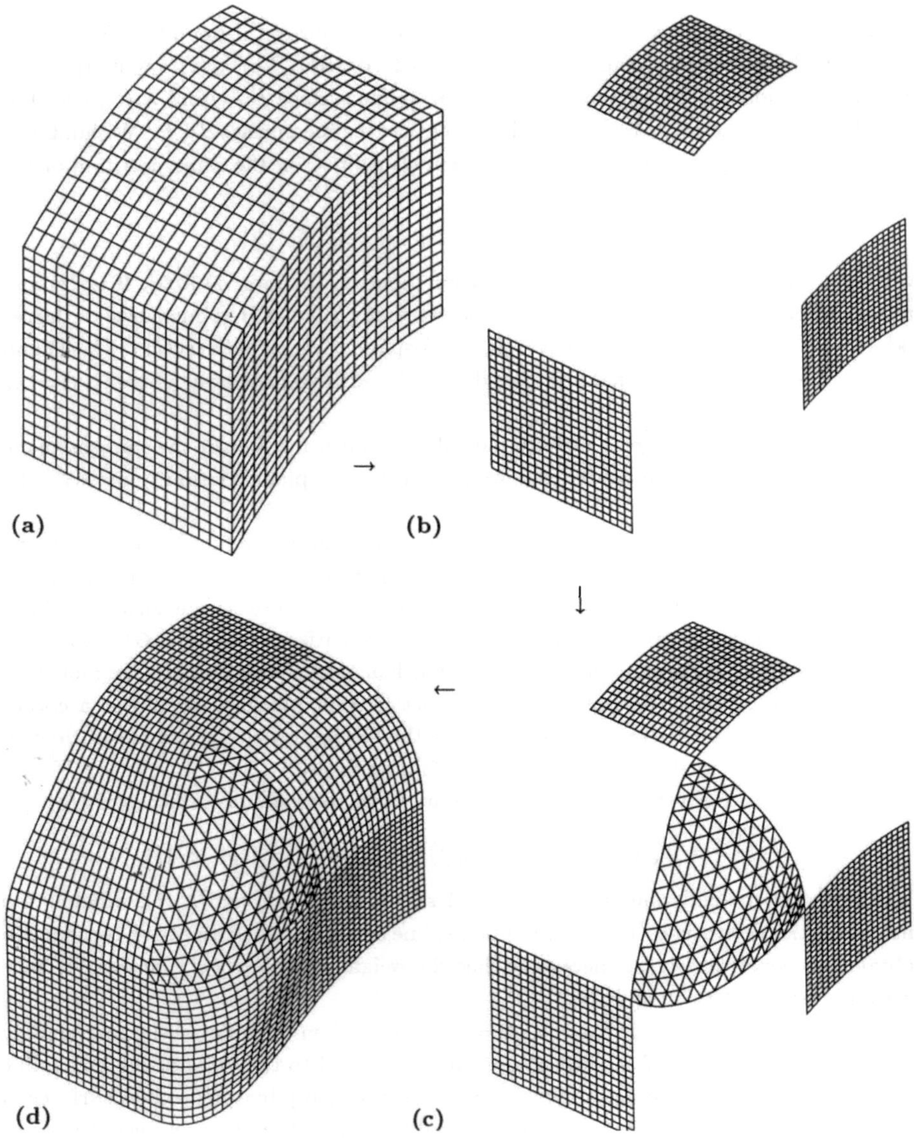

Figure 7.1. **(a)** Vertex region **(b)** Trimmed of vertex region with corner conditions **(c)** Vertex region with box spline surface in $S_{\mathbf{k}}(\Omega)$, $\mathbf{k} = (2, 2, 2)$ **(d)** Smoothly blended vertex region

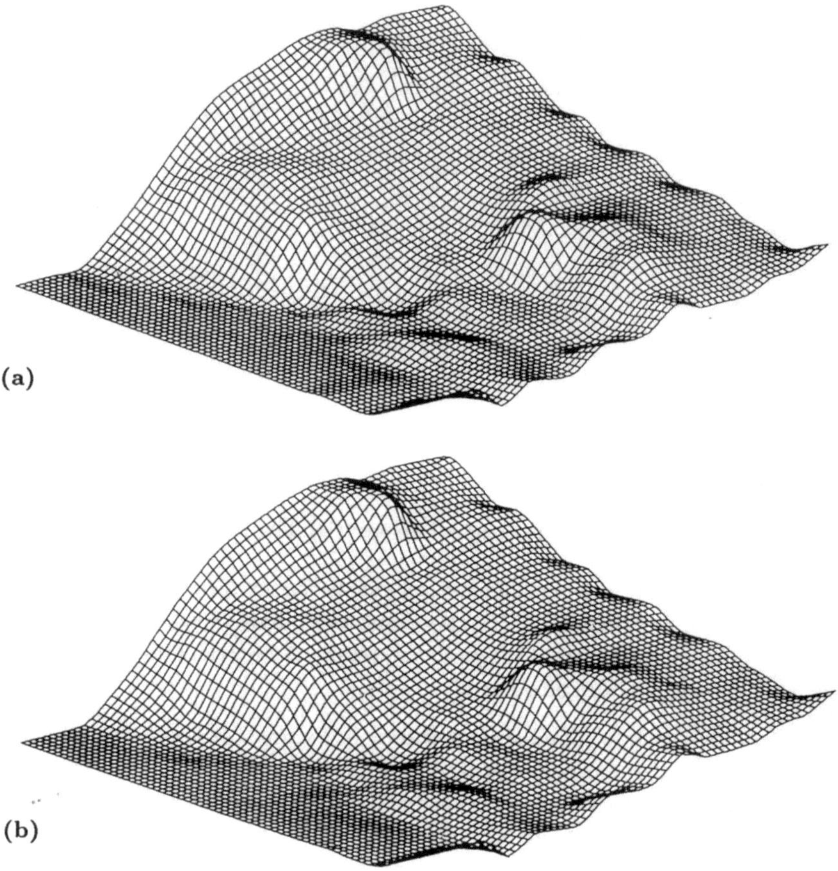

(a)

(b)

Figure 8.1. **(a)** Original terrain **(b)** The box spline reconstruction

The construction of a smooth blend at the vertex shown in Figure 7.1(a) will consist of a three-sided box spline surface at the vertex, surrounded by tensor product B–spline surfaces extending the box spline surface along the edges connected to the vertex. By constructing the box spline surface at the vertex region before constructing the surrounding tensor product surfaces we are able to generate a smoothly blended region which is tangent plane continuous. This is basically done in three steps.

- Positions, partial derivatives and twist derivatives are extracted at each corner where we are going to attach the box spline surface to the original tensor product surfaces, see Figure 7.1(b).
- We construct a box spline surface which match the corner conditions extracted in the previous step, see Figure 7.1(c). ([23])

(a) (b)

Figure 8.2. **(a)** Original image **(b)** The box spline reconstruction

- Finally, the box spline surface at the vertex region is surrounded by tensor product B-spline surfaces. The boundary conditions which ensure tangent plane continuity are described in the previous section. An overall tangent plane continuous blend is shown in Figure 7.1(d).

The box spline surface shown in Figure 7.1(b)–(d) are in $S_{\mathbf{k}}(\Omega)$, $\mathbf{k} = (2,2,2)$. Hence, it is a degree 4 surface, and using the method described in Section 7.1 we obtain tensor product surfaces of degree 6 in the directions along the edges connected to the box spline surface at the vertex.

8. A new Application

We conclude this paper by presenting two examples on new results for box splines. Using a special combination of knotline insertion and knotline removal we can use box splines to compress surfaces and images.

Figure 8.1(a) shows an original terrain modell. The terrain is given by 257×257 points on a regular grid. Figure 8.1(b) shows the reconstruction of the compressed box spline modell. The maximum deviation between the original and the reconstruction is less than 10 centimeters, while the difference in altitude in the terrain is approximately 25 meters. The storage requirements for the compressed modell corresponds to approximately 200 points.

Figure 8.2(a) shows an image. It is a 512×512 image with 256 graylevels (8 bits). Figure 8.2(b) shows a reconstruction from a box spline modell. The storage requirements for the box spline modell is approximately 4% of the requirements for the original image.

References

1. Arge, E. and M. Dæhlen, Grid point interpolation on finite regions by box splines, preprint.
2. Boehm, W., Subdividing multivariate splines, Comput. Aided Design **15** (1983), 345–352.
3. Boehm, W., Prautzsch, H., and P. Arner, On triangular splines, Constr. Approx. **2** (1987), 157–167.
4. de Boor, C., Splines as linear combinations of B-splines. A survey, in *Approximation Theory II*, G. G. Lorentz, C. K. Chui, and L. L. Schumaker (eds.), Academic Press, New York, 1976, 1–47.
5. de Boor, C., *A Practical Guide to Splines,* Springer-Verlag, New York, (1978).
6. de Boor, C., Multivariate approximation, in *The State of the Art in Numerical analysis,*A. Iserles and M.J.D. Powell (eds.), Claredon Press, Oxford 1987, 87–109.
7. de Boor, C. and R. DeVore, Approximations by smooth multivariate splines, Trans. Amer. Math. Soc. **276** (1983), 775–785.
8. de Boor, C.and K. Höllig, B–splines from parallelepipeds, J. Analyse Math. **42** (1982/83), 99–115.
9. de Boor, C. and K. Höllig, Bivariate box splines and pp functions on a three-direction mesh, J. Comput. Appl. Math. **9** (1983), 13–28.
10. Cavaretta, A. S. and C. A. Micchelli, Subdivision algorithms, in *Mathematical Methods in Computer Aided Geometric Design*, T. Lyche and L. Schumaker (eds.), Academic Press, N. Y., 1989, 115–153.
11. Chaikin, G. M., An algorithm for high speed curve generation, Computer Graphics and Image Processing **3** (1974), 346–349.
12. Cheney, E. W., *Multivariate approximation theory: Selected topics*, CBMS–NSF Reg. Conf. series in applied mathematics 51, SIAM, Philadelphia, 1986.
13. Chui, C. K., *Multivariate Splines,* CBMS-NSF, Reg. Conf. series in Appl. Math., SIAM, Philadelphia, (1988).
14. Chui, C. K., Diamond, H. and L. Raphael, Interpolation by multivariate splines, Math. Comp. **51**, (1988), 203–218.
15. Chui, C. K., Diamond, H. and L. Raphael, Shape-preserving quasi-interpolation and interpolation bybox spline surfaces, J. Comput. Appl. Math. **25**, (1989), 169–198.
16. Cohen, E., Lyche, T., and R. Riesenfeld, Discrete box-splines and refinement algorithms, Comput. Aided Geom. Design **1** (1984), 131–148.
17. Cohen, E., Lyche, T., and R. Riesenfeld, Cones and recurrence relations for simplex splines, Constr. Approx. **3** (1987), 131–141.
18. Cohen, E., Lyche, T., and L. L. Schumaker, Degree raising for splines, J. Approx. Theory **46** (1986), 170–181.

19. Dæhlen, M., An example of bivariate interpolation with translates of C^0-quadratic box splines on a three direction mesh, Comput. Aided Geom. Design **4** (1987), 251–255.

20. Dæhlen, M., On the evaluation of box splines, in *Mathematical Methods in Computer Aided Geometric Design*, T. Lyche and L. Schumaker (eds.), Academic Press, N. Y., 1989, 167–179.

21. Dæhlen, M., Box splines and applications of polynomial splines, dissertation, University of Oslo, Research Report 128, Institute for informatics, ISBN 82 7368 032 0, (1989).

22. Dæhlen, M., Knotline removal on box spline surfaces, preprint.

23. Dæhlen, M., Modelling with box spline surfaces, in *Curve and Surface Design*, H. Hagen (ed.), SIAM Conference Series, to appear.

24. Dæhlen, M. and T. Lyche, Bivariate interpolation with quadratic box splines, Math. Comp. **51** (1988), 219–230.

25. Dæhlen, M. and V. Skytt, Modelling non-rectangular surfaces using box splines, in *Mathematics of Surfaces III*, D. Handscomb (ed.), (1989), 285–300.

26. Dahmen, W., Multivariate B-splines — Recurrence relations and linear combinations of truncated powers, in *Multivariate Approximation Theory*, W. Schempp and K. Zeller (eds.), Birkhäuser, Basel, (1979), 64–82.

27. Dahmen, W. and C. A. Micchelli, Recent progress in multivariate splines, in *Approximation Theory IV*, C. K. Chui, L. L. Schumaker, and J. Ward, (eds.), Academic Press, New York, (1983), 27-121.

28. Dahmen, W. and C. A. Micchelli, Subdivision algorithms for generation of of box-spline surfaces, Comput. Aided Geom. Design **1** (1984), 115–129.

29. Dahmen, W., and C. A. Micchelli, On the local linear independence of translates of a box spline, Studia Math. **82** (1985), 243–262.

30. Dahmen, W., Dyn, N., and D. Levin, On the convergence rates of subdivision algorithms for box spline surfaces, Constr. Approx. **1** (1985), 305–322.

31. Goodman, T.N.T., Polyhedral splines, in *Computation of Curves and Surfaces*, W. Dahmen, M. Gasca and C.A. Micchelli (eds.), Kluwer, Dordrecht, 1990, 347–382.

32. Höllig, K., Box-splines, in *Approximation Theory V*, C. K. Chui, L. L. Schumaker, and J. Ward, (eds.), Academic Press, (1986), 71–95.

33. Höllig, K., Box-splines surfaces, in *Mathematical Methods in Computer Aided Geometric Design*, T. Lyche and L. Schumaker (eds.), Academic Press, N. Y., 1989, 385–402.

34. Jetter, K., A short survey on cardinal interpolation by box splines, in *Topics in Multivariate Approximation*, C. K. Chui, L. L. Schumaker, and F. Utreras, (eds.), Academic Press, Boston, (1987), 125–139

35. Jia, Rong-Qing, Linear independence of translates of box splines, J. Approx. Theory **40** (1984), 158–160.

36. Lane, J. M. and R. F. Riesenfeld, A theoretical development for the computer genera-

tion of piecewise polynomial surfaces, IEEE Trans. on Pattern Analysis and Machine Intelligence **2** (1980), 34–46.

37. Lane, J. M. and R. F. Riesenfeld, A geometric proof of the variation diminishing property of B-spline approximation, J. Approx. Theory **37** (1983), 1–4.
38. Lyche, T., Discrete B-splines and conversion problems, in *Computation of Curves and Surfaces*, W. Dahmen, M. Gasca and C.A. Micchelli (eds.), Kluwer, Dordrecht, 1990, 117–134.
39. Mueller, T. I., Geometric Modelling with multivariate B–splines, dissertation, Dept. of Comp. Science, Univ. of Utah, (1986).
40. Micchelli, C. A., Algebraic aspects of interpolation, in *Approximation Theory*, AMS Symposium in Applied Mathematics 36, Providence, R. I., 1986, 81–102.
41. Prautzsch, H., Unterteilungsalgorithmen für multivariate splines, – Ein geometrische zugang, dissertation, Technische Universität Braunschweig, (1984).
42. Prautzsch, H., Degree elevation of B-spline curves, Comput. Aided Geom. Design **1** (1984), 193–198.
43. Ron, A., Exponential box splines, Constr. Approx. **4**, (1988), 357–378.
44. Sabin, M. A., The use of piecewise forms for the numerical representation of shapes, dissertation, Hungarian National Academy of Sciences, 1977.
45. Sablonnière, P., A catalog of B-splines of degree ≤ 10 on a three direction mesh, Report ANO-132, Université de Lille, France, 1984.
46. Schoenberg, I. J., Contributions to the problem of approximation of equidistant data by analytic functions, Quart. Appl. Math. **4** (1946), 45–99, 112–141.
47. Schoenberg, I. J., letter to Philip J. Davis dated May 31, 1965.
48. Schoenberg, I. J., *Cardinal Spline Interpolation*, Reg. Conf. series in Applied Mathematics **12** SIAM, Philadelphia, 1973.
49. Schumaker, L. L., *Spline functions: Basic Theory,* Whiley & Sons, New York, (1981).

Advanced Computer Graphics Techniques for Volume Visualization

J. L. Encarnaçao M. Frühauf M. Göbel K. Karlsson

Fraunhofer-Arbeitsgruppe Graphische Datenverarbeitung
Wilhelminenstraße 7, D-6100 Darmstadt, F.R. Germany

telephone: ++ 49 6151 1000 29
telefax: ++ 49 6151 1000 99
e-mail: jle@agd.fhg.de

1 Introduction

Computer Graphics provides techniques for visualizing graphical primitives which are derived from geometric elements. Data in real life - gathered by sensors - may represent a continuum. Data in real life does not correspond directly to abstract geometric elements.

Special rendering methods have been developed in Computer Graphics to make data visible, and manifold data conversion techniques are applied to the original data samples.

Modern risc-performance workstations fairly well equipped with main memory allow users to handle massive data sets and to make a particular data set visible sample by sample. Depending on the quality of the visual representation of the data, the image generating time takes some seconds or even minutes - still far from an interactive scientific analysis of data.

This paper describes the various steps to be executed in the visualization process of volume data, i.e., to make data values in a 3-dimensional space visible. Tools are discussed that provide user assistance in allowing a more comfortable interactive handling of this data.

2 Volume Data

Volume data is produced by many different sources in many application areas. The data sets are the result of a simulation or observation process. Typical application areas according to [1,2] are:

- molecular modeling,
- medical imaging (diagnostic medicine, radiation treatment planning, orthopedic prosthesis, etc.),
- mathematics,
- geosciences (cartography, geology, meteorology),
- space exploration,
- astrophysics,
- computational fluid dynamics,
- finite element analysis,
- urban design (simulation of landscape architectures) and
- environment engineering.

The dimension of the data set is equal to the smallest number of independent variables which span the set. The data set can depend on one dimension only, e.g., the amplitude of a signal versus time, or on two dimensions, e.g., topographical maps. Data sets that depend on three spatial dimensions are called volume data. As we live in three-dimensional space these data sets are very common. Time appears often as the fourth dimension. In many cases there may be more independent variables. In material strength analysis, e.g., the material properties and the type of load on the test specimen are non-spatial independent variables. In these multi-dimensional cases, it is appropriate to restrict the visualized data to a lower-dimensional subset in order to obtain a good understanding of the data.

The dependent variables of the data can be scalar, vector or tensor fields. For example, the temperature in a room is a scalar field depending on spatial location. Wind velocity is a vector field, as its quantity has a direction as well as a magnitude. Further examples of volume data sets are stacks of two-dimensional images, e.g., CT or MRI scans. The field values in volume data are generally available at a finite number of locations in space. These may be organized into a regular grid with constant distances between the grid points, or in an irregular mesh.

3 Converting Data

The volume visualization process consists of different data transformations:

- Filtering the basic data from the simulation or measurement into another form which provides more information, perhaps less voluminous and more suitable for the chosen rendering technique (filtering data into data).
- Mapping the resulting data into geometric primitives which can be rendered (mapping data into geometry).
- Rendering the geometric data into pictures (rendering geometry into images).

A main task in the volume visualization process is the mapping of the scientific data into a visual representation [3]. A data set can be visualized in many different ways. Each representation technique illustrates another aspect of the data, and the user must be able to select the form of presentation. It is necessary to display different data in the same manner, e.g., to compare simulated data with measurements, as well as the same data in different ways simultaneously.

Computation of surface representations from volume data is an example of filtering data into data. In general we distinguish surface-oriented and volume-oriented representations of volume data. The information content of surface-oriented representations is lower than that of volume-oriented representations, since the former neglect the inhomogenity of the volumes' interior. Surface computation is performed by defining iso-contours and iso-surfaces in the volume data set. Mapping to geometric primitives is carried out by triangulating the surfaces.

Converting velocity vector fields to flow lines is performed by integration. Data conversion depends on the primitive type chosen for visual presentation of the data. In using volume elements, the size of the achieved digital image defines the necessary granularity of the volume data. A specific granularity is obtained by interpolation.

4 Primitives for Volume Visualization

Primitives used in volume visualization are dots, lines (polylines, splines), surfaces (meshes, NURBS), raster images, volume elements (voxels) and CSG. A field in three-dimensional space can be mapped to these primitives in various ways.

Zero-dimensional primitives, dots, can be used to visualize a scattered data grid where the values are mapped to marker type or color. A marker type often used to visualize vector fields is the arrow, where the length, direction and color of the arrow indicate the field values. Dots can also build up clouds and surfaces, indicating electron density or particle traces, for example.

One-dimensional primitives, curves, can be used to show contours, either the geometric contours of the object under study as a wire-frame, or contours in the data values as iso-curves. In fluid dynamics curves are used to visualize flow lines.

With two-dimensional primitives, surfaces, more complete displays of the object and its values can be created. The field values can be mapped into colors and textures, indicating the values on the surface of the object, which is rendered using polygons. The field can also be mapped into geometry through the construction of iso-surfaces. The third spatial dimension is normally indicated by shading. However, when the field values are mapped into colors, the intensity variation in the colors due to shading can make it difficult to interpret the image. Color itself represents three dimensions (e.g., hue, intensity and saturation), but it is not clear if and how those three dimensions can be transformed to a perceptible extension of dimensionality for the human viewer.

The surface representation has the disadvantage that only a small portion of the data is being displayed. The amount of information is reduced to only a surface in the volume. One technique to get more information out of the image is transparent display of iso-surfaces in the volume. An iso-surface calculated at a higher field value is completely contained within those at lower values. For each surface a unique color has to be assigned, in order to distinguish between the surfaces in the image. The color of each pixel in the image is a blend of the colors of all surfaces that are mapped into this pixel, weighted with a transparency factor and the geometrical position of each surface. The quality of the resulting image depends mainly on the algorithm used to blend the colors [4,5].

Another way to display the interior of an object is to cut away parts of the volume. The user walks through the volume by interactive definition of cutting planes, which can be combined with each other to define complex regions.

For the visualization of volume data, three-dimensional primitives have been introduced. If the field values are organized in a regular grid, the volume can be intersected into small cubic volume elements (voxels). Volume data is located at their vertices. These volume primitives implicitly define the geometry of the recorded object, i.e., the location of voxels with corresponding data values may define the shape of an object (e.g., a bone in medical applications). No surfaces are explicitly defined in the volume data set. Thus, normals for shading cannot be calculated using the vector product. Special voxel rendering techniques are mainly used in medical visualization but are appropriate also in other disciplines. The data have to be transformed into the regular voxel grid using interpolation. In order not to lose information, the resolution in the voxel grid must at least correspond to the finest grid in the original data mesh. As the volume primitives contain complete information on the volume under study, the mapping of data into geometric primitives, which is normally a computing-intensive task, is not needed. The volume can be directly displayed using volume rendering techniques.

As rendering of zero- to two-dimensional primitives is well known from other disciplines of computer graphics, we will emphazise volume rendering methods in the following section.

5 Volume Rendering

Volume rendering is defined as volume visualization directly from volume primitives and not via surface primitives. This provides the advantage of minimal precomputation or user interaction for finding and computing object surfaces in volume data. Volume rendering is the only method that preserves the entire data set and thus allows a detailed exploration of the volume. Shaded pseudo-3D representations of volume data give the most realistic impression to the scientist. As a matter of fact, such picture production is the most costly volume rendering technique.

Volume rendering, i.e., producing pictures from volume data, is one aspect of scientific visualization. Many techniques and algorithms for volume rendering are being developed, and this can be confusing to the researcher trying to apply them [6,7,8,9,10]. There are no absolutely optimal techniques for volume rendering, not only because techniques and algorithms are still in a stage of intensive development, but because the volumetric data produced by different applications are not at all similar to each other, scientists from different disciplines have different requirements concerning interactivity or details in produced pictures, and they use computers of different performances.

Two primary goals are set up for volume rendering. One is to provide optimal visual recognition of the displayed objects. Therefore, shading is necessary for visual recognition of 3D objects. The other goal is to provide the natural appearance of a real object. Due to the discreteness of the individual cubic voxels in a voxel representation, projections have a rough appearance with many aliasing effects. Volume rendering should provide the illusion of a real, smooth object, while retaining the minute details of the original data set or object.

Even with volume rendering techniques, iso-surfaces in volume data sets are visualized, although they are not geometrically modeled. Many discrete shading methods that overcome the lack of surface-normal information in the voxel representation have been developed for volume rendering. The appropriate method to recover the surface's structure to estimate the surface's normal depends mainly on the type and the properties of the data source (e.g., images, electron density maps, finite element grids or analytically defined objects). But it depends also on the time that can be spent for image generation.

A simple volume visualization technique direct from volume data is line-of-sight-integration or additive reprojection. This method does not emphasize surfaces in the volume data set and no shading is calculated. Thus, X-ray-like images are produced by that technique.

6 Discrete shading techniques

In volume rendering, the shading method and the accuracy of computing the surface normals is essential for the image quality. In volume data, an object is represented as a number of voxels with certain properties. Therefore no object surface is explicitly defined or represented as, e.g., a triangular mesh. Chen et al. [11] have introduced the following terminology: "We refer to a physical object's actual surface as an object surface; we refer to the surface made up of faces of voxels as a boundary surface".

The simplest method is depth-only (z-buffer) shading. The color of an object surface is a linear function of the distance of the boundary surface from the observer. This method produces unacceptable results, since it suppresses many details that are essential in medical applications, for instance. Other, more sophisticated shading techniques make use of the voxel's neighborhood to estimate the surface orientation. They are the z-buffer gradient shading [12], contextual shading [11] and gray-level gradient shading [13] as well as some derivates [14]. The z-buffer gradient of a voxel is calculated from the z-coordinate of its corresponding neighbors.

Other approaches in calculating surface normals are the Marching Cubes [15] and the biquadratic local surface interpolation [16].

Marching Cubes creates triangle models of constant density surfaces from 3D medical data. It uses a divide-and-conquer approach to locate the surface in a logical cube created from eight pixels. Fourteen patterns of up to four triangles are used to form the surface. The gray-level gradient is calculated to define the normals at the triangles' vertices. The triangles are diplayed using traditional rendering algorithms. Marching Cubes produces 'smooth' surfaces consisting of a huge number of triangles that have to be rendered.

In biquadratic local surface interpolation, the 3x3x3 neighborhood of a voxel provides sufficient constraints on the integral of a biquadratic function over the column's base to solve its specific coefficients. The locally interpolated surfaces are used in a ray tracer.

In the following we introduce the z-buffer gradient as well as the gray-level gradient in detail and describe our experiences using both of them in volume rendering applications.

The inclination of the object surface is estimated from the boundary surface either in the object space or in the image space. For surface normal calculation in the object space we use the gray-level gradient. This method is based on the fact that in the image data (CT, MR, ultrasound) the gray value of a pixel is an average of the material properties (e.g., material density) in a (small) region or volume. The gray-level gradient for a voxel at the location x, y, z can be computed from 6 neighbors as

$$G_x = f(x-1, y, z) - f(x+1, y, z)$$

$$G_y = f(x, y-1, z) - f(x, y+1, z)$$

$$G_z = f(x, y, z-1) - f(x, y, z+1)$$

or from all 26 neighbors as

$$G_x = \sum_{j=-1}^{1} \sum_{i=-1}^{1} (f(x-1, y+i, z+j) - f(x+1, y+i, z+j))$$

$$G_y = \sum_{j=-1}^{1} \sum_{i=-1}^{1} (f(x+i, y-1, z+j) - f(x+i, y+1, z+j))$$

$$G_z = \sum_{j=-1}^{1} \sum_{i=-1}^{1} (f(x+i, y+j, z-1) - f(x+i, y+j, z+1))$$

For volume data computed from finite element data, calculating the gray-level gradient fails. In that case the surface normal has to be calculated from the position of adjacent voxels belonging to the object or the object's surface. Thus the z-buffer gradient works only in segmented volumes [17]. The z-buffer gradient is calculated as the central difference at location x, y, z, by using the z-buffer created while projecting the data in the image space.

$$G_x = \frac{z(x+dx, y) - z(x-dx, y)}{2dx}$$

$$G_y = \frac{z(x, y+dy) - z(x, y-dy)}{2dy}$$

$$G_z = -1$$

Computing the surface normal in the image space using the z-buffer gradient with dx=dy=1 implies less computing costs than using the gray-level gradient in the object space. But gradient calculation using the z-buffer is restricted to a small number of different gradients, especially if the surface inclination is less than 45°. Thus, little details of the object's geometry can be visualized using that method. If we had the choice, as we do in medical imaging, we would use the gray-level gradient because of the capability to visualize more details [18]. However, there are applications in volume visualization where we are more interested in material properties than in the object's shape. In that case the z-buffer gradient works well.

7 Projection strategies

Most of the volume rendering techniques are based on scalar data fields arranged in a regular grid of three spatial dimensions. If data is not available in that format, it can be transformed into a suitable format in preprocessing, e.g., transforming finite element data into a regular grid consisting of cubes. A scalar value at a node in the volume represents either the relative average of data over a volume element (voxel), or is a discrete data value at that location. Volume rendering techniques can be divided in two main projection strategies. The first class is known as back-to-front (BTF) or front-to-back (FTB) strategies [19]. In back-to-front strategy, the volume space is traversed from the voxel furthest from the observer to the voxel nearest to the observer. When a voxel is visited, it is projected into the image space and its distance from the observer is recorded in the z-buffer.

The second class comprises raycasting techniques. Figure 1 shows the order of the main tasks in volume rendering by BTF and raycasting.

We have used the z-buffer gradient with BTF projection and the gray-level gradient with raycasting, although the gray-level gradient can be used with BTF projection. But the computation of the z-buffer gradient is ineffective in raycasting, since it may depend on a large region of the volume which is not determined in advance.

The calculated surface normal is used in Lambert, Gouraud, or Phong shading [20] for image production.

In raycasting techniques, rays are sent into the data volume according to the viewing direction [5]. At first we want to explain the following terminology: data location, data color, data opacity, voxel color, voxel opacity, sample location, sample color, sample opacity, and pixel color (Fig. 3).

Data is located at integer positions in the regular grid; those positions are referred to as data locations. Data color and data opacity assignment is performed using

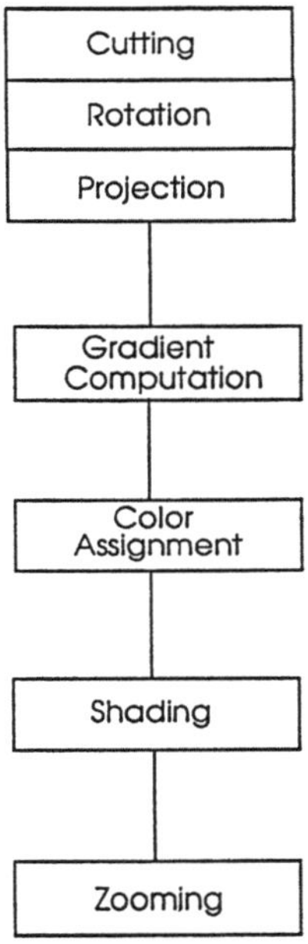

Fig. 1a: Volume rendering pipeline (BTF)

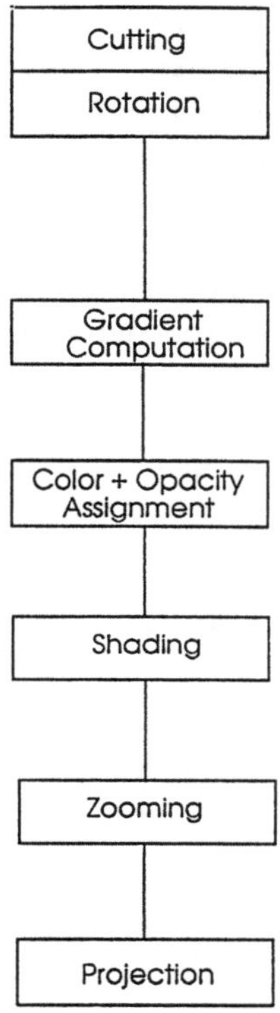

Fig. 1b: Volume rendering pipeline (Raycasting)

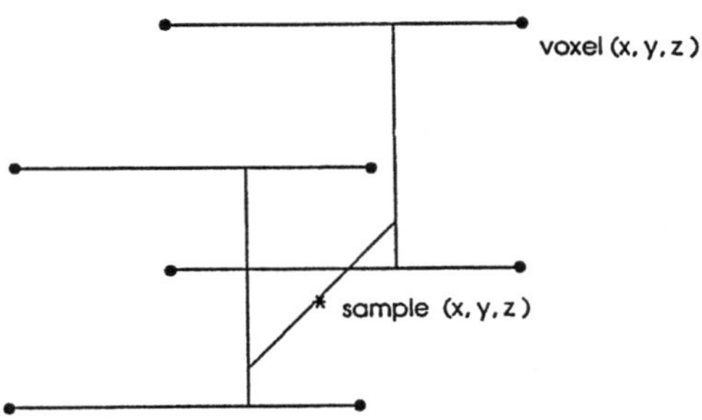

Fig. 2: Trilinear interpolation

look-up-tables (LUT). These LUTs contain colors and opacities for each value in the domain of the data. A voxel color is computed using the voxel's gradient and data color in Gouraud or Phong shading. The voxel opacity is the voxel's data opacity. Sample colors and sample opacities are computed at evenly spaced sample locations along the rays. In principle, sample locations are independent of integer data locations in the volume. If the sample location is a data location, sample color and sample opacity are equal to voxel color and voxel opacity (Fig. 3a). In any other case, the sample color and sample opacity can be computed by trilinear interpolation (Fig. 2) from the color and opacity of the eight voxels which are closest to the sample location (Fig. 3b). Instead of trilinear interpolation the nearest neighbor can be chosen. This, unfortunately, reduces the image quality as well as the computation effort. Pixel color is computed by different methods from sample colors and sample opacities.

If the data volume is rotated in advance, so that rays follow the lines of the volume, computing the sample location is simplified, and only three layers of the data volume are required for the computation of one image line at the same time. This reduces the size of the required main memory of the workstation. The three layers are required for the calculation of the gray-level gradient.

In the simplest raycasting version, rays are sent and sample colors are computed exclusively at integer data locations of the volume (Fig. 3a). In that case, an image in the size of the data volume is generated. In raycasting, the image resolution and the number of samples can be defined independently from the data resolution. In that case, sample color is computed by trilinear interpolation of colors of eight surrounding voxels (Fig. 3b). Antialiasing is performed this way, but the image generation time increases with image resolution.

Raycasting provides the capability of generating opaque or semi-transparent views of the data.

For opaque views, visibility is mostly defined by a threshold; for special applications the maximum data value at each ray is defined to be visible. If visibility is defined by a threshold, the ray can be stopped when a data value at a sample location greater than the threshold is detected or computed. For image quality, it is very important that sample color is computed at the location in the volume where trilinear interpolated data is equal to the threshold. This increases the number of samples to be computed. To accelerate this, only one sample per voxel is computed along the ray. If a sample greater than the threshold is detected, a binary sample search between the two last computed samples is performed in the object space. At the calculated location the gradient and the sample color, which is the pixel color, are computed.

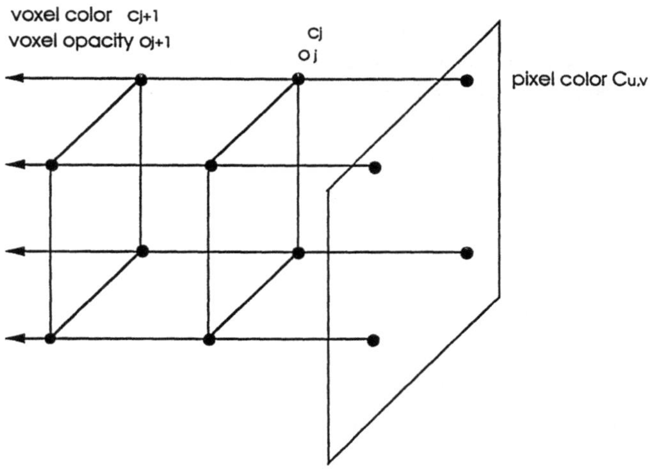

Fig. 3a: Fixed-resolution raycasting

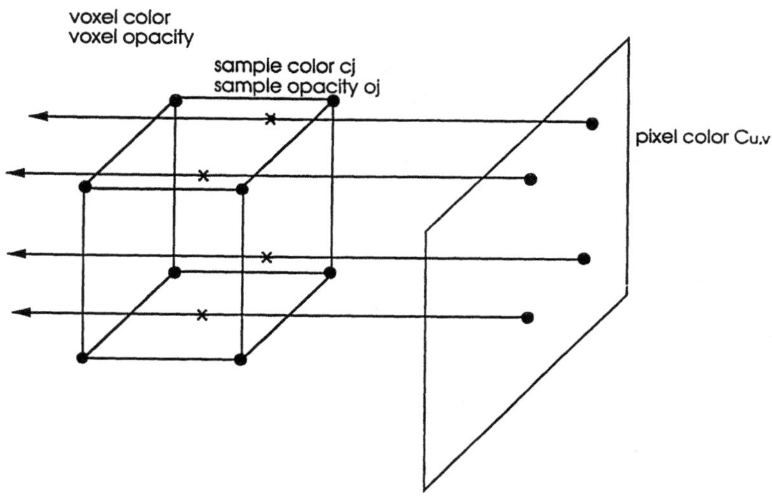

Fig. 3b: Variable-resolution raycasting

For semi-transparent views, visibility is defined by an opacity assigned to each voxel. Levoy has introduced a method for assigning opacities depending on data value and gradient magnitude in medical imaging [5]. In the semi-transparent mode, sample colors and opacities have to be computed along the entire ray. Sample colors and sample opacities of the current ray are stored in two arrays. Pixel color is then computed as

$$C_{u,v} = \sum_{j=0}^{K} c_j \, o_j \prod_{i=0}^{j-1} (1-o_i)$$

with $c_K = $ *background color* and $o_K = 1$.

Since with trilinear interpolation voxel colors and opacities at data locations are multiply used by adjacent rays, voxel colors and opacities are computed in advance and stored in a table for the three required layers. This does not work, if rays are cast with an arbitrary angle through the volume. The semi-transparent mode can be accelerated by stopping tracing each ray, if the accumulated opacity reaches a defined threshold, i.e., if further samples do not contribute to the pixel color.

8 Visualization of Scalar Data in Finite Element Grids

Finite element data is usually arranged in an irregular mesh. Normally, the mesh is rather loose in the areas of low gradients and denser in the areas of high gradients.

There are various methods to visualize finite element data [21]. The simplest form is the wire-frame display. It is mainly used to detect mesh generation errors, as the complete geometry of the element mesh is presented in a comprehensive manner.

In the available tools for the visualization of finite element data, the field values of the data set are displayed with surface-oriented techniques. The surface of the object under study is rendered using polygons, where the values on the surface are mapped onto colors. The third geometric dimension is indicated by shading or, in order not to distort the colors, by overlaying the wire-frame mesh onto the image.

The data from finite element analysis are often vector fields. In these cases arrows are used to indicate the direction as well as the magnitude of the field.

At FhG-AGD we visualize finite element data with volume rendering techniques. The data in the coarse element mesh have to be transformed to the regular voxel grid, using three-dimensional scan-conversion and trilinear interpolation of the field values. There are many algorithms for two-dimensional scan-conversion used in raster systems. These algorithms can be extended to the three-dimensional domain

[22,23]. Afterwards the voxel data is displayed with standard voxel rendering techniques. The main advantage of this method is that it preserves the entire complexity and inhomogenity of the data set.

One disadvantage with this method is the great number of data points, also in the regions where the data is varying slowly. There are millions of voxels containing practically no information for the viewer. This overhead also has implications on the performance. The amount of data to be processed is higher than for the surface-oriented display, and only the simple, low-quality volume rendering algorithms allow interactive manipulation of the data.

9 Animation

Different reasons require the generation of image sequences in volume visualization. As mentioned above, time appears often as the fourth dimension of a data set. This implies image sequences for visualization. This is a mapping of time to time.

In interactive visualization systems, data set browsers are used. A stack of computer tomograms, for instance, is visualized image by image, mapping the third spatial domain to time.

As 3D objects are projected to 2D images, the entire geometry of an object can only be visualized by changing viewing parameters in animated sequences. Therefore, for efficient computation of animated sequences the rendering process has to be split up into time-independent and time-dependent tasks [24,25]. Using some of the simple volume rendering methods, artifacts arise while rotating an object. This is caused by the discreteness of the data set and the fact that the estimation of the object's surface inclination is not independent of the angle of rotation of the data set in these algorithms.

10 Interaction

To date, a number of systems that visualize volume data, such as post-processors of finite element programs, have been developed. But they are all lacking in interactivity. Visualization is not a one-way communication. The possibility to "play around" with the data as well as with their visual representation is crucial for the understanding of the process under study. Interaction with the data must be possible on all levels of the visualization process, from the data sources to the final images.

Scientists do not just want to analyse the result of computation, they also want to know what is happening to the data during the computation. This implies a close connection between the data sources and the display of the data. Intermediate results are presented and the scientist steers the calculations by changing parameters, etc. The visualization system not only displays data, but allows interaction with the data. This connection of data sources is rarely found today, mainly, of course, because most computations cannot be performed in real time.

Further, users have to be able to enhance the data, to transform the data into another representation which is more informative or less voluminous by filtering, interpolation, image processing, etc. The selection and combination of parameters as well as the construction of derived values are functions needed for the preparation of the data.

A variety of presentation forms must be available. The user has to be able to choose between text, tables, diagrams, wire-frames, iso-surfaces, volumes, etc., to combine all of them arbitrarily and smoothly vary color, textures, transparency, shading.

Further, it must be possible to change some parameters of the object under study and perform the computations again iteratively, e.g., to see how the stress in a steel shaft is changed when the geometry of the shaft is changed. Only in this way is the scientist able to interpret the data, analyse their correlation and understand the process behind them.

Volume rendering of huge volume data sets in scientific visualization is very computing-intensive. High-quality images from those data sets cannot be computed in or near real time on general purpose graphic workstations, not even on the super-workstations. Thus a special tool for interactive specification of viewing parameters for volume rendering is required. Viewing parameters in this case are the viewpoint and the location of cut planes through the data set. The tool must provide scientists with the opportunity of orientation even in huge data sets. The echo of every user interaction must be computed in real time. Moreover, a user interface for volume visualization has to provide tools for graphical interactive definition of the parameters required in the visualization process (e.g., data transformation, color and opacity assignment, specification of light source and viewing parameters).

In animation systems, for instance, wire-frame representations are used to define the motion of objects interactively, while the final frames are rendered using these motion parameters afterwards. Wire-frame representations are also used in CAD systems while constructing objects, and shaded representations are computed afterwards. In volume rendering, the use of a wire-frame representation is not possible for different reasons. The first reason is the lack of any explicit surface representation of objects in the data set. The second reason is that the interior of 'objects' in the data set is not homogeneous. The neglect of that inhomogeneity as in wire-frame representation would complicate the orientation in the data set for the scientist. The third reason is that the structure and thus the surface of 'objects' created by the interpretation of the data set is very complex. Therefore a wire-frame representation is difficult to compute and would in most cases consist of many vectors. Furthermore, surface representations of volume data have to be re-computed after slicing the data set. For these reasons we have developed a special tool for the interactive definition of viewing parameters and we are using this tool with different volume renderers.

In the following, we describe the concepts and the implementation of a tool for interactive definition of point of view to scientific volume data and cut planes in such data, i.e., rotation and cutting of volume data in real time [26]. Volume data is mostly arranged in a regular grid, i.e., a data cube. The orientation of the cube is perceived by the user from the location of its vertices and edges. Back and front, left and right, bottom and top can be distinguished by the interior structure of the cube surfaces. Therefore we project 2D pixmaps from the volume data set onto the six surfaces of the cube. This is known as multiplanar reprojection of volume data. 2D pixmaps on a cube are sufficient for orientation because most scientists are now accustomed to evaluating their data sets with the aid of 2D images, and 2D images are the source of many scientific volume data sets.

The simplest version is to map the data from the outer layers to the cubes surfaces. In cases where these layers do not contain any data, a threshold depending on the user interpretation of the data set above the data noise is specified. Data above this threshold is then orthogonally projected onto the cube surfaces. These six projections are performed in preprocessing steps. Only one new surface is computed at a time after a cutting operation through the data set, because cut planes are perpendicular to the coordinate axes of the volume space.

To accelerate the rendering of the echo, only the six faces of the volume are rendered, not the inner parts. The faces are kept in main memory as two-dimensional pictures in the form of pixmaps. These are normally the outer voxel layer of the corresponding faces of the volume. In case this will result in empty pictures the pixmaps contain the parallel projections of an inner object on the faces.

For the rotation of the volume a rotation matrix is used. To accelerate the process, only the eight vertices of the volume are rotated. For each rotation, i.e., each time a mouse button is pressed, the matrix is updated and the new positions of the vertices are calculated by multiplication with the matrix.

From any given viewing direction only one, two or at most three faces are visible. Thus at most three of the six faces must be drawn. This, together with the fact that the volume is convex, eliminates the hidden surface problem. The visible faces are determined from their normal vectors. The faces with a positive z-coordinate (Fig. 4) of the normal vector are visible.

The volume is displayed with a parallel projection. This is done simply by skipping the z-coordinates of the vertices of the volume, and no calculations are necessary. Each visible face appears as a parallelogram on the screen. Through shearing and scaling of the pixmap the face is mapped onto the parallelogram (Fig. 5). The method used for the mapping is a scan-line based fill algorithm similar to the one presented in [27].

11 Parallel architectures

To be really interactive, a visualization system has to produce several images per second. The processing of large data sets with that performance can only be achieved by parallel processing. The discrete topology and repetitive nature of the voxel data representation and the fact that, normally, all voxels in the volume are independent of each other make parallel processing appropriate.

Some special hardware architectures have been developed for voxel-based imaging, e.g., the Voxel Processor architecture [28,29], which is a hierarchical, pipelined hardware organization, in which the volume is divided into equal subcubes, each processed by a separate processor.

The Cube architecture [30,31] is a hardware architecture centered around a three-dimensional cubic frame buffer of voxels. A voxel data set can be directly loaded into the cubic frame buffer. Alternatively, a geometric model is scan-converted into its discrete voxel representation. The three-dimensional images in the frame buffer are directly interactively manipulated and finally projected and rendered onto a two-dimensional display.

Nevertheless, parallalization can be realized using multiple general purpose workstations or multiprocessor superworkstations communicating via UNIX sockets and remote procedure calls [32].

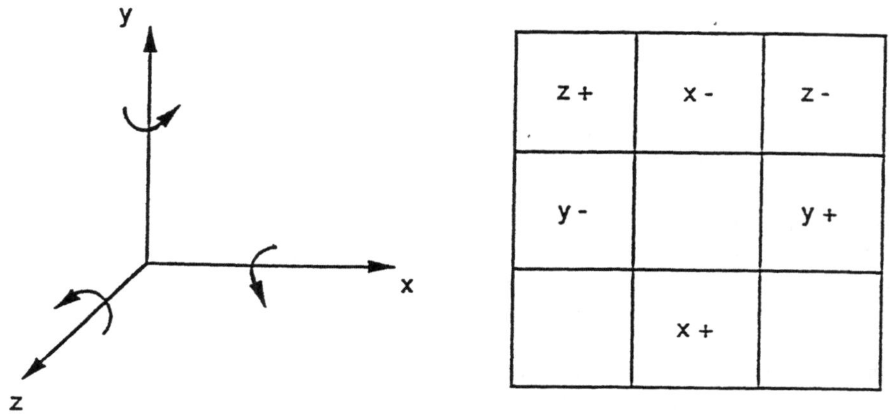

Fig. 4: Division of the window into rotation-sensitive areas

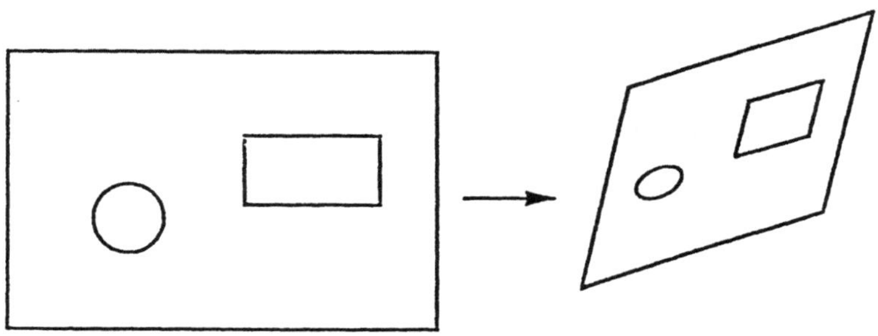

Fig. 5: Mapping of pixmaps to cube surfaces

12 Conclusion

The visualization of volume data sets using advanced rendering methods provides tools for many scientists from various disciplines to obtain insight from their measurements or simulations.

Rendering methods for the integration of different primitives in combined visualizations have to be developed. One method for combined polygon and volume rendering is described in [33]. Other combinations are volumes and lines or dots as well as heterogenous volumes and analytically defined CSG.

The development of convenient user interfaces to steer the simulation and visualization processes interactively will increase the profitability of applying such systems. The interactivity will be increased by developing advanced visualization systems on powerful parallel graphic workstations.

References

[1] McCormick, B. H.; DeFanti, T. A.; Brown, M.: Visualisation in scientific computing. Computer Graphics, Vol.21, No.6 (1987)

[2] Encarnacao, J. L.; Schönhut, J.: High performance, visualisation and integration - the computer graphics headlines for the 90's. In: Proc. IFIP TC 5 Conference on CAD/CAM Technology Transfer, Mexico City; North-Holland, Amsterdam (1988)

[3] Upson, C. et al.: The application visualization system: a computational environment for scientific visualization. IEEE Comp Graph App, July 1989, pp. 30-41 (1989)

[4] Kaneda, K. et al: Reconstruction and semi-transparent display method for observing inner structure of an object consisting of multiple surfaces. In: Proc. Computer Graphics International '87, Springer-Verlag, Berlin, Heidelberg, New York (1987)

[5] Levoy, M.: Display of surfaces from volume data. IEEE Comp Graph App, May 1988, pp. 29-37 (1988)

[6] Drebin, R.; Carpenter, L.; Hanrahan, P.: Volume rendering. Computer Graphics, Vol.22, No.4, pp.65-74 (1988)

[7] Kajiya, J. T.; Von Herzen, B. P.: Ray tracing volume densities. Computer Graphics, Vol.18, No.3, pp.165-174 (1984)

[8] Sabella, P.: Rendering algorithm for visualizing 3D scalar fields. Computer Graphics, Vol.22, No.4, pp.51-58 (1988)

[9] Upson, C.; Keeler, M.: V-BUFFER: visible volume rendering. Computer Graphics, Vol.22, No.4, pp.59-64 (1988)

[10] Johnson, E. R.; Mosher, C. E.: Integration of volume rendering and geometric graphics. In: Upson, C. (ed.): Chapel Hill Workshop on Volume Visualization, pp.1-8, Dept.Computer Science, Univ. North Carolina, Chapel Hill (1989)

[11] Chen, L. et al.: Surface shading in the Cuberille environment. IEEE Comp Graph App, Dec. 1985, pp. 33-43 (1985)

[12] Gordon, D.; Reynolds, R. A.: Image space shading of 3-dimensional objects. Comp Vision Graph and Image Proc, 29, pp. 361-376 (1985)

[13] Höhne, K. H.; Bernstein, R.: Shading 3D-images from CT using gray-level gradients. IEEE Trans Med Imag, Vol MI-5, No.1, pp.45-47 (1986)

[14] Cohen, D. et al.: Real time discrete shading. Visual Comp, 6, pp. 16-27 (1990)

[15] Lorensen, W. E.; Cline, H. E.: Marching Cubes: A high resolution 3D surface construction algorithm. Computer Graphics, Vol.21, No.4, pp. 163-169 (1987)

[16] Webber, R. E.: Ray tracing voxel data via biquadratic local surface interpolation. Visual Comp., Vol.6, pp.8-15 (1990)

[17] Back, S.; Neumann, H.; Stiehl, H. S.: On segmenting computed tomograms. In: Lemke, H.U. (ed.): Proc. Int. Symp. CAR'89, pp. 691-696 Springer-Verlag, Berlin, Heidelberg, New York (1989)

[18] Pommert, A.; Tiede, U.; Wiebecke, G.; Höhne, K. H.: Image quality in voxel-based surface shading. In: Lemke, H.U. (ed.): Proc. Int. Symp. CAR'89, pp. 737-741 Springer-Verlag, Berlin, Heidelberg, New York (1989)

[19] Frieder, G.; Gordon, D.; Reynolds, R. A.: Back-to-front display of voxel-based objects. IEEE Comp Graph App, Jan. 1985, pp. 52 (1985)

[20] Foley, J. D.; Van Dam, A: Fundamentals of interactive computer graphics. Addison-Wesley, Reading, Massachusetts (1983)

[21] Gallagher, R.S.; Nagtegaal, J.C.: An efficient 3-D visualization technique for finite element models and other coarse volumes. Computer Graphics, Vol.23, No.3 (1989)

[22] Kaufman, A.: An algorithm for 3D scan-conversion of polygons. In: Proc. Eurographics '87, Elsevier, Amsterdam (1987)

[23] Kaufman, A.: Efficient algorithms for 3D scan-conversion of parametric curves, surfaces and volumes. Computer Graphics, Vol. 21, No.4 (1987)

[24] Farrell, E. J. et al.: Animated 3D CT imaging. IEEE Comp Graph App, Dec. 1985, pp. 26-32 (1985)

[25] Levoy, M.: Design for a real-time high-quality volume rendering workstation. In: Upson, C. (ed.): Chapel Hill Workshop on Volume Visualization, pp.85-92, Dept.Computer Science, Univ. North Carolina, Chapel Hill (1989)

[26] Frühauf, M.; Karlsson, K.: The rotating cube: interactive specification of viewing for volume visualization.
Eurographics Workshop on Visualization in Scientific Computing, Clamart, France (1990)

[27] Hofmann, G. R.: Non-planar polygons and photographic components for naturalism in computer graphics. In: Hansmann, W.; Hopgood, F. R. A.; Strasser, W. (eds.): Eurographics '89, North-Holland, Amsterdam (1989)

[28] Goldwasser, S.M. et al.: Physician's workstation with real-time performance.
IEEE Comp Graph App, Vol.5, No.12 (1985)

[29] Goldwasser, S.M. et al.: High performance graphics processors for medical imaging applications.
Proc. International Conference on Parallel Processing for Computer Vision and Display, Leeds, UK (1988)

[30] Kaufman, A.; Bakalash, R.: Memory and processing architecture for 3D voxel-based imagery.
IEEE Comp Graph App, Vol.8, No.6 (1988)

[31] Kaufman, A.; Bakalash, R.: Parallel processing for 3D voxel-based graphics.
Proc. International Conference on Parallel Processing for Computer Vision and Display, Leeds, UK (1988)

[32] Frühauf, M.; Karlsson, K.: Visualisierung von Volumendaten in verteilten Systemen.
In: Proc. GI-Fachtagung Visualisierung von Umweltdaten in Supercomputersystemen, Springer-Verlag, Berlin, Heidelberg, New York (in German 1989)

[33] Levoy, M.: A hybrid ray tracer for rendering polygon and volume data.
IEEE Comp Graph App (March 1990)

Rational B-splines

Gerald Farin
Arizona State University
Computer Science Dept. Tempe, Arizona 85287-5406, USA

Abstract

We give an introduction to rational B-splines together with a critical evaluation of their potential for industrial applications.

Keywords: Conic sections, rational Bézier curves, reparametrization, rectangular and triangular surfaces, the sphere.

1 Introduction

Ask anyone in the CAD/CAM industry or in Graphics about the most promising curve or surface form – the "in" answer is invariably "NURBS", or non-uniform rational B-splines. In these Course Notes, we shall describe the main features of this curve and surface representation. It is assumed that the reader is familiar with the concepts of integral (i.e., nonrational) Bézier and B-spline curves and surfaces, as described in Farin [17]. More material on rational curve and surface schemes is in Piegl and Tiller [37], and Tiller [42].

The development of rational B-splines is nicely exemplified by the development at Boeing. That company, like may other aircraft companies, relied heavily on constructions that involved the use of conics – e.g., for the definition of the fuselage. This use of conics goes back to the book by R. Liming [29]. Based on it, Boeing had a large number of mathematical tools for fuselage design, and when the first computers arrived, it was natural to program those methods up, thereby entering the phase of "Computer Adied Design".

In the late fifties, a research group became interested in the use of spline curves (C^2 piecewise cubic interpolants), culminating in the pioneering paper by J. Ferguson [20] in 1964. Several successful design methods were developed using the spline paradigm.

Eventually, it became clear that two incompatible design systems were being used by one company: the output of the conic-based system could not be represented in spline form and vice-versa.[1] B. Blomgren attempted to unify the two systems into one, namely by incorporating both into the rational B-spline format, which is capable of representing conics as well as piecewise polynomials.

On a final historical note, rational bicubics were first considered by S. Coons [12]; rational B-splines first appear in K. Vesprille's thesis [43].

The rest of the paper is structured as follows:

- conics in rational quadratic form,
- rational Bézier curves,

[1]There is an exception, of course: a *parabola* is both a conic section and a quadratic spline curve.

- rational B-splines,
- rational surfaces,
- comments on rational B-splines.

2 Conics

As mentioned above, conic sections, the oldest known curve form, still are essential to many CAD systems. ManyA number of equivalent ways exist to define a conic section; for our purposes the following one is very useful: *A conic section in* $\mathbb{E}^2$ *is the projection of a parabola in* $\mathbb{E}^3$ *into a plane.*[2]

When it comes to the formulation of conics as rational curves, one typically chooses the center of the projection to be the origin $\mathbf{0}$ of a 3D cartesian coordinate system. The plane into which one projects is taken to be the plane $z = 1$. Since we will study planar curves in this section, we may think of this plane as a copy of $\mathbb{E}^2$, thus identifying points $\begin{bmatrix} x & y \end{bmatrix}^{\mathrm{T}}$ with $\begin{bmatrix} x & y & 1 \end{bmatrix}^{\mathrm{T}}$. Our special projection is characterized by

$$\begin{bmatrix} x \\ y \\ z \end{bmatrix} \rightarrow \begin{bmatrix} x/z \\ y/z \\ 1 \end{bmatrix}.$$

Note that a point $\begin{bmatrix} x & y \end{bmatrix}^{\mathrm{T}}$ is the projection of a whole family of points: every point on the straight line $\begin{bmatrix} wx & wy & w \end{bmatrix}^{\mathrm{T}}$ projects to $\begin{bmatrix} x & y \end{bmatrix}^{\mathrm{T}}$. In the following, we will use the shorthand notation $\begin{bmatrix} w\mathbf{x} & w \end{bmatrix}^{\mathrm{T}}$ with $\mathbf{x} \in \mathbb{E}^2$ for $\begin{bmatrix} wx & wy & w \end{bmatrix}^{\mathrm{T}}$.[3] Figure 1 gives an example of how to obtain a conic using this projection method.

Let $\mathbf{c}(t) \in \mathbb{E}^2$ be a point on a conic. One can show that there exist numbers $w_0, w_1, w_2 \in \mathbb{R}$ and points $\mathbf{b}_0, \mathbf{b}_1, \mathbf{b}_2 \in \mathbb{E}^2$ such that

$$\mathbf{c}(t) = \frac{w_0\mathbf{b}_0 B_0^2(t) + w_1\mathbf{b}_1 B_1^2(t) + w_2\mathbf{b}_2 B_2^2(t)}{w_0 B_0^2(t) + w_1 B_1^2(t) + w_2 B_2^2(t)}, \tag{1}$$

i.e. $\mathbf{c}$ may be expressed as a *parametric rational quadratic curve.* For a proof, see Patterson [31] or Farin [17].

We call the points $\mathbf{b}_i$ the *control polygon* of the conic $\mathbf{c}$; the numbers w_i are called *weights* of the corresponding control polygon vertices. Thus the conic control polygon is the projection of the control polygon with vertices $\begin{bmatrix} w_i\mathbf{b}_i & w_i \end{bmatrix}^{\mathrm{T}}$, which is the control polygon of the 3D parabola that we projected onto the conic $\mathbf{c}$.

The form (1) is called the *rational quadratic form* of a conic section. If all weights are equal, we recover nonrational quadratics, i.e., parabolas. As w_1 becomes larger,i.e., as $[w_1\mathbf{b}_1, w_1]$ moves "up" parallel to the $z-$axis, the conic is "pulled" towards $\mathbf{b}_1$.

[2]Here, $\mathbb{E}^3$ denotes Euclidean 3-space etc.

[3]Sometimes the set of all points $\begin{bmatrix} wx & wy & w \end{bmatrix}^{\mathrm{T}}$ is called the *homogeneous form* or *homogeneous coordinates* of $\begin{bmatrix} x & y \end{bmatrix}^{\mathrm{T}}$.

We may evaluate (3) at the endpoint $t = 0$:

$$\dot{\mathbf{c}}(0) = \frac{2}{w_0}[w_1\mathbf{b}_1 - w_0\mathbf{b}_0 - (w_1 - w_0)\mathbf{b}_0].$$

After some simplifications we obtain

$$\dot{\mathbf{c}}(0) = \frac{2w_1}{w_0}\Delta\mathbf{b}_0. \tag{4}$$

Similarly, we obtain

$$\dot{\mathbf{c}}(1) = \frac{2w_1}{w_2}\Delta\mathbf{b}_1. \tag{5}$$

In both formulas, $\Delta\mathbf{b}_i = \mathbf{b}_{i+1} - \mathbf{b}_i$.

Let us now consider two conics, one defined over the interval $[u_0, u_1]$ with control polygon $\mathbf{b}_0, \mathbf{b}_1, \mathbf{b}_2$ and weights w_0, w_1, w_2 and the other defined over the interval $[u_1, u_2]$ with control polygon $\mathbf{b}_2, \mathbf{b}_3, \mathbf{b}_4$ and weights w_2, w_3, w_4. Both segments form a C^1 curve if

$$\frac{w_1}{u_1 - u_0}\Delta\mathbf{b}_1 = \frac{w_3}{u_2 - u_1}\Delta\mathbf{b}_2. \tag{6}$$

The appearance of the interval lengths is due to the application of the chain rule, which is necessary since we now consider a composite curve with a global parameter u. Thus there exist C^1 piecewise conics that are not the projections of C^1 piecewise parabolas! One may also combine conics to *conic spline curves* with curvature continuity, see Farin [16].

3 Osculatory Interpolation

With rational cubics, it is easy to solve an interesting kind of interpolation problem: given a Bézier polygon $\mathbf{b}_0, \mathbf{b}_1, \mathbf{b}_2, \mathbf{b}_3$ and a curvature value at each endpoint, find a set of weights w_0, w_1, w_2, w_3 such that the corresponding rational cubic assumes the given curvatures at $\mathbf{b}_0$ and $\mathbf{b}_1$. The following method is very similar to one developed by T. Goodman in 1988, see [23]. We assume without loss of generality that $w_0 = w_3 = 1$.[4] The given curvatures κ_0 and κ_3 are then related to the unknown weights by

$$\kappa_0 = \frac{4}{3}\frac{w_2}{w_1^2}c_0, \quad \kappa_3 = \frac{4}{3}\frac{w_1}{w_2^2}c_1, \tag{7}$$

where

$$c_0 = \frac{\text{area}[\mathbf{b}_0, \mathbf{b}_1, \mathbf{b}_2]}{\text{dist}^3[\mathbf{b}_0, \mathbf{b}_1]}, \quad c_1 = \frac{\text{area}[\mathbf{b}_1, \mathbf{b}_2, \mathbf{b}_3]}{\text{dist}^3[\mathbf{b}_2, \mathbf{b}_3]}.$$

Equations(7) decouple nicely, so that we can determine our unknowns w_1 and w_2:

$$w_1 = \frac{4}{3}[\frac{c_0^2}{\kappa_0^2}\frac{c_1}{\kappa_1}]^{\frac{1}{3}}, \quad w_2 = \frac{4}{3}[\frac{c_0}{\kappa_0}\frac{c_1^2}{\kappa_1^2}]^{\frac{1}{3}}. \tag{8}$$

A similar interpolation problem was addressed by Klass [28] and de Boor, Hollig, and Sabin for the nonrational case: they prescribe two points and corresponding tangent

[4]Goodman [23] assumes that $w_1 = w_2 = 1$.

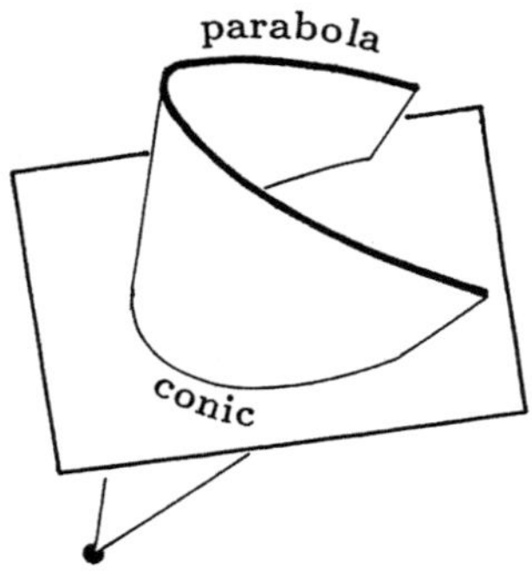

Figure 1: Conic sections: a parabolic arc in 3-space is projected into the plane $z = 1$; the result is a conic section.

2.1 Reparametrization

Note that a common nonzero factor in the w_i does not affect the conic at all. If $w_0 \neq 0$, one may therefore always achieve $w_0 = 1$ by a simple scaling of all w_i. There are other changes of the weights that leave the curve shape unchanged: these correspond to *rational linear parameter transformations* $t = t(\hat{t})$. Let us set

$$t = \frac{\hat{t}}{\hat{\rho}(1-\hat{t}) + \hat{t}}, \quad (1-t) = \frac{\hat{\rho}(1-\hat{t})}{\hat{\rho}(1-\hat{t}) + \hat{t}}$$

for some $\rho \in I\!R$. We may insert this into (1) and obtain:

$$\mathbf{c}(\hat{t}) = \frac{\hat{\rho}^2 w_0 \mathbf{b}_0 B_0^2(\hat{t}) + \hat{\rho} w_1 \mathbf{b}_1 B_1^2(\hat{t}) + w_2 \mathbf{b}_2 B_2^2(\hat{t})}{\hat{\rho}^2 w_0 B_0^2(\hat{t}) + \hat{\rho} w_1 B_1^2(\hat{t}) + w_2 B_2^2(\hat{t})}. \tag{2}$$

Thus the curve shape is not changed if each weight w_i is replaced by $\hat{w}_i = \hat{\rho}^{2-i} w_i$. If, for a given set of weights w_i, we select

$$\hat{\rho} = \sqrt{\frac{w_2}{w_0}},$$

then we obtain $\hat{w}_0 = w_2$, and, after dividing all three weights through by w_2, we even have $\hat{w}_0 = \hat{w}_2 = 1$. A conic that satisfies this condition is said to be in *standard form.* All conics with $w_0, w_2 \neq 0$ may be rewritten in standard form with the above choice of $\hat{\rho}$. If in standard form,i.e., $w_0 = w_2 = 1$, the point $\mathbf{s} = \mathbf{c}(\frac{1}{2})$ is called *shoulder point.* The shoulder point tangent is parallel to $\mathbf{b}_0\mathbf{b}_2$.

2.2 Derivatives

The derivative of a conic section, i.e., the vector $\dot{\mathbf{c}}(t) = \mathrm{d}\mathbf{c}/\mathrm{d}t$, is given by (see Farin [17]):

$$\dot{\mathbf{c}}(t) = \frac{1}{w(t)}[\dot{\mathbf{p}}(t) - \dot{w}(t)\mathbf{c}(t)]. \tag{3}$$

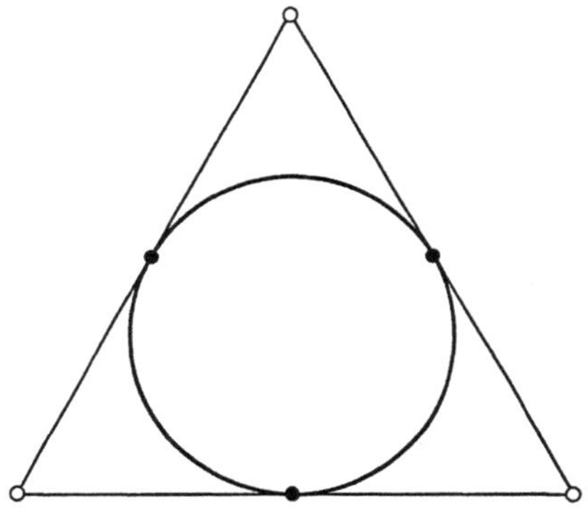

Figure 2: Circles: an arc of a circle may be written as three rational Bézier quadratics.

directions and curvatures [14]. The solution (when it exists) can only be obtained using an iterative method.

This method should be viewed as a building block for other interpolation methods – one would need algorithms to determine $\mathbf{b}_1$ and $\mathbf{b}_2$ from endpoints and tangent directions. Research for this is currently under way by the author.

3.1 The Circle

Let a rational quadratic (with $w_1 < 1$) describe an arc of a circle. Because of the symmetry properties of the circle, the control polygon must form an isosceles triangle. If we know the angle $\alpha = \angle(\mathbf{b}_2, \mathbf{b}_0, \mathbf{b}_1)$, we should be able to determine the weight w_1.[5] Some elementary trigonometry readily yields

$$w_1 = \cos\alpha.$$

A whole circle is representable by piecing several such arcs together. For example, we might choose to represent a circle by three equal arcs, resulting in a configuration as shown in Figure 2. The angles α equal 60 degrees, and so the weights of the inner Bézier points is 1/2, whereas the junction Bézier points have weights of unity, since each arc is in standard form.

Note that we have a representation of the circle that is C^1, assuming uniform parameter intervals. It is not C^2, however! Still we have an *exact* representation of the circle, not an approximation. Thus this particular representation of the circle is an example of a curve that is not twice differentiable yet is curvature continuous; such curves are called *geometrically continuous* and are discussed in Boehm [7], Farin [17], and Gregory [24].

[5]The actual size of the control polygon does not matter, of course: it can be changed by a scaling to any size we want, and scalings do not affect the weights!

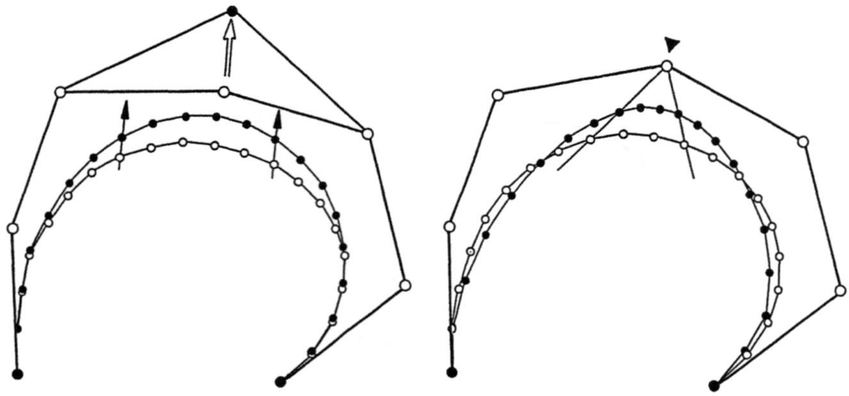

Figure 3: Influence of the weights: Left, a nonrational curve with a change in one control point. Right, with one weight changed.

How can rational Bézier curves in non-standard form arise? A common case occurs in connection with rational Bézier surfaces, as discussed in Section 6: the end weights of an isoparametric curve will in general not be unity. Such curves are often "extracted" from a surface and then treated as entities in their own right.

We may perform *degree elevation* by degree elevating the 4D polygon with control vertices $\begin{bmatrix} w_i\mathbf{b}_i & w_i \end{bmatrix}^{\mathrm{T}}$ and projecting the resulting control vertices into the hyperplane $w = 1$. Let us denote the control vertices of the degree elevated curve by $\mathbf{b}_i^{(1)}$; they are given by

$$\mathbf{b}_i^{(1)} = \frac{w_{i-1}\alpha_i\mathbf{b}_{i-1} + w_i(1-\alpha_i)\mathbf{b}_i}{w_{i-1}\alpha_i + w_i(1-\alpha_i)}; \quad i = 0, \ldots, n+1 \tag{10}$$

and $\alpha_i = i/(n+1)$. The weights $w_i^{(1)}$ of the new control vertices are given by

$$w_i^{(1)} = w_{i-1}\alpha_i + w_i(1-\alpha_i); \quad i = 0, \ldots, n+1.$$

The connection of reparametrization and degree elevation may lead to surprising situations. Consider the following procedure: take any rational Bézier curve in standard form and degree elevate it. Next, take the original curve, reparametrize it, then degree elevate it and bring it to standard form. We end up with two different polygons (and two different sets of standardized weights) that both describe the same rational curve. This situation is very different from the nonrational case! It is illustrated in Figure 5.

For the sake of completeness, we should mention that other ways exist to reparametrize rational curves than just by rational linear reparametrizations. For example, the reparametrization $t \leftarrow \alpha t(1-t)$ does not change the curve, but it raises the degree

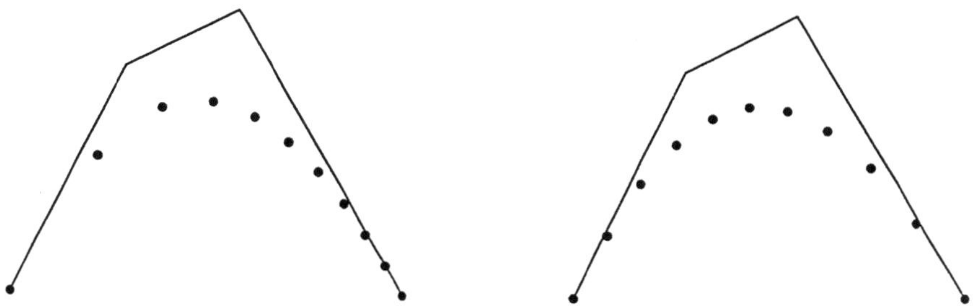

Figure 4: Reparametrizations: left, a rational Bézier curve evaluated at parameter values $0, 0.1, 0.2, \ldots, 1$. Right, the same curve and parameter values but after a reparametrization with $c = 3$.

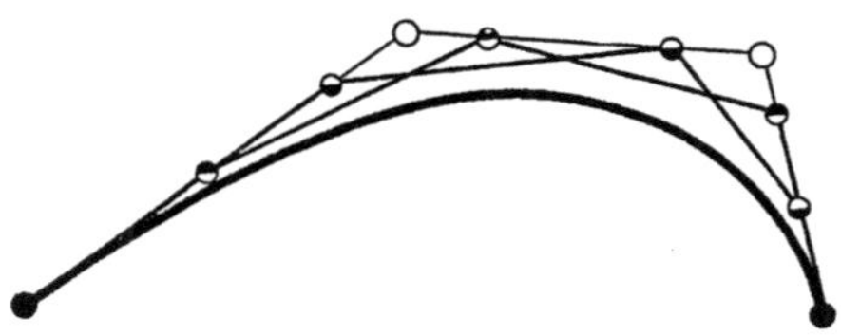

Figure 5: Ambiguous curve representations: the two heavy polygons represent the same rational quartic. Also indicated is the rational cubic representation that they were both obtained from.

from n to $2n$. As long as the reparametrization is of the form $t \leftarrow r(t)$, where $r(t)$ is a rational polynomial, we do not leave the class of rational curves. It is an interesting task to determine if a rational curve has been "improperly parametrized" in this way. For a solution, consult Sederberg [39].

4 Rational Cubic B-spline Curves

A 3D rational cubic B-spline curve is the projection through the origin of a 4D nonrational cubic B-spline curve into the hyperplane $w = 1$. The control polygon of the rational B-spline curve is given by vertices $\mathbf{d}_{-1}, \ldots, \mathbf{d}_{L+1}$; each vertex $\mathbf{d}_i \in \mathbb{E}^3$ has a corresponding weight w_i. The rational B-spline curve has a piecewise rational cubic Bézier representation. It may be obtained by projecting the corresponding 4D Bézier points into the hyperplane $w = 1$. Thus we obtain

$$\mathbf{b}_{3i-2} = \frac{w_{i-1}(1-\alpha_i)\mathbf{d}_{i-1} + w_i\alpha_i\mathbf{d}_i}{v_{3i-2}} \tag{11}$$

$$\mathbf{b}_{3i-1} = \frac{w_{i-1}\beta_i\mathbf{d}_{i-1} + w_i(1-\beta_i)\mathbf{d}_i}{v_{3i-1}}, \tag{12}$$

where all points $\mathbf{b}_j, \mathbf{d}_k$ are in $\mathbb{E}^3$ and

$$\Delta = \Delta_{i-2} + \Delta_{i-1} + \Delta_i,$$
$$\alpha_i = \frac{\Delta_{i-2}}{\Delta},$$
$$\beta_i = \frac{\Delta_i}{\Delta}.$$

The weights of these Bézier points are given by

$$v_{3i-2} = w_{i-1}(1-\alpha_i) + w_i\alpha_i, \tag{13}$$
$$v_{3i-1} = w_{i-1}\beta_i + w_i(1-\beta_i). \tag{14}$$

For the junction points, we obtain

$$\mathbf{b}_{3i} = \frac{\gamma_i v_{3i-1}\mathbf{b}_{3i-1} + (1-\gamma_i)v_{3i+1}\mathbf{b}_{3i+1}}{v_{3i}}. \tag{15}$$

where

$$\gamma_i = \frac{\Delta_i}{\Delta_{i-1} + \Delta_i}$$

and

$$v_{3i} = \gamma_i v_{3i-1} + (1-\gamma_i)v_{3i+1}$$

is the weight of the junction point $\mathbf{b}_{3i}$.

Another way to generate the piecewise rational Bézier polygon is by taking the control polygon $[w_i\mathbf{d}_i, w_i]^{\mathrm{T}}$, converting it to Bézier form, and then dividing through by the Bézier weights. This is less geometric, but it is certainly more efficient.

Designing with rational B-spline curves is not very different from designing with their nonrational counterparts. We now have the added freedom of being able to change weights. A change of only one weight affects a rational B-spline curve only locally, as shown in Figure 6.

Let us close this section with a result that somewhat limits the use of C^2 rational B-spline curves. *There is no symmetric periodic representation of a circle as a C^2 rational cubic B-spline curve.*

If such a representation existed, it would be of the form

$$\mathbf{x}(u) = \sum w_i\mathbf{d}_i N_i^3(u) / \sum w_i N_i^3(u),$$

where all w_i are equal by symmetry. Then the w_i cancel out, leaving us with an integral B-spline curve, which is not capable of representing a circle. Note, however, that we can represent any *open* circular arc by C^2 rational cubics.

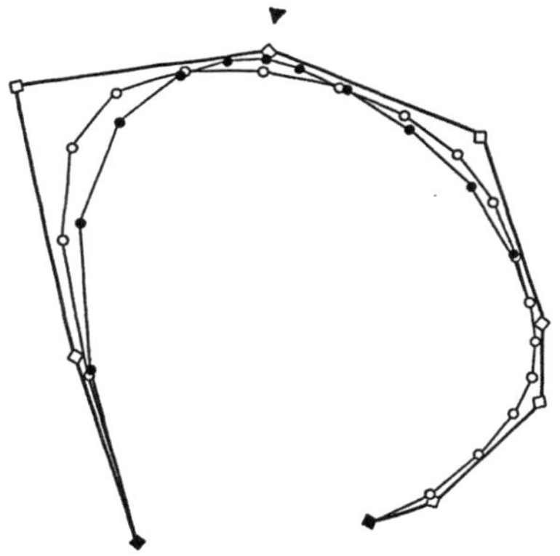

Figure 6: Rational B-splines: the weight of the indicated control point is changed. The curve is only affected locally.

5 Rational Bézier and B-spline Surfaces

We can generalize Bézier and B-spline surfaces to their rational counterparts in much the same way as we did for the curve cases. In other words, we define a rational Bézier or B-spline surface as the projection of a 4D tensor product Bézier or B-spline surface. Thus, the rational Bézier patch takes the form

$$\mathbf{x}(u,v) = \frac{\sum_i \sum_j w_{i,j}\mathbf{b}_{i,j}B_i^m(u)B_j^n(v)}{\sum_i \sum_j w_{i,j}B_i^m(u)B_j^n(v)}, \tag{16}$$

and a rational B-spline surface is written as

$$\mathbf{s}(u,v) = \frac{\sum_i \sum_j w_{i,j}\mathbf{d}_{i,j}N_i^m(u)N_j^n(v)}{\sum_i \sum_j w_{i,j}N_i^m(u)N_j^n(v)}. \tag{17}$$

Figure 8 shows an example of a rational B-spline surface. Its control net, shown in Figure 7, may be used to define a bicubic spline surface. With weights as shown in Figure 8, the "dip" became more pronounced, as well as the "vertical ridge".

Rational surfaces are obtained as the projections of tensor product patches – but they are not tensor product patches themselves. Recall that a tensor product surface is of the form $\mathbf{x}(u,v) = \sum_i \sum_j \mathbf{c}_{i,j}F_{i,j}(u,v)$, where the basis functions $F_{i,j}$ may be expressed as products $F_{i,j}(u,v) = A_i(u)B_j(v)$. The basis functions for (17) are of the form

$$F_{i,j}(u,v) = \frac{w_{i,j}N_i^m(u)N_j^n(v)}{\sum_i \sum_j w_{i,j}N_i^m(u)N_j^n(v)}.$$

Because of the structure of the denominator, this may in general not be factored into the required form $F_{i,j}(u,v) = A_i(u)B_j(v)$.

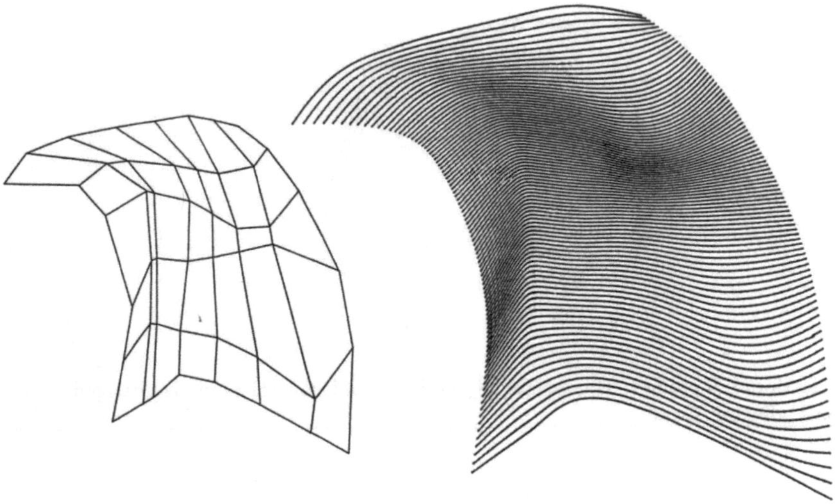

Figure 7: Bicubic B-spline surfaces: a control net and the (nonrational) surface generated by it.

But even though rational surfaces do not possess a tensor product structure, we may utilize many tensor product algorithms for their manipulation. Consider, for example, the problem of finding the piecewise rational bicubic Bézier form of a rational bicubic B-spline surface. All we have to do is to convert each row of the B-spline control net into piecewise rational Bézier cubics (according to Section 5). Then we repeat this process for each column of the resulting net (and the resulting weights!), simply following the standard recipe for tensor product surfaces (consult de Boor [13] or Farin [17]).

As another example, consider the problem of extracting an isoparametric curve from a rational Bézier surface. Suppose the curve corresponds to $v = \hat{v}$. We simply interpret all columns of the control net as control polygons and evaluate each at $\hat{v}$, using a rational version of the de Casteljau algorithm, for example. Keep in mind that we also have to compute a weight in each case. We can now interpret all obtained points together with their weights as the Bézier control polygon of the desired isoparametric curve. In general, its end weights will not be unity, i.e., the curve will not be in standard form (as described in Section 4.1). This situation may be remedied by the use of the reparametrization algorithm which is also described in that section.

Currently, rational B-spline surfaces are used for two reasons: they allow the exact representation of *surfaces of revolution* and of *quadric surfaces*. We will briefly describe surfaces of revolution in rational B-spline form here; for quadric surfaces see Farin [17].

1	1	1	1	1	1	1	1
1	1.5	1.5	1	1	1	1	1
1	2	2	1	1	1.5	1.5	1
1	1.5	1.5	1	1	1	1	1
1	1.5	1.5	1	1	1	1	1
1	1	1	1	1	1	1	1

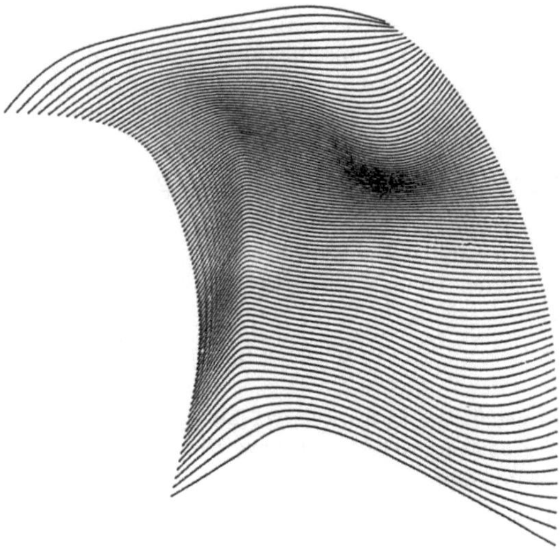

Figure 8: Rational bicubic B-spline surfaces: the control net is the same as in the previous figure. The weights for the rational form are indicated on the left.

5.1 Surfaces of Revolution

A surface of revolution is given by

$$\mathbf{x}(u,v) = \begin{bmatrix} r(v)\cos u \\ r(v)\sin u \\ z(v) \end{bmatrix}.$$

For fixed v, an isoparametric line $v = const$ traces out a circle of radius $r(v)$, called a *meridian*. Since a circle may be exactly represented by rational quadratic arcs, we may find an exact rational representation of a surface of revolution provided we can represent $r(v), z(v)$ in rational form.

The most convenient way to define a surface of revolution is to prescribe the (planar) generating curve, or generatrix, given by

$$\mathbf{g}(v) = [r(v), 0, z(v)]^{\mathrm{T}}$$

and by the axis of revolution, in the same plane as $\mathbf{g}$. Suppose $\mathbf{g}$ is given by its control polygon, knot sequence, and weight sequence. We can construct a surface of revolution such that each meridian consists of three rational quadratic arcs, as shown in Figure 2. For each vertex of the generating polygon, construct an equilateral triangle (perpendicular to the axis of revolution) as in Figure 2. Assign the given weights of the generatrix to the three polygons corresponding to the triangle edge midpoints; assign half those weights to the three control polygons corresponding to the triangle vertices. In this way, we represent *exactly* "classical" surfaces such as cylinders, spheres, or tori.

Instead of breaking down each meridian into three arcs, we might have used four or more. This would have resulted in a larger number of patches; for an illustration, see Piegl and Tiller [37].

Note that although the generatrix may be defined over a knot sequence $\{v_j\}$ with only simple knots, this is not possible for the knots of the meridian circles: we have to use double knots, thereby essentially reducing it to the piecewise Bézier form.

6 Rational Bézier Triangles

Following the familiar theme of generating rational curve and surface schemes, we define a rational Bézier triangle to be the projection of a nonrational 4D Bézier triangle. We thus have:

$$\mathbf{b}^n(\mathbf{u}) = \mathbf{b}_{\mathbf{0}}^n(\mathbf{u}) = \frac{\sum_{|\mathbf{i}|=n} w_{\mathbf{i}} \mathbf{b}_{\mathbf{i}} B_{\mathbf{i}}^n(\mathbf{u})}{\sum_{|\mathbf{i}|=n} w_{\mathbf{i}} B_{\mathbf{i}}^n(\mathbf{u})}, \tag{18}$$

where, as usual, the $w_{\mathbf{i}}$ are the weights associated with the control vertices $\mathbf{b}_{\mathbf{i}}$. Needless to say, for positive weights we have the convex hull property, and we have affine and projective invariance.

Rational Bézier triangles may be evaluated by a de Casteljau algorithm in a not too surprising way:

Rational de Casteljau algorithm

Given: a triangular array of points $\mathbf{b}_{\mathbf{i}} \in \mathbb{E}^3; |\mathbf{i}| = n$, corresponding weights $w_{\mathbf{i}}$, and a point in a domain triangle with barycentric coordinates $\mathbf{u}$.

Set:

$$\mathbf{b}_{\mathbf{i}}^r(\mathbf{u}) = \frac{u w_{\mathbf{i}+\mathbf{e}1}^{r-1}(\mathbf{u}) \mathbf{b}_{\mathbf{i}+\mathbf{e}1}^{r-1}(\mathbf{u}) + v w_{\mathbf{i}+\mathbf{e}2}^{r-1}(\mathbf{u}) \mathbf{b}_{\mathbf{i}+\mathbf{e}2}^{r-1}(\mathbf{u}) + w w_{\mathbf{i}+\mathbf{e}3}^{r-1}(\mathbf{u}) \mathbf{b}_{\mathbf{i}+\mathbf{e}3}^{r-1}(\mathbf{u})}{w_{\mathbf{i}}^r(\mathbf{u})} \tag{19}$$

where

$$w_{\mathbf{i}}^r(\mathbf{u}) = u w_{\mathbf{i}+\mathbf{e}1}^{r-1}(\mathbf{u}) + v w_{\mathbf{i}+\mathbf{e}2}^{r-1}(\mathbf{u}) + w w_{\mathbf{i}+\mathbf{e}3}^{r-1}(\mathbf{u})$$

and

$$r = 1, \ldots, n \quad \text{and} \quad |\mathbf{i}| = n - r$$

and $\mathbf{b}_{\mathbf{i}}^0(\mathbf{u}) = \mathbf{b}_{\mathbf{i}}, w_{\mathbf{i}}^0 = w_{\mathbf{i}}$. Then $\mathbf{b}_{\mathbf{0}}^n(\mathbf{u})$ is the point with parameter value $\mathbf{u}$ on the rational Bézier triangle $\mathbf{b}^n$.

This algorithm works since we can interpret each intermediate $\mathbf{b}_{\mathbf{i}}^r$ as the projection of the corresponding point in the de Casteljau algorithm of the nonrational 4D preimage of our patch.

6.1 Derivatives

We now give a formula for the *directional derivative* of a rational Bézier triangular patch. Let $\mathbf{d}$ denote a direction in the domain triangle, expressed in barycentric coordinates. We are interested in the directional derivative $D_{\mathbf{d}}$ of a rational triangular Bézier patch $\mathbf{b}^n(\mathbf{u})$. Proceeding exactly as in the curve case (see Section 2.2), we obtain:

$$D_{\mathbf{d}}\mathbf{b}^n(\mathbf{u}) = \frac{1}{w(\mathbf{u})}[\dot{\mathbf{p}}(\mathbf{u}) - D_{\mathbf{d}}(\mathbf{u})\mathbf{b}^n(\mathbf{u})],$$

where we have set

$$\mathbf{p}(\mathbf{u}) = w(\mathbf{u})\mathbf{b}^n(\mathbf{u}) = \sum_{|\mathbf{i}|=n} w_\mathbf{i}\mathbf{b}_\mathbf{i}B_\mathbf{i}^n(\mathbf{u}).$$

Higher derivatives follow the pattern outlined in Section 2.2, i.e.

$$D_\mathbf{d}^r\mathbf{b}^n(\mathbf{u}) = \frac{1}{w(\mathbf{u})}[D_\mathbf{d}^r\mathbf{p}(\mathbf{u}) - \sum_{j=1}^{r} D_\mathbf{d}^j w(\mathbf{u}) D_\mathbf{d}^{r-j}(\mathbf{u})].$$

6.2 The Sphere

Rational Bézier triangles may be used to represent an octant of a sphere – as it turns out, this representation has to be rational quartic and not, as one might guess intuitively, rational quadratic. For the exact form of the representation, see [19]. In order to represent the whole sphere, we would assemble eight copies of this octant patch. Other representations are also possible: each octant may be written as a rational biquadratic patch (introducing singularities at the north and south poles), see [37]. A representation of the whole sphere as two rational bicubics (Piegl [36]) turned out to be incorrect (see Cobb [10]). Quite a different way of representing the sphere is also due to J. Cobb: he covers it with six rational bicubics having a cube-like connectivity [11].

7 Concluding Remarks

Presently, many CAD systems use rational B-splines to represent "exactly" surfaces of revolution or conic sections. Since there is no such thing as "exact" on a digital computer, this is not quite as advantageous as it may sound. These curve and surface types may equally well be approximated by piecewise polynomial schemes – even with the rational form, we cannot do better than machine accuracy. If we cannot gain on accuracy, then what about storage? Here, the rational form wins out over the piecewise polynomial form. Consider an arc of a circle. If we were to approximate it by a piecewise cubic, we would definitely need many more segments than the two or three (depending on how large the arc is) that are required for its rational quadratic representation.

It does not seem to be a viable option to introduce NURBS just in order to be able to handle conic sections and surfaces of revolution "exactly". The overhead that is thus created does not justify the alleged payoff. It seems therefore that the present popularity of NURBS is more of a trend than a real necessity.

Should we conclude that we should return to integral splines and abandon the rational ones? Probably not – rational curves and surfaces have a lot more to offer than just conics and surfaces of revolution. They speed up realistic perspective maps in graphics or they offer designers more flexibility through the use of weights. Also, they are as of now the most general surface description, thus offering themselves as a flexible tool in the conversion of geometry data from one CAD system to another.

Typically, C^k NURBS are conceived as the projection of a C^k integral 4D curve or surface. As we have seen, not every C^k piecewise rational curve or surface can be obtained that way. While the description of curve schemes that are are derived from projections is now straightforward, discussion of piecewise rational schemes that are not generated in that way is more involved. As an example of the first kind of curve schemes, we cite the projective generation of beta-splines by Barsky [5]; as an example of the second kind, we mention the more general derivation of rational gamma-splines by Boehm [6].

8 Acknowledgements

This research was supported in part by NSF grant DMC-8807747 and by DoE grant DE-FG02-87ER25041 to Arizona State University.

References

[1] A. Ball. Consurf I: introduction of the conic lofting tile. *Computer Aided Design*, 6(4):243–249, 1974.

[2] A. Ball. Consurf II: description of the algorithms. *Computer Aided Design*, 7(4):237–242, 1975.

[3] A. Ball. Consurf III: how the program is used. *Computer Aided Design*, 9(1):9–12, 1977.

[4] A. Ball. The parametric representation of curves and surfaces using rational polynomial functions. In R. Martin, editor, *The Mathematics of Surfaces II*, pages 39–62, Oxford University Press, 1987.

[5] B. Barsky. Introducing the rational beta-spline. In *Proceedings of the third Int. Conf. on Engineering Graphics and Descriptive Geometry*, 1988. Vienna, Austria.

[6] W. Boehm. Rational geometric splines. *Computer Aided Geometric Design*, 4(1-2):67–78, 1987.

[7] W. Boehm. Visual continuity. *Computer Aided Design*, 20(6):307–311, 1988.

[8] W. Boehm, G. Farin, and J. Kahmann. A survey of curve and surface methods in CAGD. *Computer Aided Geometric Design*, 1(1):1–60, 1984.

[9] F. Bookstein. Fitting conic sections to scattered data. *IEEE Computer Graphics and Applications*, 9:56–71, 1979.

[10] J. Cobb. Letter to the editor. *Computer Aided Geometric Design*, 6:85, 1989. Concerning Piegl's sphere approximation.

[11] J. Cobb. *A rational bicubic representation of the sphere.* Technical Report, Computer science, U. of Utah, 1988.

[12] S. Coons. *rational bicubic surface patches.* Technical Report, MIT, 1968. Project MAC.

[13] C. de Boor. *A Practical Guide to Splines.* Springer, 1978.

[14] C. de Boor, K. Hollig, and M. Sabin. High accuracy geometric Hermite interpolation. *Computer Aided Geometric Design*, 4(4):269–278, 1988.

[15] G. Farin. Algorithms for rational Bézier curves. *Computer Aided Design*, 15(2):73–77, 1983.

[16] G. Farin. Curvature continuity and offsets for piecewise conics. *ACM Transactions on Graphics*, 8(2):89–99, 1989.

[17] G. Farin. *Curves and Surfaces for Computer Aided Geometric Design.* Academic Press, second edition, 1989.

[18] G. Farin. Rational curves and surfaces. In T. Lyche and L. Schumaker, editors, *Mathematical Aspects in CAGD*, pages 215–238, Academic Press, 1989.

[19] G. Farin, B. Piper, and A. Worsey. The octant of a sphere as a nondegenerate triangular Bézier patch. *Computer Aided Geometric Design*, 4(4):329–332, 1988.

[20] J. Ferguson. Multivariable curve interpolation. *JACM*, II/2:221–228, 1964.

[21] A. Forrest. The twisted cubic curve: a computer-aided geometric design approach. *Computer Aided Design*, 12(4):165–172, 1980.

[22] R. Goldman. The method of resolvents: a technique for the implicitization, inversion, and intersection of non-planar, parametric, rational cubic curves. *Computer Aided Geometric Design*, 2(4):237–255, 1985.

[23] T. Goodman. *Shape preserving interpolation by parametric rational cubic splines.* Technical Report, University of Dundee, 1988. Department of Mathematics and Computer Scinece.

[24] John A. Gregory. Geometric continuity. In T. Lyche and L. Schumaker, editors, *Mathematical Methods in Computer Aided Geometric Design*, pages 353–372, Academic Press, 1989.

[25] J. Hands. Reparametrisation of rational surfaces. In R. Martin, editor, *The Mathematics of Surfaces II*, pages 87–100, Oxford University Press, 1987.

[26] B. Joe. Multiple-knot and rational cubic beta-splines. *ACM Transactions on Graphics*, 8(2):100–120, 1989.

[27] S. Katz and T. Sederberg. Genus of the intersection curve of two rational surface patches. *Computer Aided Geometric Design*, 5(4):253–258, 1988.

[28] R. Klass. An offset spline approximation for plane cubics. *Computer Aided Design*, 15(5):296–299, 1983.

[29] R. Liming. *Practical Analytical Geometry with Applications to Aircraft.* Macmillan, 1944.

[30] N. Patrikalakis. Approximate conversion of rational splines. *Computer Aided Geometric Design*, 6(2):155–166, 1989.

[31] R. Patterson. Projective transformations of the parameter of a rational Bernstein-Bézier curve. *ACM Transactions on Graphics*, 4:276–290, 1986.

[32] T. Pavlidis. Curve fitting with conic splines. *ACM Transactions on Graphics*, 2:1–31, 1983.

[33] L. Piegl. A geometric investigation of the rational Bézier scheme in computer aided geometric design. *Computers in Industry*, 7:401–410, 1987.

[34] L. Piegl. Interactive data interpolation by rational Bézier curves. *IEEE Computer Graphics and Applications*, 7(4):45–58, 1987.

[35] L. Piegl. On the use of infinite control points in CAGD. *Computer Aided Geometric Design*, 4:155–166, 1987.

[36] L. Piegl. The sphere as a rational Bézier surface. *Computer Aided Geometric Design*, 3(1):45–52, 1986.

[37] L. Piegl and W. Tiller. Curve and surface constructions using rational B-splines. *Computer Aided Design*, 19(9):485–98, 1987.

[38] V. Pratt. Techniques for conic splines. In *SIGGRAPH '85 Proceedings*, pages 151–159, 1985.

[39] T. Sederberg. Improperly parametrized rational curves. *Computer Aided Geometric Design*, 1(3):67–75, 1986.

[40] T. Sederberg and X. Wang. Rational hodographs. *Computer Aided Geometric Design*, 4(4):333–336, 1988.

[41] M. Shantz and S. Chang. Rendering trimmed NURBS with adaptive forward differencing. *Computer Graphics*, 22(4):189–198, 1988.

[42] W. Tiller. Rational B-splines for curve and surface representation. *IEEE Computer Graphics and Applications*, 3(6), 1983.

[43] K. Vesprille. *Computer aided design application of the rational B-spline approximation form*. PhD thesis, Syracuse University, 1975.

[44] A. Vinacua and P. Brunet. A construction for VC^1 continuity for rational Bezier patches. In T. Lyche and L. Schumaker, editors, *Mathematical Methods in Computer Aided Geometric Design*, pages 601–611, Academic Press, 1989.

Scattered Data Interpolation and Applications: A Tutorial and Survey

Richard Franke
Department of Mathematics
Naval Postgraduate School
Monterey, California 93943
0083P@cc.nps.navy.mil

Gregory M. Nielson
Computer Science Deartment
Arizona State University
Tempe, Arizona 85287-5406
nielson@enuxva.eas.asu.edu

Abstract. *The multivariate scattered data interpolation problem is introduced and the reasons for the difficulty of the problem compared to the one dimensional case are discussed. Basic ideas for interpolation (or approximation) of scattered data are introduced. Various types of data sets and some strategies for dealing with some of them are given. Readily available algorithms for the solution of the problem are discussed and suitability for various types of data, along with discussion of situations where they have been useful is given. Some related ideas are briefly mentioned. Throughout there are bountiful references to the existing literature.*

1. Introduction

This paper discusses the problem of constructing an interpolating (or approximating) function for scattered data in more than one independent variable. The problem is easily stated, and we do so for the case of two independent variables. Given data points $(x_k, y_k, f_k), k = 1, \dots, N$, scattered in the sense that there are no assumptions about the disposition of the independent data, $(x_k, y_k), k = 1, \dots, N$ except that there are no repeated points, the problem is to construct a smooth (at least C^1) function $F(x,y)$ which takes on the value f_k at $(x_k, y_k), k = 1, \dots, N$; that is, $F(x_k, y_k) = f_k, k = 1, \dots, N$. It may be more desirable, for various reasons such as errors in the data or possibly too much data, to solve the more general problem of approximating the data. While we primarily concentrate on the case of two independent variables, some of the methods have immediate and obvious extensions to more independent variables, while others extend with some difficulty.

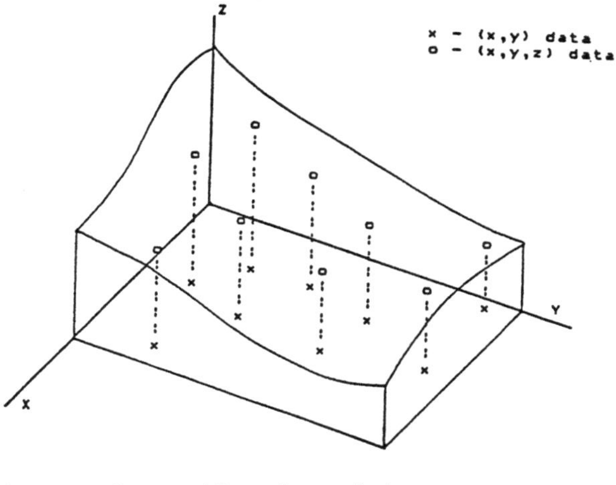

Figure 1. Scattered Data Interpolation

Many authors have treated the topic of scattered data interpolation from various points of view. Some have only been interested in the construction of surfaces which satisfy the basic criteria. Others have attempted to obtain some kind of "optimum" surface or to develop methods especially suited to a particular kind of problem. Some authors have attempted to assess the usefulness of various methods in terms of computational cost, accuracy, and other quantitative and qualitative criteria. We will make reference to many of these works here. We further point out the existence of two bibliographies, each over a broader field than we discuss: A Bibliography of Multivariate Approximation (Franke and and Schumaker 1987), and A CAGD Bibliography (Farin and Luscher 1990). For earlier surveys on this topic, the reader is referred to (Schumaker 1976) and (Barnhill 1977), each covering broader areas. A minisymposium on scattered data called "Surfaces" was held at Stanford in 1982 and the proceedings, which was edited by R. E. Barnhill and G. M. Nielson, appeared as a special issue of the Rocky Mountain Journal of Mathematics (Barnhill and Nielson 1984). A survey and results of evaluations of many methods for scattered data interpolation is contained in (Franke 1982b). There have been numerous assessments of limited sets of methods appearing in the literature for various scientific fields, with the application being to data from those fields. References to some of these can be found in the aforementioned bibliographies. A more recent survey, (Alfeld 1989) covers methods for more than two independent variables. The book by Lancaster and Salkauskas (Lancaster and Salkauskas 1986) discusses some methods for the problem and gives a number of applications and examples.

While the statement of the problem is simple, and the analog in one dimension has viable solutions for many applications, the multidimensional problem is fundamentally harder. This is indicated by a theorem due to Haar (see Davis 1963). We must first make a definition.

Definition. *A set of functions, $f_i, i = 1, \ldots, N$ defined on a point set S is called unisolvent on S if $det(f_i(p_j)) \neq 0$ for every set of distinct points $p_j, j = 1, \ldots, N$ lying in S.*

Theorem. *Let S be a point set in Euclidian space of n-dimension, $R_n, n \geq 2$. Suppose that S contains an interior point p. Let $f_1, f_2, \ldots, f_N, N > 1$ be defined on S and continuous in a neighborhood of p. Then this set of functions cannot be unisolvent on S.*

The property of unisolvence is exactly what is required to solve the interpolation problem for a general configuration of independent data by using a linear combination of the basis functions. The practical importance of this theorem is that no fixed set of basis functions can be guaranteed to lead to a solvable set of equations. Even in cases where the data points do lead to a solvable system, the problem may be arbitrarily badly conditioned. While it is possible to circumvent this problem by using explicit methods which do not require the solution of equations or by the blending of local approximations based on small subsets of points guaranteed to lead to solvable systems; by far the most popular way to get around it is through the use of basis functions which depend on the data points themselves and are guaranteed to lead to solvable systems of equations for the coefficients. For example, the set of basis functions $b_j(p) = ||p - p_j||, j = 1, \ldots, N$ leads to a solvable system.

The plan of this tutorial is to first give some structure to the subject by a giving survey of several classes of important methods along with a bit of the history and references to the initial (so far as we can tell) developments of these basic ideas. Following this will be a section detailing the difficulty of two dimensional data sets from the viewpoint of the variety of data configurations that may occur, and how some problems that arise may be alleviated by scaling in one or both variables. We then give more detailed discussions of some algorithms which have proven useful in treating data from various sources, and an indication of how the computer programs can be obtained. Little emphasis is put on commercially available programs which are often embedded in libraries of mathematical or plotting software. Finally, we give a brief survey of related topics.

2. Basic Ideas for Scattered Data Interpolation

There are a number of basic ideas for construction of formulas for smooth interpolation of scattered data. We will discuss the most important of these as preliminary to more detailed developments.

2.1 Inverse Distance Weighted Methods

One of the first ideas to be used is that of inverse distance weighting of data. In the mathematical literature, the scheme has become known as Shepard's method (Shepard 1968), although the method and variations of it were in use before Shepard's original publication (see e.g. Crain and Bhattacharyya 1967 and Cressman 1959). The basic method is to take the interpolant to be a weighted average of the value of the ordinates, the weight being a power of inverse distance:

$$F(x,y) = \frac{\sum_{k=1}^{N} \frac{f_k}{[d_k(x,y)]^\mu}}{\sum_{k=1}^{N} \frac{1}{[d_k(x,y)]^\mu}},$$

where $d_k(x,y) = \sqrt{(x-x_k)^2 + (y-y_k)^2}$ and μ is commonly taken as 2. In this form, it may not be so easy to verify that the basic interpolation properties hold. If we rewrite the expression as

$$F(x,y) = \sum_{k=1}^{N} f_k W_k(x,y)$$

where

$$W_k(x,y) = \frac{\prod_{j\neq k} d_j(x,y)^\mu}{\sum_{k=1}^{N} \prod_{j\neq k}^{N} d_j(x,y)^\mu}$$

then it is easier to see that $F(x_k, y_k) = f_k, k = 1, \ldots, N$, since $W_i(x_j, y_j) = \delta_{ij}$. It is well-known that the basic method has serious deficiencies such as "flat spots" at the data points for $\mu > 1$ and "corners" for $\mu = 1$, and undue influence of points which are far away (especially for $\mu = 2$). The example of Figure 2 with the data from Table 1 points out some of this behavior. This example was given by Gordon and Wixom (Gordon and Wixom 1974) and actually covers the more general case where the $\mu's$ are allowed to vary from data site to data site.

Table 1. Data for Shepard's Interpolant of Figure 2.

p_i	f_i	μ_i
(0.0,0.0)	4.0	2.5
(1.0,1.0)	0.0	2.5
(1.2,0.2)	3.0	3.0
(0.0,0.5)	1.0	4.0
(1.0,0.5)	1.0	4.0

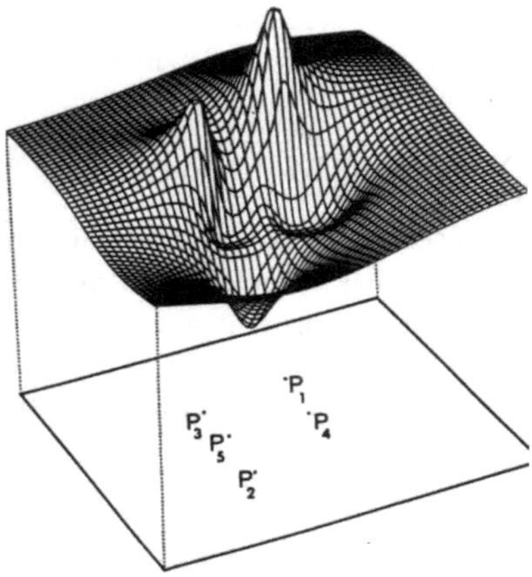

Figure 2. Shepard's method, from (Gordon and Wixom 1974)

These deficiencies can be largely overcome by replacing d_k by a local weight function and f_k by a local approximation to the surface. Methods of this type will be discused in Section 4.1. Another way to generalize and modify Shepard's method is to view it as an inverse distance weighted least squares approximation by a constant (see e.g. Franke and Nielson 1980 or Schumaker 1976). This viewpoint immediately leads to inverse distance weighted approximations by higher order polynomials. The coefficients in the approximation are functions of the point (x,y), and thus a least squares problem must be solved for each point at which F is to be evaluated. This type of approximation is rather expensive (see McLain 1976 and Franke 1982b), especially since at least quadratic approximations must be used.

2.2 Finite Element Methods

Finite element methods derive their name from the fact that the underlying approximation used is commonly associated with finite element methods for the solution of partial differential equations. The first step in these methods is the triangulation of the convex hull of the data points, $(x_k, y_k), k = 1, \ldots, N$. The topic of triangulating planar data sets has been treated quite extensively in the literature (see e.g. Lawson 1977 and Schumaker 1987). Usually some type of "optimal" triangulation of the convex hull is sought so as to avoid (insofar as possible; not always possible, especially around the boundary of the convex hull) triangles with small angles.

The obvious piecewise linear approximation over the triangulation is not smooth. To achieve continuous partial derivatives, it is necessary to use a piecewise interpolant with more flexibility that simple linear. This will usually necessitate the estimation of some partial derivatives. The first published scheme of this sort is due to Akima (Akima 1978). His scheme uses the reduced quintic element (Bell 1969). It is no surprise that a crucial part of these finite element schemes is the need for good estimates of the derivatives and empirical studies have substantiated this (see Nielson and Franke 1983). Akima's scheme does not do a very good job of estimating derivatives. This is particularly serious, since the reduced quintic element requires estimates of both first and second derivatives. Later methods used elements requiring only first partial derivatives at the vertices. The most common C^1 elements used are either the Clough-Tocher (Clough and Tocher 1965) piecewise cubic split of each triangle or a rationally corrected cubic element (see e.g. Nielson 1981 or Herron 1985). The first of these later methods was due to Lawson (Lawson 1977), followed by Nielson (Nielson 1983) and others. Some details of the elements and derivative estimation used in these methods will be discussed in Section 4.2.

2.3 Radial Basis Function Methods

Some of the most elegant schemes from a mathematical point of view, and which often work very well, belong to a class of methods which have become known as radial basis function methods. The name derives from the fact that the basis functions are constructed from a function of one variable, $h(s)$, and the basis function associated with each point is of the form $h_k(x,y) = h(d_k)$, where d_k is distance from (x,y) to the kth data point (x_k, y_k), as above. In general these methods have no polynomial precision, but precision can be built in at small cost and sometimes occurs naturally in the development. The form of the approximating function is

$$F(x,y) = \sum_{k=1}^{N} a_k h_k(x,y) + \sum_{k=1}^{M} b_k q_k(x,y)$$

where the set $< q_k(x,y) >$ is a basis for the polynomials of degree $< m$. The coefficients a_k and b_k satisfy the system of equations

$$\sum_{j=1}^{N} a_j h_j(x_i, y_i) + \sum_{j=1}^{M} b_j q_j(x_i, y_i) = f_i, \quad i = 1, \ldots, N$$

$$\sum_{i=1}^{N} a_i q_j(x_i, y_i) = 0, j = 1, \ldots, M$$

The first N equations are the interpolation requirements and the the last M equations guarantee polynomial precision. Since a system of at least N linear equations in as many unknowns, with a full matrix must be solved to obtain the coefficients in the approximating function, these methods are only suitable for data sets with up to a few hundred points. Some work has been done on iterative methods (see e.g. Dyn, Levin and Rippa 1986) for the solution of the equations, perhaps allowing larger data sets to be treated. Other ideas allowing treatment of large sets of data based on radial basis function approximation are discussed in Section 2.4

The very first radial basis function method was introduced by Hardy (Hardy 1971) and was applied to data from various sources. The basis function for Hardy's method is the multiquadric function $h_k(x,y) = \sqrt{d_k^2 + c^2}$, or the reciprocal of this function, each with $m = 0$ (i.e., no polynomial precision). The value of c is a constant, although the original motivation for it arose from the application, and represented a uniform distance in a third independent variable. Until a few years ago, no mathematical theory was available to support the method, although in practice it often worked very well. In the same time frame, Harder and Desmarais (Harder and Desmarais 1972) developed what they called "surface splines" as a two dimensional analog of the cubic spline. They proceeded from an operational engineering point of view. Their method has become known as thin plate splines, with the theoretical work showing the optimality of the approximation in a Hilbert space setting being done by Duchon (Duchon 1975a). The basis functions for surface splines are $h_k(x,y) = d_k^2 log d_k$, and $m = 2$ arises naturally in the development. Other schemes, some with special features, will be discussed in Section 4.3.

2.4 Blended Local Methods

If a set of data is too large to be treated globally, one must treat local subsets of the data. Depending on the application, it may be desirable to blend local approximations together in overlapping regions, although in other applications (not requiring a strictly smooth approximation), only the local treatment is necessary.

One type of blending of local approximations in a way that achieves a smooth approximation is a localized version of Shepard's method, and will be discussed in some detail in Section 4.1.

Because some of the finite element methods are local (depending on the derivative estimation scheme), these may be applied over subsets of the data. Care must be taken to ensure that the triangulation over a subset is a subset of the global triangulation, not necessarily a trivial task near the boundary of the convex hull, since boundary triangles may involve data very far away.

The main idea for discussion here is the idea of achieving local approximations by blending local interpolants together in such a way as to achieve a smooth global interpolant. Let $w_i(x,y), i = 1, \ldots, K$ be a set of functions which are nonnegative and sum to one for all (x,y). Such a set of functions is said to form a partition of unity. Let $\mathbf{R}_i = \{(x,y) : w_i(x,y) \neq 0\}$. If the "local interpolant" $Q_i(x,y)$ has the property that $Q_i(x_k, y_k) = f_k$ whenever $(x_k, y_k) \in R_i$, then the function

$$F(x,y) = \sum_{i=1}^{K} w_i(x,y) Q_i(x,y)$$

will interpolate the data (see Franke 1977). For suitable $w_i(x,y)$ only a few terms will be nonzero in the sum, so evaluation of the function $F(x,y)$ will be fast, provided the nonzero terms can be determined easily. A scheme using overlapping rectangles for the regions $\mathbf{R}_i$ with product Hermite cubic functions for the $w_i(x,y)$ and thin plate splines for the local interpolation functions $Q_i(x,y)$ is given in (Franke 1982a). A similar method using least squares approximations in place of the local interpolants was developed in (Jancaitus and Junkins 1974).

3. Sources of Data

Data requiring approximation by scattered data methods occur in virtually every field of science and engineering. Sources include both experimental results (experiments in chemistry, physics, engineering) and measured values of physical quantities (meteorology, oceanography, optics, engineering, geodetics, mining, geology, geography, cartography), and computational values (e.g., output from finite element solutions of partial differential equations).

3.1 Variety

The many sources of data and the myriad patterns that can occur in the x-y placement of the data points almost certainly guarantee that no one method will successfully treat all sets of data. A early guide to the type of data sets and the possible modes of treatment is given in (Sabin 1980) and (Sabin 1985).

The best configuration (gridded data is clearly the most desirable; here we mean in the actually scattered case) is a reasonably uniform distribution of data points. This seems to almost never occur in real problems, although it certainly may occur locally, for example in weather measurements taken by radiosonde over large land masses (see Franke 1985a). In such cases, depending on whether the data is dense enough to give a reasonable sampling of the behavior of the underlying function, any of the better schemes (many of the methods discussed here have been compared and ranked by Franke 1982b) will give reasonable results.

Data may be clustered into groups which are then scattered around the domain, leaving sparsely populated regions, or voids. Even though such voids are in the convex hull of the data set, they may be further away from data points than the region near but outside the convex hull. Thus, even though extrapolation is usually considered to be serious, voids are also a serious problem. As with any configuration of data, the clusters and voids may represent regions in which the surface is undergoing rapid change and very little, respectively. It is also possible, however, that the configuration represents more about the nature in which the data was taken or accessibility of data in various regions than behavior of the surface. In any case, it is probably best to use a method which is stable in the sense that it will not "misbehave" in the voids; unfortunately such methods (e.g., Shepard's method) have other deficiencies which may not be acceptable.

Tracked data occurs in many instances, but especially when data is obtained from a moving platform such as a ship taking hydrographic data or an aircraft taking magnetometer data. If the tracks do not cross, it may be possible to use lofting techniques across the one dimensional fits to the tracks. In some cases this may not be attractive because the the tracks themselves are ragged. Frequently it is the case that the distance between data points on the tracks are much closer together than the distance between tracks. In such cases methods based on distances will likely perform poorly since the closest data to a given point probably all lies on one track, hence is essentially one dimensional information. Misbehavior of the interpolant between tracks is nearly a certainty. Acceptable solutions to this problem can be obtained by scaling in one direction, and this is discussed briefly in the next section. If tracks cross, measured data subject to otherwise acceptably small errors may lead to large gradients due to inconsistencies when the tracks cross. Modification of the data by replacing the conflicting set of two or more points with one is a possibility and has been used (in another context) in meteorology.

Many examples of the variety of data sets that can be encountered in practice and in simulations are given in (Ripley 1981). The sources of his data include most of the above, plus some from forestry, biology, crytallography, ecology, among others. It is an interesting source for x-y scattered data (in some instances there are no z coordinates, e.g. for the location of eagle's nests).

3.2 Scaling and Affine Invariance

When one proceeds to provide input data to a subroutine for scattered data interpolation, the question arises as to what units or what numerical scale should be used for the input data. Often in the use of numerical procedures, concern over numerical accuracy may be the governing factor. In the case of scattered data interpolation, there are other matters to worry about. What units or scale is used can affect the results of the method and more than just that of round-off error. This can be illustrated by a small example. Given five data points, an interpolant of the form $S(x,y) = a+bx+cy+dxy+e(x^2+y^2)$ is determined. With the data of Table 2, the interpolant is easily verified to be $S_1(x,y) = 2(x+y-x^2-y^2)$. If we think of the units for the x-variable to be feet and those of the y-variable to be minutes, then the same data in the units of inches and seconds is given in Table 3. The interpolant is, in this case, $S_2(x,y) = (12x + 60y - x^2 - y^2)/936$. Now if one were to use this interpolant to estimate the dependent argument at, say, .75 feet and .5 minutes, we get in the first case, $S_1(.75,.50) = 0.875$ and in the second case $S_2(9,30) = \frac{103}{104} \neq 0.875$.

The reason these two estimates are different is that the basis functions $1, x, y, xy, x^2 + y^2$ are not closed under affine transformations. The space of bivariate polynomials $P_m = < x^i y^j : i+j \leq m >$ is affine invariant and consequently approximations based on these functions will be translation, rotation, and scale invariant. Unfortunately, polynomial basis functions are not appropriate for scattered data interpolation due to a variety of reasons including the lack of unisolvence. As we mentioned earlier, the need for unisolvent basis functions has lead to a number of scattered data interpolation methods which use basis functions that depend upon the data. Usually this dependence is such that the span of the basis functions is **not** closed under affine transformations and so the resulting methods are not affine invariant. This lack of affine invariance was first pointed out by Nielson (see Nielson 1987b) who also proposed an automatic way to modify a number of the methods so as to make them affine invariant. In order to briefly describe these modifications, we introduce the notion that a method of scattered data interpolation is an operator that accepts independent data values $\mathbf{V} = (V_1, \ldots, V_N) = ((x_1,y_1), \ldots, (x_N,y_N))$ and associated dependent values $f_i, i = 1, \ldots, N$ and produces a function $S[\mathbf{V}](V) = S(V) = S(x,y)$ such that $S(x_i,y_i) = f_i, i = 1, \ldots, N$.

x_i	y_i	z_i
1.0	1.0	0
1.0	0.0	0
0.5	0.5	1
0.0	1.0	0
0.0	0.0	0

Table 2.

x_i	y_i	z_i
12	60	0
12	0	0
6	30	1
0	60	0
0	0	0

Table 3.

Definition. *A scattered data interpolation method is said to be affine invariant provided*

$$S[T(\mathbf{V})](T(V)) = S[\mathbf{V}](V)$$

for any affine transformation T.

It is interesting to identify certain less restrictive classes of scattered data interpolation methods. Namely, those that are *translation invariant*, those that are *rotation invariant* and those that are *scale* and *scalar* invariant. It is clear what the definition of the first two is, but the difference between scale and scalar in not so standard and so we elaborate. By scale invariance, we mean that $S[\begin{pmatrix} \alpha & 0 \\ 0 & \beta \end{pmatrix}\mathbf{V}](\begin{pmatrix} \alpha & 0 \\ 0 & \beta \end{pmatrix}V) = S[\mathbf{V}](V)$, where $\alpha, \beta \neq 0$. By scalar invariance, we mean the special case of scale invariance where $\alpha = \beta$. The reason we distinguish these two cases is that a number of methods that are scalar invariant are not scale invariant. The difference between scale invariance and scalar invariance is related to the problem of the choice of units of measurement used for the independent data.

Many of the well known and widely used methods of scattered data interpolation satisfy some of these invariant properties but not all of them. For example, Hardy's method based upon multiquadrics (Hardy 1971) is rotation and translation invariant but not scalar invariant. The delta iteration method of Foley and Nielson (Foley and Nielson 1980) is translation and scalar invariant, but not rotation invariant. Franke's local, thin plate spline method (Franke 1977) is scale invariant due to the fact that the data is always scaled to the unit square, but this scaling destroys the rotation invariant aspect of thin plate splines. A number of methods are translation, rotation and scalar invariant. This includes Shepard's method (Shepard 1968), McLain's method (McLain 1976), Akima's method (Akima 1978), the surface splines of Harder and Desmarais (Harder and Desmarais 1972) and Duchon's rotation invariant splines (Duchon 1975a, 1975b). But none of these methods are scale invariant. It is this observation that prompted Nielson to investigate this subject. His modification is based upon replacing the metric resulting from the Euclidean norm with one that is affine invariant. This requires that the norm depend on the independent data values. This dependence is indicated by the use of the notation

$$\|\ \|_{\mathbf{V}}, \quad \text{where} \quad \mathbf{V} = \begin{pmatrix} V_1 \\ V_2 \\ \vdots \\ V_N \end{pmatrix}.$$

Definition. *Assume that P and Q are any two points in the domain of the norm $\|\ \|_{\mathbf{V}}$. This norm is said to be affine invariant provided*

$$\|p - q\|_{\mathbf{V}} = \|T(p) - T(q)\|_{T(\mathbf{V})}$$

for any affine transformation T.

The following affine invariant norm, $|||\ |||_{\mathbf{V}}$ was first introduced by Nielson (Nielson 1987b).

$$|||p|||_{\mathbf{V}}^2 = (x, y) \begin{pmatrix} \frac{\Sigma_{\mathbf{Y}}^2}{\Sigma_{\mathbf{X}}^2\Sigma_{\mathbf{Y}}^2 - (\Sigma_{\mathbf{XY}})^2} & \frac{-\Sigma_{\mathbf{XY}}}{\Sigma_{\mathbf{X}}^2\Sigma_{\mathbf{Y}}^2 - (\Sigma_{\mathbf{XY}})^2} \\ \frac{-\Sigma_{\mathbf{XY}}}{\Sigma_{\mathbf{X}}^2\Sigma_{\mathbf{Y}}^2 - (\Sigma_{\mathbf{XY}})^2} & \frac{\Sigma_{\mathbf{X}}^2}{\Sigma_{\mathbf{X}}^2\Sigma_{\mathbf{Y}}^2 - (\Sigma_{\mathbf{XY}})^2} \end{pmatrix} \begin{pmatrix} x \\ y \end{pmatrix}$$

where

$$\Sigma^2_{\mathbf{X}} = \frac{\sum_{i=1}^N (x_i - \bar{x})^2}{N}, \qquad \bar{x} = \frac{\sum_{i=1}^N x_i}{N}$$

$$\Sigma^2_{\mathbf{Y}} = \frac{\sum_{i=1}^N (y_i - \bar{y})^2}{N}, \qquad \bar{y} = \frac{\sum_{i=1}^N y_i}{N}$$

and

$$\Sigma_{\mathbf{XY}} = \frac{\sum_{i=1}^N (x_i - \bar{x})(y_i - \bar{y})}{N}.$$

It is quite easy to verify that this norm is actually affine invariant once it is observed that

$$|||p|||^2_{\mathbf{V}} = (\,x, y\,)\, N[\bar{\mathbf{V}}^* \bar{\mathbf{V}}]^{-1} \begin{pmatrix} x \\ y \end{pmatrix},$$

where

$$\bar{V} = \begin{pmatrix} x_1 - \bar{x}, & y_1 - \bar{y} \\ x_2 - \bar{x}, & y_2 - \bar{y} \\ \vdots & \vdots \\ x_N - \bar{x}, & y_N - \bar{y} \end{pmatrix}.$$

This observation also makes it obvious what the generalization of this norm is for points of E^n.

The question now arises as to what effect this modification has on the performance of the method. One way to assess this is to compare the ability of these two methods, the original and the modified, to reproduce a given function based upon sampled values. In the examples that follow,the basic method is the surface spline of Harder and Desmarais (Harder and Desmarais 1972) (or equivalently, the thin plate splines of Duchon 1976). The errors of the first row of Table 4 are based upon the difference between a function, F, and the spline approximation obtained by evaluating this function at the independent data sites. The function is (Franke's function)

$$\begin{aligned} F(x,y) &= .75\exp\left[-\frac{(9x-2)^2+(9y-2)^2}{4}\right] + .75\exp\left[-\frac{(9x+1)^2}{49} - \frac{9y+1}{10}\right] \\ &+ .5\exp\left[-\frac{(9x-7)^2+(9y-3)^2}{4}\right] - .2\exp\left[-(9x-4)^2-(9y-7)^2\right] \end{aligned} \tag{3.1}$$

and the data sets are shown in Figure 3.

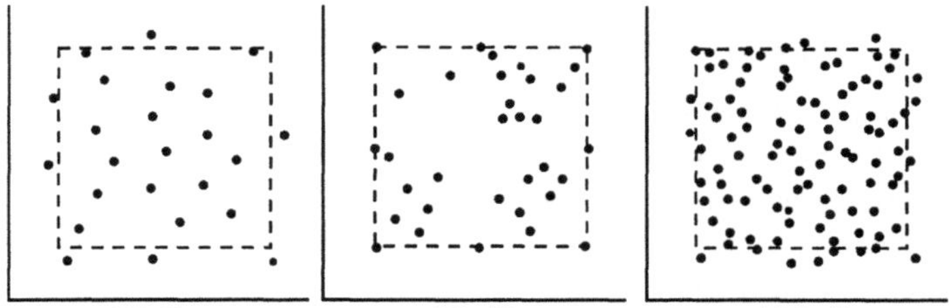

Figure 3. Data sets (25 Pt., 33 Pt. and 100 Pt.)

Table 4. RMS errors for three data sets

	25 Pt. Data Set	33 Pt. Data Set	100 Pt. Data Set
Original	.03480	.04210	.00947
Aff. Inv.	.03380	.04150	.00942

Also shown in Table 4 are the errors for the modified, affine invariant method. The errors are not very different in this case because the methods are not very different. This is due to the fact that the affine invariant norm is not much different than the Euclidean norm for these data sets. The unit disks of the affine invariant norm are very close to circular. This is often the case for a large number of data sets that one encounters in applications. But it is possible for the results to be quite different; particularly if the units of measurement of the data are very different. In order to illustrate this, we show in Table 4 the results of applying the original method to three more data sets where we have artificially scaled the x component by a factor of 10. The results are significantly larger compared to the errors for the original method.

Table 5. RMS errors for scaled data sets ($x \leftarrow 10x$)

	Scaled 25 Pt.	Scaled 33 Pt.	Scaled 100 Pt.
Original	.14609	.15041	.07887
Aff. Inv.	.03380	.04150	.00942

In Table 6 we show the results of further comparisons. Here the the ratio of the RMS errors of the original method and the affine invariant modification for three data sets and three functions are given. The three data sets are shown in Figure 4 and the three functions are

$$F_1(x,y) = \frac{1}{3}\exp\left[-\frac{81}{16}((x-\frac{1}{2})^2 + (y-\frac{1}{2})^2\right],$$

$$F_2(x,y) = \frac{1.25 + \cos(5.4y)}{6[1 + (3x - 1)^2]},$$

and

$$F_3(x,y) = \text{same as in equation (3.1)}$$

The graphs of these three functions are shown in Figure 5. The results of this table indicate that for typical data sets and typical functions, the affine invariant modification often gives slightly better results than the original method. But this is not always the case and in a given situation it may not yield the best performance of a method. Thus it may be better to retain the parameter(s) of scaling differently in different independent variables in order to achieve more desirable results.

We cite an example where poor results were obtained using several methods before scaling in one variable, while satisfactory results were obtained using two methods when suitable scaling was carried out. The data consisted of 64 sea surface temperature measurements off the Big Sur coast of California, taken during a hydrographic survey. The data is tracked, with 5 tracks running more or less perpendicular to the coast; the distances between tracks is about 25 kilometers, while distances between points on the grid vary from less than three kilometers near the coast to more than ten kilometers further away. The data was considered by (Breaker 1984) and required interpolation from the scattered points to a rectangular grid in order to enable processing by algorithms that required a grid of data. Breaker found that this disparity in the distances along and between tracks caused difficulties for methods based on distance because the closest points were nearly on a straight line. By scaling the distance along the tracks by a factor of about three, very satisfactory results were obtained using the Modified Quadratic Shepard method (see Section 4.1 and Franke and Nielson 1980) and local thin plate splines method (see Section 2.4). Breaker's ultimate choice was the former because of its relative simplicity. While these studies were carried out before the development of a affine invariant metric, we note that application of the affine invariant metric results in scaling by about a factor of 1.5 in a direction nearly equal to that of the tracks, not as large a scaling as was found desirable by empirical means. The temperature data is also treated in (Foley 1986) and (Foley 1987) where it is referred to as Montery Coast data, with a table of the data being listed in the former.

Table 6. $\dfrac{\text{RMS error surface spline}}{\text{RMS error affine invariant surface spline}}$

	Data Set 1	Data Set 2	Data Set 3
Function F_1	1.05	1.28	1.08
Function F_2	0.98	1.13	1.01
Function F_3	0.96	1.04	0.99

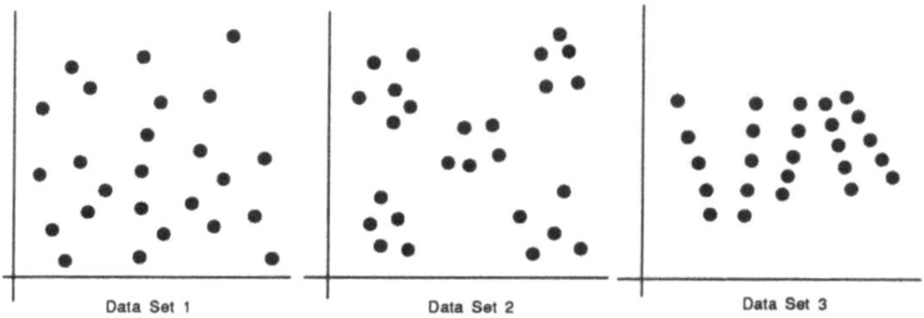

Figure 4. Three data sets

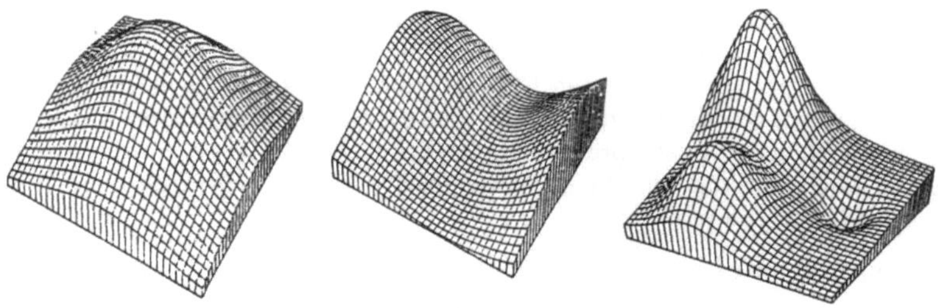

Figure 5. Graphs of F_1, F_2 and F_3

4. Available Algorithms That (sometimes) Work

The purpose of this section is to discuss some algorithms that are widely available and which have proven to be useful for solving some scattered data interpolation problems. Our emphasis is primarily on algorithms in the public domain and other schemes which are easy to code. The ACM Algorithms and many others are available by electronic mail via NETLIB (see Dongarra and Grosse 1878, but note that the address has changed; send the message "send index" to netlib@ornl.gov). Further information is also given in (Grosse 1990). Various commercial libraries and plotting packages also have routines for solution of the problem, and these are generally, but not always, based on published ideas or algorithms. A few comments will be made about such software.

4.1 Modified Quadratic Shepard

The deficiencies of the basic Shepard's method were indicated in Section 2. The primary modifications required involve modifying the weight function d_k to localize the overall approximation, and replacing f_k with a suitable "local approximation", $Q_k(x, y)$, to the surface. The method was developed by Franke and Nielson (Franke and Nielson 1980) where the local weight functions were of the form $W_k(x, y) = (r_w - d_k)_+^2/(r_w d_k)^2$, for some constant r_w. Here the sub+ denotes the truncated power function, hence the weight is zero at distances greater than r_w from the data point. The $Q_k(x, y)$ were taken to be quadratic polynomials, obtained by a weighted least squares fit, the quadratic constrained to take on the value f_k at (x_k, y_k). The weights in the least square process were taken of the same form as the weight functions, but with r_w replaced by another value, r_q. The approximating function is of the form

$$F(x, y) = \frac{\sum_{k=1}^{N} W_k(x, y) Q_k(x, y)}{\sum_{k=1}^{N} W_k(x, y)}.$$

Note that the function is locally determined, the influence of any point not extending further than a distance $(r_w + r_q)$ from each data point.

Assuming somewhat uniform density of data, the constant values for r_w and r_q are appropriate, and Franke and Nielson suggested values for which about 9 and 18 data points would be in the disks of radius r_w and r_q, respectively. Fitting data outside the region in which the weight function for the point goes to zero seems to be a prudent idea as the transition from one region to another (and one local quadratic to another) would tend to give less abrupt changes in the overall surface. Many copies of the Franke and Nielson program have been distributed, however the more recent ACM implementation by Renka is readily available and has been extensively tested.

If the data are not of reasonably uniform density, it then may be desirable to let the radii r_w and r_q depend on k. Renka (Renka 1988) reports that this idea is generally beneficial, and further that good results are obtained over many data sets by taking the radii just big enough to include 26 and 13 points, respectively, in the disks associated with the weight functions for the kth data point. Note that the relative sizes of the regions is contrary to the thoughts in the previous paragraph. The Renka program is also available in a version to handle data in three independent variables, where different criteria are used for selection of the radii.

4.2 Minimum Norm Network and Related Methods

There are three steps to Nielson's minimum norm network (MNN) method:

(i) *Triangulation.* The convex hull, D, of the planar points $V_i = (x_i, y_i), i = 1, \ldots, N$ is triangulated using only these points as vertices.

(ii) *Curve Network.* Initially, the interpolant,S, and its derivatives are defined over the union of all edges of the triangulation.

(iii) *Blending.* The network of (ii) is extended to all of D by means of triangular patches which interpolate to the boundary curves provided by the network and specified boundary derivatives.

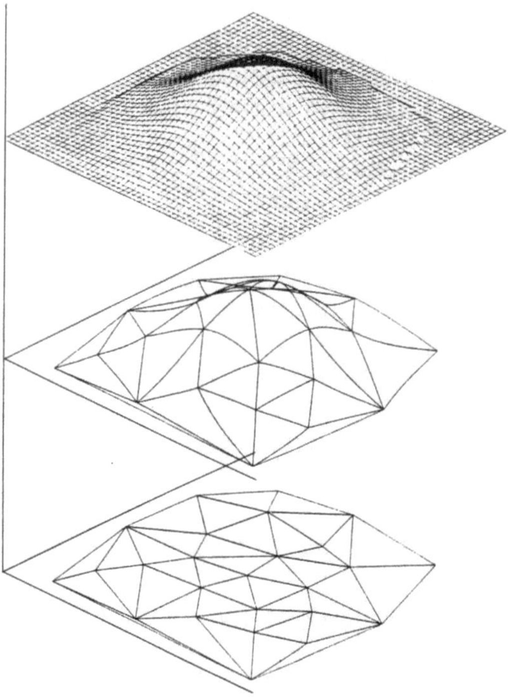

Figure 6. Steps of Minimun Norm Network Method.

As we mentioned above, the problem of triangulating planar data sets has been treated quite well in the literature. If one uses a method which involves a triangulation and also wants the method to be affine invariant, then a method for choosing an affine invariant triangulation must be used. Using concepts related to the affine invariant norm discussed above, Nielson has developed and studied a method for characterizing "optimal" affine invariant triangulations (see Nielson 1987b).

The main result of the second step is to produce the first order partial derivatives at each of the data sites. Nielson does this by defining a network which is analogous to the univariate cubic spline in that it has a minimum norm characterization. Given the data $(t_i, S_i), i = 1, \ldots, N$, the univariate natural spline of interpolation is characterized as the unique solution to the problem

$$\frac{MIN}{f \in H[t_1, t_N]} \int_{t_1}^{t_N} [f''(t)]^2 dt, \text{ subject to : } f(t_i) = s_i,$$

where $H[t_1, t_n] = \{f : f \in C[t_1, t_N],\ f'$is absolutely continuous, $f'' \in L^2[t_1, t_N]\}$ As a result of this minimization, it can be shown that s is a piecewise cubic polynomial which has a continuous second derivative and satisfies the "natural" end conditions, $s''(t_1) = s''(t_N) = 0$. Analogous to this, Nielson defines the functional

$$\sigma(F) = \sum_{ij \in N_e} \int_{e_{ij}} [\frac{\partial^2 F}{\partial e_{ij}^2}]^2 ds_{ij}$$

and proves the following characterization theorem:

Theorem. (Nielson) *Let $S \in C[F]$ be the unique piecewise cubic network with the properties that $S(V_i) = f_i, i = 1, \ldots, N$ and*

$$\sum_{ij \in N_i} \frac{(x_j - x_i)}{||e_{ij}||^3} [(x_j - x_i) S_x(V_i) + (y_j - y_i) S_y(V_i) + \frac{1}{2}(x_j - x_i) S_x(V_j) + \frac{1}{2}(y_j - y_i) S_y(V_j) + \frac{3}{2}(z_i - z_j)\Big] = 0,$$

$$\sum_{ij \in N_i} \frac{(y_j - y_i)}{||e_{ij}||^3} [(x_j - x_i) S_x(V_i) + (y_j - y_i) S_y(V_i) + \frac{1}{2}(x_j - x_i) S_x(V_j) + \frac{1}{2}(y_j - y_i) S_y(V_j) + \frac{3}{2}(z_i - z_j)\Big] = 0, \tag{4.1}$$

Then, among all functions $F \in C[E], F(V_i) = f_i, i = 1, \ldots, N$, the function S uniquely minimizes $\sigma(F)$.
Here,

$$N_e = \{ij : e_{ij} \text{is an edge of the triangulation }\},$$

$$N_i = \{ij : e_{ij} \text{is the edge of the triangulation with the endpoint} V_i\},$$

and $C[E] = \{F : F$ is the restriction to the set of edges of some C^1 function defined on D and the univariate function function obtained as the restriction of F to e_{ij} is a element of $H[e_{ij}]\}$.

In order to implement this method, the system of equations (4.1) has to be solved. The system is sparse and easily solved for the first order derivatives by an iterative method.

Once the first order partials have been computed, the interpolating curve network is extended to the entire domain by means of a triangular patch interpolant. Nielson uses a 9-parameter, C^1 triangular interpolant which is itself characterized on the basis of minimizing a certain norm (see Nielson 1981). Renka has programmed the MNN method of estimating derivatives and this Fortran program is available from ACM (see Renka 1984b). Renka and Cline (Renka 1984a) use the Clough-Tocher element for triangle interpolation, but as Nielson and Franke have reported (Nielson and Franke 1983), the choice of the particular triangle interpolant is not so critical; it is the method of estimating the partial derivatives that really affects the performance of these types of methods.

Motivated by the problems of ringing and overshoot often experienced with rapidly varying data, Nielson and Franke (Nielson and Franke 1984) developed a generalization of the MNN method which allows the user to specify "tension" values along the segments of the curve network. If the tension values are zero then this method reduces to a method very similar to the conventional MNN. This program has been successful used in terrain modelling algorithm developed by Kelley, Malin and Nielson (see Kelley, Malin and Nielson 1988). The Fortran implementation of this tension version of the MNN method has been distributed by Franke. The program was written for developmental and experimental purposes and so it is distributed with the usual caveats for this type of program.

4.3 Radial Basis Function Methods

The original radial basis function method, due to Hardy (Hardy 1971) was mentioned above. From the initial ideas, probably in the late 1960's, into the early 1980's, no mathematical theory existed for the method, not even so simple and crucial an idea as the assured existence of a solution to the equations in all possible point configurations. The method was touted by Hardy, and verified by others (see e.g. Franke 1982b) as being capable of fitting the underlying surface very accurately, and the calculations are not too onerous provided the data set contained no more than (say) a few hundred points. Another investigation using a variable value of c was used to obtain functions with steep gradients that were then used as basis functions in the solution of partial differential equations (Kansa 1990).

Recently there has been an explosion of theory, yielding some explanation of why the method works well. The first theoretical results were due to Micchelli (Micchelli 1986), the most important being that the system of equations is nonsingular, and that the multiquadric function $h_k(x,y) = \sqrt{d_k^2 + c^2}$ is conditionally positive definite of order one. As a practical matter, this means that the inclusion of a constant in the approximation (with the resulting side condition that the sum of the coefficients of the multiquadrics sum to zero) is the natural strategy and results in precision for constant functions.

Further results which apply to the multiquadric method are those of Madych and Nelson (Madych and Nelson 1988). This paper shows that the multiquadric method minimizes a certain pseudonorm and thus is the optimum approximation in a certain Sobolev space with a weighted norm.

In another vein, recent work at Cambridge University by Powell, Jackson, and Buhmann (see e.g., Powell 1987a, Powell 1987b, Jackson 1988, and Buhmann 1988) has considered the problem of degree of approximation by multiquadric (and other radial basis) functions. While they have primarily been concerned with cases where the data lies on a regular grid in n dimensions, their surprising results give some information about the degree of approximation possible. In particular, with an infinite amount of data, low degree polynomials are reproduced, the degree (and the order of approximation) increasing with the dimension of the space in which the approximation is being carried out. Further interesting information concerns the rate of decay of the cardinal functions for multiquadric and thin plate spline approximation (Buhmann and Powell 1988).

Finally, a recent history of the multiquadric method and its applications is given in (Hardy 1990).

The system of equations for the coefficients in the multiquadric method seems to have a rather mild condition number, examples being known where the condition number for 500 points in the unit square is less than 10^6, yielding adequate accuracy in single precision on some machines and in double precision on others. Two recent papers on the condition numbers of such coefficient matrices are (Narcowich and Ward 1989) and (Narcowich and Ward 1990).

The method of thin plate splines was, as mentioned, developed by Harder and Desmarais from an operational point of view. This scheme was later developed by Duchon (Duchon 1975, see also Meinguet 1979a-c, 1981, 1984) and is a special case of a family of rotation invariant schemes which minimize various seminorms in Sobolev spaces. In the case of thin plate splines, the name arises from the fact that the thin plate functional,

$$\int_{R^2} (F_{xx}^2 + 2F_{xy}^2 + F_{yy}^2) dy dx$$

is minimized over all functions which interpolate the data and belong to a certain Sobolev space. The equations represent the physical situation considered by Harder and Desmarais, a thin plate with point loads at the data points that enforce the interpolation conditions. The coefficients of the $d_k^2 \log d_k$ terms differ by a constant factor from the required loads. The thin plate spline includes linear terms and hence has precision for linear functions. In practice the method performs very well, nearly as well as the multiquadrics. The system of equations tends to be somewhat less well-conditioned, hence cannot be used as a global approximation for as many points as the multiquadric method.

A variation of thin plate splines is the thin plate spline with tension (see Franke 1985b). These functions model a thin plate with tension applied and loads which force the plate to pass through the given points. The radial basis function in this case is

$$g(r) = \frac{1}{2\pi\alpha^2}[K_0(\alpha r) + \log(\frac{\alpha r}{2}) + \gamma]$$

Here α is the tension parameter, K_0 is the modified Bessel function, and the natural value of m is 1. Of course, in practice the leading constant in the function can be dropped (incorporated into the coefficient, in effect), as well as the constant γ (Euler's constant) because the sum of the coefficients will be constrained to be zero.

The method has been applied to meteorological problems (interpolating wind fields) in Hong Kong and found to be useful (see Hickernell 1989). An initial data transformation is used to account for the effects of topography. The parameters in the approximation are determined by cross-validation.

Another type of interpolation which is in fairly wide use in some fields is a scheme based on statistical considerations. It is called "Kriging" in the mining and geology literature, and "Statistical Interpolation" in the meteorological literature. Details vary slightly, but the schemes are equivalent. Initial developments in meteorology and geology were by (Gandin 1965) and (Matheron 1970), respectively. It is assumed that the data to be approximated (usually not interpolated in the sense of exact matching) represents values from a function selected from an ensemble of stationary (the parameters of the probability distribution do not depend on (x, y)) random functions with mean of known form. The approximation is chosen to minimize the mean squared error over all functions in the ensemble, under the assumption that the ensemble of functions is normally distributed so that the optimal approximation is linear. Carrying out this process, it is found that the approximation is a radial basis function method with the basis function being the spatial covariance function for the ensemble, and the part of the polynomial terms being taken by the assumed form of the spatially dependent mean value over the ensemble, which could be zero, a constant or linear polynomial, or a linear combination of any other set of functions. The estimated mean-squared error is easily available and is mentioned by practitioners as one of the attractive features of the method. The methods are generally approximation methods which account for the error in measured data by including an additive delta function with the radial basis function for purposes of calculating the coefficients in the approximation from the interpolation equation; the delta function is then dropped from the approximating function.

A major difficulty with statistical methods is that data from real sources is not known to satisfy the assumptions, and more seriously, the spatial covariance function is unknown, and hence must be estimated from the data. In meteorology (see Thiébaux and Pedder 1987) the amount of historical data available ranges from more than merely adequate to staggering. Thus, while the form of the covariance function to be assumed is somewhat arbitrary, estimation of the parameters can be done in reasonable fashion. Various authors have found that a special form of the second order autoregressive correlation function, plus a constant,

$$g(r) = A + (1 - A)(1 + br)\exp^{-br}$$

fits correlation data from meteorological sources remarkably well (see Franke, Barker and Goerss 1988).

The problem of estimating the spatial covariance function for rather small sets of data such as occur in mining and geology is a major problem. Further, the statistical assumptions seem less likely to be valid. Nonetheless, practitioners of Kriging spend considerable amounts of computational effort to determine suitable covariance functions (see Dubrule 1984 for the flavor of some of the work). This part of the process is called "structural identification", and identifies the order of the trend (the degree of the polynomial mean value, a function of position), and the generalized covariance function K(h) of the data after the mean has been removed. The generalized covariance is frequently assumed to be a linear combination of h, h^3, h^5 and $h^2 \log h$. Inclusion of a delta function (called a nugget in this context) allows for errors in the data. Because the process is often applied to relatively small sets of data, the success of the method may depend as much on the fact that radial basis function methods based on h, h^3, h^5 and $h^2 \log h$ can be explained in functional analysis terms as well as in statistical terms. Any one of the radial basis functions often leads to a reasonable approximation. The calculation of the resulting estimated mean squared errors can be misleading, since these are much more highly dependent on proper estimation of parameters than the actual errors (see Franke 1985). A related approach (in some cases identical) is the method of Laplacian Smoothing Splines (LSS) developed by Wahba and Wendelberger (Wahba and Wendelberger 1980). The scheme is a generalization of thin plate splines to approximation (by including a delta function in the radial basis function for calculation of the coefficients; again, it is dropped from the approximating function) and to more dimensions and seminorms involving higher order derivatives. The method was mentioned by Duchon in its simplest form, as well as by Harder and Desmarais. The method involves a smoothing parameter governing the fidelity to the data, and this can be chosen by generalized cross-validation. The functional minimized is a linear combination of a seminorm, the integral of sums of the squares of all possible derivatives of order m, and a weighted sum of the squares of the deviations from the data. In dimension d, the radial basis functions are of the form r^{2m-d} for d odd, and $r^{2m-d} \log r$ for d even, with the polynomial part of the approximation having degree $m - 1$. The algorithm is quite robust and handles reasonably large sets of data (up to 200-300 points, even repeated independent coordinates). The Fortran code can be obtained from the University of Wisconsin Computer Center for a modest fee. Documentation of the code is in (Wendelberger 1981).

When the size of the data set is very large, it is necessary to use local schemes. The previously mentioned (Section 2.4) method of Franke (Franke 1982a) uses thin plate splines as the underlying approximation (any other radial basis function method could be easily incorporated). The method uses weighting functions which are Hermite bicubics giving derivative continuity across grid lines. If the underlying approximation has more smoothness (as in the multiquadric method, for example) higher order weight functions could be used. It is probably not useful to do this for cosmetic purposes since the higher order Hermite functions tend to force rapid transitions into the region near one-half giving more rapid transitions between local approximations.

4.4 Cautionary Advice

Finally, we have mentioned that there are a number of commercial codes available that give solutions to the scattered data interpolation/approximation problem. Some of these are imbedded in graphics packages, while others are marketed as being especially designed to handle data from particular sources, and are especially prevalent for problems in the petroleum and related industries. Despite any claims for performance, use of these codes should be approached in the same manner as any other for the problem. No one algorithm will handle all data presented to it in a satisfactory manner, hence it is unlikely there will ever be a "black box" for scattered data approximation. Quality varies widely: In some cases fairly primitive schemes are proposed, while in others there has clearly been a lot of work and thought put into the algorithm. Some investigation of the ideas on which the code is based (not always easy for commercial packages) and how it works for data similar to that the user wants to approximate is a necessary step toward obtaining acceptable results. Since there is usually no quantitative way of measuring the quality of the approximation, final judgements must be made on the basis of subjective measures. For this an important ingredient is visual representation of the interpolant, a subject beyond the scope of this article.

5. Other Problems

In this section we will discuss briefly some other problems relevant to the scattered data interpolation problem. Since the topic of data lying on manifolds or other surfaces is dealt with elsewhere in this volume (Barnhill and Foley 1990), we only touch on it briefly. In many situations modern data gathering techniques have given rise to a surfeit of data, and we discuss some ways of dealing with this, and we also mention an approximation scheme that can be very effective in dealing with some problems. Finally, we discuss the problem of constructing a surface which passes through a collection of $3D$ points. Often the techniques here are used for surface design rather that data analysis, but the subject is sufficiently related to warrant discussion here.

5.1 Surfaces on Surfaces

The "surfaces on surfaces" problem assumes that the independent data sites $p_i, i = 1, \ldots, N$ are given on some surface domain D and the problem is to construct a function F, defined on **all** of D which interpolates the data, i.e. $F(p_i) = f_i, i = 1, \ldots, N$. For example, one may have temperature measurements at certain locations on the surface of a spacecraft and wish to interpolate temperature on the entire surface; or one may have a sequence of measurements from various locations on the human skull (as is the case for an EKG) and wish to fill in and display this information for the entire surface. An important special case is where D is a sphere. This situation not only arises in many applications (where the domain is the surface of the earth) but it is also a key component in the very general ,"domain mapping" method of surfaces on surfaces recently developed by Foley, Lane, Nielson, Franke and Hagen (Foley, Lane, Nielson, Franke and Hagen 1990). We now describe some of the methods available for the spherical domain.

Lawson has extended his basic planar method to a spherical domain (Lawson 1984). First, a triangulation of the spherical domain is computed. Spherical triangles are comprised of edges which are geodesic curves. The next step requires the estmation of derivatives at the data sites. Lawson uses techniques similar to those used in his planar method. The final step involves filling in the surface with spherical triangular patches. Lawson uses a spherical modification of Nielson's side-vertex method (Nielson 1979) for this purpose. A similar method is available as an ACM algorithm (see Renka 1984a)

Ramaraj and Nielson (see Ramaraj 1986 and Nielson and Ramaraj 1987) have extended the ideas of the MNN method (see Section 4.2) to a spherical domain. Here the network is defined over the union of geodesics comprising the triangulation and similar to the planar case, a sparse linear system must be solved for the derivatives at each data site. The final step extends the network to the entire sphere by means of spherical triangular patches. The details of this method along with some code for implementation can be found in Ramaraj's thesis (Ramaraj 1986) and the subsequent paper (Nielson and Ramaraj 1987). An additional important component of the work of Nielson are Ramaraj is the development of techniques for displaying functions defined over a spherical domain. These techniques are covered quite thoroughly in Ramaraj's thesis. They have been finely tuned and improved upon recently and these results can be found in (Foley, Lane, Nielson and Ramaraj 1990).

The idea of extending Hardy's method to a spherical domain is appealing. Rather than the usual notion of Euclidean distance, one could use geodesic distance. The procedure is the same as the planar case where one solves the system of equations

$$\sum_{i=1}^{N} a_i \sqrt{d(p_j, p_i) + c^2} = f_j, \quad j = 1, \ldots, N$$

where $d(p, q)$ is the distance along the great circle from p to q, i.e. it is simply the inner-product (p, q). The problem with this method is that the basis functions, $\sqrt{d(p, p_j) + c^2}$ have discontinuities in the first derivatives at the antipodal points. One can smooth these off, say with piecewise cubics (used in a radial sense), in order to get a C^1 or C^2 basis function. Foley has done this and reported the results in (Foley 1989). The multiquadric function can be viewed as the length of a hypotenuse. Pottmann and Eck (Pottman and EcK 1989) have used this observation to develope a different type of extension of Hardy's method to a sphere.

5.2 Subset Selection

When there is a great deal of data available, two choices seem to be available to the investigator. Either a subset of the data must be chosen, to which one of the methods we have already mentioned can be applied, or some type of approximation to all of the data that involves fewer coefficients might be used. We first discuss the subset selection problem.

In subset selection the problem is choose a "representative" subset of the data from which the approximating function is constructed, probably by interpolation although an approximation method could be used. This can be done a priori, using some local computations to assess how important various points are to the definition of the surface, or adaptively by making a choice of subset and then adjusting the subset based on how good the approximation is at the remaining points.

The first strategy is the subject of (Bozzini, deTisi and Lenarduzzi 1986) where the local behavior of the surface is assessed and a figure of "importance" assigned to each data point. This is done by computing a local average deviation of the surface from its value at a given data point, and then comparing its variation from nearby data points to that. Larger relative variations result in a larger figure of "importance". A subset is then selected based on these values. According to the authors the process is not too sensitive to deletion of a data point, so the subset can be selected based on one pass of the algorithm, although multiple passes could be made as well, the selected points being deleted from consideration on later passes.

Another approach was taken by Pickrell (Pickrell 1979), and later improved on by Schiro and Williams (Schiro and Williams 1984). The application for their ideas was to the representation of ocean bottom topography. This problem has some special considerations; for example, navigational maps have to have good accuracy in shallow areas, and especially in the vicinity of a sharp underwater hill. Here the idea was to adaptively choose points by starting with a small (somewhat arbitrary) subset. The multiquadric interpolant for this subset was computed, and errors at the remaining points were computed. More points were added to the subset (and perhaps some deleted) based on these errors, and the process repeated. Some care was necessary to prevent adding points too close together on any given iteration. Schiro improved on Pickerell's ideas by subdividing the region into rectangular areas which represent somewhat homogeneous behavior of the surface. These regions were considered individually, then blended together to obtain a smooth approximation in a manner somewhat like that suggested earlier in (Jancaitus and Junkins 1974) and (Franke 1977).

Another approach to subset selection is from the other direction. The removal of "redundant" data has been suggested by LeMehaute and Lafrance (LeMehaute and Lafranche 1989), inspired by work of Lyche and coworkers in the univariate and gridded data case (see e.g. Lyche and Mørken 1988). Here the idea is to use a finite element method for the approximation (they suggest the use of Bell's interpolant (Bell 1969), but others could be used; it is assumed that all necessary data for construction of the interpolant is known), and then assess the effect of removing points from the set. Points are removed if the reduced point set varies by less than ϵ from the surface constructed from all of the data. The process can proceed adaptively, with several "nonadjacent" points being removed in any given iteration and nearby ones being removed on successive iterations.

5.3 Least Squares and Knot Selection

The algorithms of the previous section are distinguished from those of this section by the use of interpolation at a subset of the data points, the subset being chosen to obtain a surface that approximates the other data points in some reasonable fashion also. Here we will not require interpolation at any particular point, but attempt to approximate all points in some reasonable way by ultimately computing a least squares approximation for the data.

The efficacy of radial basis function methods seem to make them attractive for use as least squares approximants. Schmidt (Schmidt 1985) has done some initial work on the use of least squares thin plate splines with the knot locations (points at which the radial basis functions are centered) being parameters to be determined as part of the (nonlinear) least squares process. If n basis functions are used, this results in a system of about $3n$ equations. Her results appear to indicate that multiple local minima for the objective function may be a serious problem. This led to the idea of decoupling the knot selection process from that of the least squares problem (which then is linear, and hence relatively uninteresting since it is reliably and quickly solved using the QR decomposition).

The selection of knots on an a priori basis necessarily involves some assumptions about the data points, and the assumption made in (McMahon and Franke 1987) (see also McMahon 1986) was that the "Venezia Criterion" should prevail which is that data density indicates something about the behavior of the surface, in particular that regions with greater density indicate greater variation in the surface. This assumption led to the idea that each point had the same relative importance, and thus each point should be "close", in some sense, to a knot point. The criteria adopted were (1) each knot point should have the same number of data points which were closer to it than any other knot point, and (2) that the sum of the squares of the distances of the data points from the nearest knot point should be a (local) minimum. The latter function has very many local minima, and McMahon's work involved an algorithm for searching for a knot configuration which corresponded to a local minimum for the second criterion in which the first criterion was nearly satisfied. The algorithm was capable of generating knot sets that reflected varying densities of data sets quite well.

5.4 Two Stage Methods

Because of the computational attractiveness of tensor product methods, several investigations into the use of bicubic splines and other bicubic functions to fit scattered data have been made. This idea was apparently first mentioned by Schumaker (Schumaker 1976) as a two stage process. The first stage is to construct a local approximation using a method for scattered data interpolation (or approximation) to generate the needed data for a grid of points. For bicubic splines, only function values are necessary, while for piecewise Hermite bicubics the values of first partial derivatives and the cross partial (or twist) would also be estimated. The second stage would be the tensor product approximation on the grid. This approximation does not, in general, interpolate to the original data. It this is deemed desirable, a third stage could be used to achieve interpolation. One way to accomplish this is to use the technique of delta iteration developed by Foley and Nielson (Foley and Nielson 1980). In most instances it is the case that interpolation is not required. Further studies of this type of interpolant have been made by Foley (Foley 1984) using the tensor product Hermite bicubic functions.

Weighted cubic splines arose from an investigation by Salkauskas (Salkauskas 1984) where sampling of contours along straight lines led to data with rapidly changing slopes. By minimizing a weighted seminorm involving the second derivative the ringing effect common in spline approximations to such data were controlled. Further investigations by Foley then treated the rapidly variation in the bivariate case through the use of a seminorm incorporating a weight function (Foley 1987b). The computer programs written by Foley are discussed in (Foley 1987b), with details of their application to several different type of data, including the Big Sur coast water temperature data (mentioned in Section 3.2, above). This type of method can be very effective, especially in cases where the data is somewhat regularly distributed. In situations where there are clumps or voids, care must be taken to ensure the function behaves properly in void regions.

5.5 Surface Interpolation to Scattered Data

Another related problem is that of constructing a surface which passes through a set of specified control points. Given $(x_i, y_i, z_i), i = 1, \ldots, N$ we want to determine a surface, S, which passes through these points. S can be defined implicitly or parametrically. There has been more attention given to the parametric case (see Farin 1983). For this case, the first thing that must be assumed is that there has been assigned some interconnection topology joining the control points. An easy way to satisfy this is to require that a triangulation of the points be given. This gives a polyhedral surface which can serve as a rough and none smooth approximation to the surface. An example is shown in Figure 7.

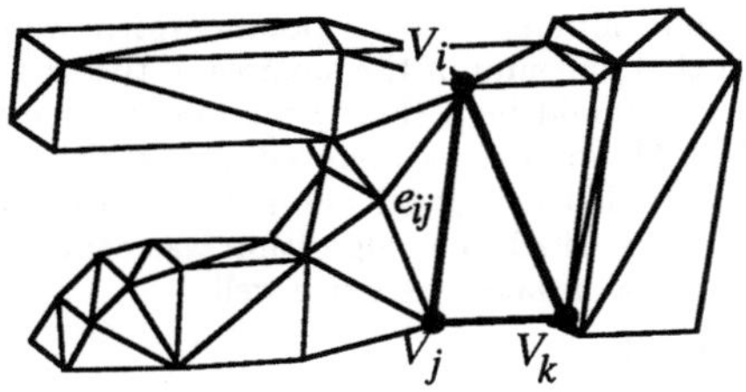

Figure 7. Polyhedron

One approach to getting a smooth surface of interpolation is to use triangular patches which pass through the vertices and then join them up smoothly along their edges. A possible condidate is the G^1, six parameter patch developed by Nielson (Nielson 1987a). This patch, G, is defined over some barycentric coordinates, $b_1, b_2, b_3; b_1+b_2+b_3+1$ and has the property that it interpolates the given the vertices, $V_i = (x_i, y_i, z_i)$, $V_j = (x_j, y_j, z_j)$ and $V_k = (x_k, y_k, z_k)$ and the corresponding normal vectors, N_i, N_j, and N_k. That is

$$G(1,0,0) = V_i, \quad G(0,1,0) = V_j, \quad G(0,0,1) = V_k)$$

and

$$\text{Normal of G at } (1,0,0) \text{ is } N_i$$
$$\text{Normal of G at } (0,1,0) \text{ is } N_j$$
$$\text{Normal of G at } (0,0,1) \text{ is } N_k$$

It also has the property that when two curve boundaries meet, there is a G^1 join, i.e. the tangent plane is continuous across patch boundaries. In order to turn this approach into an actually method, someway of estimating the normals, $N_i, i = 1, \ldots, N$ must be specified. This is probably the most crucial step in an approach of this type. It is akin to estimating the first order partial derivative of one of the finite element methods (see Section 2.2) which we know is extremely important to the success of these method. As was pointed out in (Nielson 1987a), using normals which are the averages of the normals of adjoining triangles does not work very well. A more involved approach is to extend the ideas of the MNN network to this context. This has been done (see Nielson 1988a), but the method has not been completely tested. We also mention that recently Hamann has developed a G^1 patch which may be better that the Nielson G^1 patch in some cases. This patch only applies to a restricted data set (convex), but there may be ways to circumvent this.

References

1. Akima, H.(1978), Algorithm 526: bivariate interpolation and smooth surface fitting for irregularly distributed data points, ACM TOMS **4**,160-164.
2. Alfeld, P.(1989), Scattered data interpolation in three and more variables, in: Lyche, T., Schumaker, L. L. (eds), Mathematical Methods in Computer Aided Geometric Design, Academic Press, New York, pp. 1-33.
3. Barnhill, R. E.(1977), Representation and approximation of surfaces, in: Rice, J. R., (ed) Mathematical Software III, Academic Press, New York, pp.68-119.
4. Barnhill, R. E., Foley, T. A.(1990), Surfaces on Surfaces, this volume.
5. Barnhill, R. E., Nielson, G. M. (1984), "Surfaces", A special issue of the Rocky Mountain Journal of Mathematics, Vol. **14**, No. 1.
6. Bell, K. (1969), A refined triangular plate bending element, Internat. J. Numer. Methods Engrg. **1**,101-122.
7. Bozzini, M., deTisi, F., Lenarduzzi, L.(1986), A new method in order to determine the most significant members within a large sample, SIAM J. Sci. Stat. Comp. **7**,98-104.
8. Breaker, L. C.(1983), The space-time scales of variability in oceanic thermal structure off of the central California coast, TR NPS-68-84-001, Naval Postgraduate School, Montery, CA.
9. Buhmann, M. D.(1988), Multivariate interpolation in odd dimensional Euclidean spaces using multiquadrics, Rpt NA/6, University of Cambridge, England.
10. Buhmann, M. D.(1989), Convergence of univariate quasi-interpolation using multiquadrics, IMA J. of Numer. Anal. **13**, 69-78.
11. Buhmann, M. D., Powell, M.J.D.(1988), Radial basis function interpolation on an infinite regular grid, Rpt DAMTP NA/13, University of Cambridge, England.
12. Clough, R. W., Tocher, J. L.(1965), Finite element stiffness matrices for analysis of plates in bending, in Proc. Conf. Matrix Methods in Struct. Mech., Air Force Inst. of Tech., Wright-Patterson AFB, Ohio.
13. Crain, I. K., Bhattacharyya (1967), Treatment of nonequispaced two- dimensional data with a digital computer, Geoexploration **5**, 173-194.
14. Cressman, G. P.(1959), An operational objective analysis system, Mon. Wea. Rev. **87**, 367-374.
15. Davis, P. J.(1963), Interpolation and Approximation, Blaisdell, Waltham, MA.
16. Dongarra, J. J., Grosse, E. (1987), Distribution of mathematical software via electronic mail, CACM **30** ,403-407.
17. Dubrule, O.(1984), Comparing splines and Kriging, Computers and Geosciences **10**, 327-338.

18. Duchon, J.(1975), Fonctions-spline du type plaque mince en dimesion 2, Rpt 231, Grenoble.
19. Duchon, J.(1975), Splines minimizing rotation - invariant semi - norms in Sobolev spaces, in: Schempp, W., Zeller, K. (eds), Multivariate Approximation Theory, Birkhauser, Basel, 85-100.
20. Duchon, J.(1976), Interpolation des fonctions de deus variables suivant le principe de la flexion des plaques minces, RAIRO Anal. Numer. **10**, 5-12.
21. Dyn, N. , Levin, D., and Rippa, S.(1986), Numerical procedures for global surface fitting of scattered data by radial functions, SIAM J. Sci. Stat. Comp. **7**, 639-659.
22. Farin, G., (1983), Smooth interpolation to scattered 3D data, in: Barnhill, R. E. , Boehm , W. (eds), Surfaces in Computer Aided Design, North Holland, Amsterdam, 43-64.
23. Farin, G., Luscher, N.(1990), Bibliography on CAGD, (to appear)
24. Foley, T. A.(1984), Three stage interpolation to scattered data, Rocky Mtn. J. Math. **14**, 141-149.
25. Foley, T. A.(1986), Scattered data interpolation and approximation with error bounds, CAGD **3**, 163-177.
26. Foley, T. A.(1987), Interpolation and approximation of three and four- dimensional scattered data, Comp. Maths. Appls. **13**, 711-740.
27. Foley, T. A.(1987), Weighted bicubic spline interpolation to rapidly varying data, ACM TOGS **6**, 1-18.
28. Foley, T. A.(1990), Interpolation to scattered data on a spherical domain, in: Cox, M., Mason, J., (eds), Algorithms for Approximation II, Chapman and Hall, London.
29. Foley, T. A., Lane, D., Nielson, G. M., Ramaraj, R., Visualizing functions over a sphere, IEEE Computer Graphics and Applications **10**,32-40.
30. Foley, T. A.,Lane, D., Nielson, G. M., Franke, R.,Hagen, H. (1990), Interpolation of scattered data on closed surfaces, CAGD, to appear.
31. Foley, T. A., Nielson, G. M.(1980), Multivariate interpolation to scattered data using delta iteration, in: Cheney, E. W. (ed), Approximation Theory III, Academic Press, New York, 419-424.
32. Franke, R.(1977), Locally determined smooth interpolation at irregularly spaced points in several variables, J. Inst. Math. Appl **19**, 471-482.
33. Franke, R.(1982), Smooth interpolation of scattered data by local thin plate splines, Comp. Maths. Appls. **8**, 273-281.
34. Franke, R.(1982), Scattered data interpolation: Tests of some methods, Math. Comp. **38**, 181-200.
35. Franke, R.(1985), Sources of error in objective analysis, Mon. Wea. Rev. **113**, 260-270.
36. Franke, R.(1985), Thin plate splines with tension, CAGD **2**, 87-95.
37. Franke, R., Barker, E., Goerss, J.(1988), The use of observed data for the initial-value problem in numerical weather prediction, Comp. Maths. Applic. **16**, 169-184.
38. Franke, R., Nielson, G. M.(1980), Smooth interpolation to large sets of scattered data, Intern. J. Numer Meth. Engrg. **15**, 1691-1704.
39. Franke, R., Schumaker, L. L.(1987), A bibliography of multivariate approximation, pp. 275-335 in: C. K. Chui, L. L. Schumaker, and F. I. Utreras (eds), Topics in Multivariate Approximation, Academic Press.
40. Gandin, L. S.(1965), Objective Analysis of Meteorological Fields, Translated from Russian by Israel Program for Scientific Translations, 242 pp. (NTIS TT65-50007).

41. Gordon, W. J. and Wixom, J. (1978), On Shepard's method of metric interpolation to scattered bivariate and multivariate data interpolation, Math. Comp. **32**, 253-264.
42. Grosse, E.(1990), A catalog of algorithms for approximation, in: M. G. Cox and J. C. Mason (eds), Algorithms for Approximation II,Chapman and Hall, London
43. Harder, R. L., and Desmarais, R. N.(1972), Interpolation using surface splines, Journal of Aircraft 9,189-197.
44. Hardy, R. L.(1971), Multiquadric equation of topography and other irregular surfaces, J. Geophysical Res. **76**,1905-1915.
45. Hardy, R. L.(1990), Theory and applications of the multiquadric - biharmonic method : 20 years of discovery 1968-1988, in: Advances in Partial Differential Equations (to appear).
46. Herron, G. J.(1985), A characterization of certain C^1 discrete triangluar interpolants, SIAM J. Numer. Anal. **22**, 811-819.
47. Hickernell, F. J., Ho, K. H., Kong, P. K., Lau, W. W.(1989), Spline approximation of the surface wind field from scattered data: an operational model for Hong Kong, TR, Dept. Math., Hong Kong Baptist College, Hong Kong.
48. Jackson, I.R.H.(1988), Convergence properties of radial basis functions, Constr. Approx. **4**, 243-264.
49. Jancaitus, J. R., Junkins, J. L.(1974), Modeling in n-dimensions using a weighting function approach, J. Geophys. Res. **79**, 3361- 3366.
50. Kansa, E. J.(1990), Multiquadrics - a scattered data approximation scheme with applications to computational fluid dynamics, I. Surface approximations and derivative estimates, in Advances in Partial Differential Equations, (to appear)
51. Kansa, E. J.(1990), Multiquadrics - a scattered data approximation scheme with applications to computational fluid dynamics, II. Solutions to parabolic, hyperbolic, and elliptic partial differential equations, in Advances in Partial Differential Equations, (to appear)
52. Kelley, A. D. , Malin, M. C., Nielson, G. M. (1988), Terrain simulation using a model of stream erosion, Computer Graphics **22**, 263-268.
53. Lancaster, P., Salkauskas, K. (1986), Curve and Surface Fitting: An Introduction, Academic Press, New York.
54. Lawson, C. L.(1977), Software for C^1 surface interpolation, in Mathematical Software III, J. R. Rice (ed), Academic Press, New York, 161-194.
55. Lawson, C. L.(1984), C^1 Surface interpolation for scattered data on a sphere, Rocky Mountain J. Math. **14**, 177-202.
56. LeMehaute, A., Lafranche, Y.(1989), A knot removal strategy for scattered data in R^2, Lyche, T., and Schumaker, L. L. (eds), in: Mathematical Methods in CAGD, Academic Press, 419-426.
57. Lyche, T.,Mørken, K.(1968), A data-reduction strategy for splines with applications to the approximation of functions and data, IMA J. Numer. Anal **8**, 185-208.
58. Madych, W. R., Nelson, S. A.(1988), Multivariate interpolation and conditionally positive definite functions, J. Approx. Theory and its Applic. **4**, 77-89.
59. Matheron, G.(1970), Random functions and their applications in geology, in: Computer Application in the Earth Sciences, Plenum Press, New York, 70-87.
60. McLain, D. H.(1974), Drawing contours from arbitrary data points, The Comp. J. **17**, 318-324.
61. McLain, D. H.(1976), Two dimensional interpolation from random data, Comput J. **19**, 178-181 (see errata, p. 384).

62. McMahon, J. R.(1986), Knot selection for least squares approximation using thin plate splines, M.S. Thesis, Naval Postgraduate School.
63. McMahon, J. R., Franke, R.(1987), Knot selection for least squares thin plate splines, Naval Postgraduate School Rpt NPS-53-87-005, Montery.
64. Meinguet, J.(1979), Multivariate interpolation at arbitrary points made simple, Z. Angew. Math. Phys. **30**, 292-304.
65. Meinguet, J.(1979), An intrinsic approach to multivariate spline interpolation at arbitrary points, in: Sahney, B. N. (ed), Polynomial and Spline Approximations, Reidel, Dordrecht, 163-190.
66. Meinguet, J.(1979), A convolution approach to multivariate representation formulas, in: Schempp, W., Zeller, K. (eds), Multivariate Approximation Theory, Birkhauser, Basel, 198-210.
67. Meinguet, J.(1981), From Dirac distributions to multivariate representation formulas, in: Ziegler, Z. (ed), Approximation Theory and Applications, Academic Press, 225-248.
68. Meinguet, J. (1984), Surface spline interpolation: basic theory and computational aspects, in: Singh, S. P., Burry, J. H. W., Watson, B. (eds), Approximation Theory and Spline Functions, Reidel, Dordrecht, 127-142.
69. Micchelli, C. A.(1986), Interpolation of scattered data: Distance matrices and conditionally positive definite functions, Constr. Approx. **2**,11-22
70. Narcowich, F. J., Ward, J. D.(1989), Norms of inverses and condition numbers for matrices associated with scattered data, CAT Rpt No.196, Texas A and M University, College Station.
71. Narcowich, F. J., and Ward, J. D.(1990), Norm estimates for inverses of scattered-data interpolation matrices associated with completely monotonic radial functions, Texas A and M University, College Station.
72. Nielson, G. M. (1979), The side-vertex method for interpolating in triangles, J. Approximation Theory **25**,318-336.
73. Nielson, G. M.(1981), Minimum norm interpolation in triangles, SIAM J. Numer. Anal. **17**, 44-62.
74. Nielson, G. M.(1983), A method for interpolation of scattered data based upon a minimum norm network, Math. Comp. **40**, 253-271.
75. Nielson, G. M.(1987), A transfinite, visually continuous, triangular patch, in: Farin, G. (ed), Geometric Modelling, SIAM, Philadelphia, 235-246.
76. Nielson, G.M.(1987), Coordinate free scattered data interpolation, in: Chui, C., Schumaker, L. L., Utreras, F., (eds), Topics in Multivariate Approximation, Academic Press, New York, 175-184.
77. Nielson, G. M.(1988), Interactive surface design using triangular network splines, in: Slaby, S. M., Stachel, H. (eds), Proceedings of the Third Internations Conference on Engineering Graphics and Descriptive Geometry, Vienna.
78. Nielson, G. M.(1988), Affine invariant triangualtions and tessellations, Technical Report TR88-023, Arizona State University, Tempe.
79. Nielson, G.M., Foley,T. A. (1989), A survey of applications of an affine invariant norm, in: Lyche, T., Schumaker, L. L., (eds), Mathematical Methods in CAGD, Academic Press, New York.
80. Nielson, G. M., Franke, R.(1983), Surface construction based upon triangulations, in: Barnhill, R., Boehm, W. (eds), Surfaces in Computer Aided Design, North Holland, Amsterdam, 163-177.

81. Nielson, G. M., Franke, R.(1984), A method for construction of surfaces under tension, Rocky Mt. J. Math. **14**, 203-221.
82. Nielson, G. M., Ramaraj, R.(1988), Interpolation over a sphere based upon a minimum norm network, CAGD **4**, 41-57.
83. Pickrell, A. J.(1979), Representation of hydrographic surveys and ocean bottom topography by analytical models, M.S. Thesis, Naval Postgraduate School, Monterey.
84. Pottmann, H., Eck, M.(1989), Modified multiquadric methods for scattered data interpolation over a sphere, manuscript, 1989.
85. Powell, M.J.D.(1987), Radial basis functions for multivariable interpolation: a review, in: Cox, M. G., Mason, J. C., (eds), Algorithms for Approximation, Oxford University Press, 143-167.
86. Powell, M.J.D.(1987), Radial basis function approximations to polynomials, Rpt DAMTP NA/6, University of Cambridge, England.
87. Ramaraj, R.(1986), Interpolation and display of scattered data over a sphere, Masters Thesis, Computer Science Department, Arizona State University, Tempe.
88. Renka, R. L.(1984), Algorithm 623: Interpolation on the surface of a sphere, ACM TOMS **10**,437-439.
89. Renka, R. L.(1984), Algorithm 624: Triangulation and interpolation of arbitarily distributed points in the plane, ACM TOMS **10**, 440-442.
90. Renka, R. L. (1988), Algorithm 660: QSHEP2D: Quadratic Shepard method for bivariate interpolation of scattered data, ACM TOMS **14**, 149-150.
91. Renka, R. L., Cline, A. K.(1984), A triangle-based C^1 interpolation method, Rocky Mt. J. Math. **14**, 223-238.
92. Ripley, B. D.(1981), Spatial Statistics, Wiley, New York.
93. Sabin, M. A.(1980), Contouring - a review of methods for scattered data, in: Brodlie, K. (ed), Mathematical Methods in Computer Graphics and Design, Academic Press, New York, 63-85.
94. Sabin, M. A.(1985), Contouring: the state of the art, in: Earnshaw, R. A.(ed), Fundamental Algorithms for Computer Graphics, Springer-Verlag, 411-482.
95. Salkauskas, K.(1984), C^1 splines for interpolation of rapidly change data, Rocky Mtn. J. Math. **14**, 239-250.
96. Schiro, R. A., and Williams, G.(1984), An adaptive application of multiquadric interpolants for numerically modelling large numbers of irregularly space hydrographic data, Surveying and Mapping **34**, 365-381.
97. Schmidt, R. M.(1985), Ein beitrag zur flachenapproximation uber unregelmassig verteilten daten, in: Schempp, W., Zeller, W. (eds), Multivariate Approximation Theory III, Birkhauser, Basel, 363-369.
98. Schumaker, L. L.(1976), Fitting surfaces to scattered data, in: Chui, C. K., Schumaker, L. L., Lorentz, G. G. (eds), Approximation Theory II, Wiley, New York, 203-268.
99. Schumaker, L. L. (1987), Triangulation methods, in : Chui, C., Schumaker, L. L., Utreras, F., (eds), Topics in Multivariate Approximation, Academic Press, New York, 219-232.
100. Shepard, D.(1968), A two dimensional interpolation function for irregularly spaced data, in Proc. 23rd Nat. Conf. ACM, 517-524.
101. Thiébaux, H. J., Pedder, M. A. (1987), Spatial Objective Analysis, Academic Press.
102. Wahba, G., Wendelberger, J. G.(1980), Some new mathematical methods for variational objective analysis using splines and cross validation, Mon. Wea. Rev. **108**, 1122-1143.

103. Wendelberger, J. G.(1981), The computation of Laplacian smoothing splines with examples, Rpt. 648, Dept. Statistics, Univ. Wisconsin.

Acknowledgements

The first author has been supported by various agencies of the U. S. Navy over a period of many years. The contributions of the second author were supported by the U.S. Dept. of Energy contract DE-FG-02-87ER25041 to Arizona State University and by NATO RG. 0097/88.

Variational Principles in Curve and Surface Design

H. Hagen und **G. Schulze**
FB Informatik
Universität Kaiserslautern

Abstract: Free-form surfaces are an essential part of sophisticated CAD-systems. A major topic is the design of smooth curves and surfaces, which are appropriate for the NC-process. In this publication we present some new techniques based on a calculus of variation approach.

0. Introduction

Free form objects are an essential part of powerful CAD systems. A major topic is the generation of smooth curves and surfaces which can be immediately supplied to the NC-process. In this paper a brief survey of well established techniques and some new methods is given which are developed using minimal principles.

The fundamental idea of the described methods is the use of modeling tools (curves or surfaces) which minimize a certain functional that can be interpreted in the sense of physics or geometry. In the case of curves a thin elastic beam serves as a model for a fair shape. Such a beam tends to take a position of least elastic energy and following D. Bernoulli (1742) the elastic energy stored in the beam is proportional to the integral

$$\int_a^b \varkappa^2(t) \mid x'(t) \mid \, dt.$$

Therefore a cubic spline gives a rough approximation for the position of a thin beam that passes through a sequence of freely rotating sleeves attached at fixed points.

In the first section of this paper the classical method of (weighted) spline interpolation is briefly summarized. In the second paragraph the main features of splines in tension and their polynomial alternatives are described. New results about non-linear spline curves are given in the third section and finally nonlinear splines in tension are treated in section 4. Further minimal principles using higher derivatives are briefly mentioned.

1. Weighted Spline Interpolation

Let Δ_n be an arbitrary partition of the interval $[a,b] \subset \mathbb{R}$

$$\Delta_n: a = t_0 < t_1 < \cdots < t_n = b \quad , \qquad n \in \mathbf{N} ,$$

and let $y_0, y_1, \cdots, y_n, y_0', y_n'$ be fixed real numbers. With F_n we denote the set of all C^k-functions on $[a,b]$ that interpolate the given data:

$$F_k = \{ f \in C^k [a,b] \mid f(t_i) = y_i, \ i = 0,\cdots, n \text{ and } f'(t_0) = y_0', \ f'(t_n) = y_n' \}.$$

The unique function $s \in F_2$ that consists of cubic polynomials defined on the subintervals $[t_i, t_{i+1}]$ $(i = 0,\ldots, n-1)$ is called the cubic **spline** with first derivative end conditions. Among all functions of F_2 the cubic spline s minimizes the functional

$$(1.1) \quad \int_a^b (f''(t))^2 \ dt$$

(see [Müller '78]).

The integral (1.1) can be interpretated as an approximation of the energy integral E

$$(1.2) \quad E = \int_a^b \varkappa^2(t) \mid x'(t) \mid \ dt$$

of a parametrized curve x with curvature function $\varkappa$. In the special case $x(t) = (t, f(t))$ E is given by

$$E = \int_a^b \frac{[f''(t)]^2}{[1+[f'(t)]^2]^{5/2}} \ dt \simeq \int_a^b [f''(t)]^2 \ dt,$$

if we use the additional assumption

$$(1.3) \quad (f''(t))^2 \ll 1 .$$

As D. Bernoulli has pointed out in 1742 the elastic energy stored in a thin beam is proportional to the quantity (1.2). Because the beam tends to take a position of least energy the cubic spline gives an approximation for the shape of an elastic beam that passes through a sequence of freely rotating sleeves attached at fixed points.

Algorithms for the efficient computing of cubic splines can be found in [Späth, '83] (see also section 2).

In the context of CAGD splines are used to represent curve (or surface) information. Therefore splines are mainly used as component functions x^i $(i = 1,\cdots, n)$ of parametrized curves $x:[a,b] \to \mathbb{R}^n$. The main problem with parametrized splines is the appropriate choice of the partition Δ_n of the parameter interval which is now not determined by the interpolation data (see [Foley / Nielson, '89]). After choosing the right knots the computation of the spline curve is simply done by computing n spline functions.

Because the elastic energy stored in a thin beam is only proportional to integral (1.2) it is obvious to introduce a weighted spline interpolation by using the extremals of the functional

$$(1.4) \sum_{i=0}^{n-1} \omega_{i+1} \int_{t_i}^{t_{i+1}} (f''(t))^2 \, dt \, .$$

Here one associates with each subinterval $[t_i, t_{i+1}]$ a weight coefficient ω_{i+1} which has the effect that high values for the weights will "tighten" the shape of the curve.

For given weights $w_j (j = 1,\ldots, n)$ there is a unique function $s_\omega \in F_1$ that minimizes (1.4) in set F_1 of all C^1 interpolating functions. s_ω is a piecewise cubic and can be computed by solving a tridiagonal system of equations whose unknowns are the first derivatives at the data points (see [Foley, '87]).

For equal weights $\omega_i = \omega_{i+1}$ in consecutive spline segments the weighted spline s_ω is even twice continuously differentiable in the knot t_i. This result follows from the fact that

$$\omega_{i+1} \, s''_\omega(t_i^+) = \omega_i \, s''_\omega(t_i^-).$$

In the case of identical weights $\omega_1 = \omega_2 = \cdots = \omega_n$ the weighted spline s_ω is again a common cubic C^2-spline.

Remark:

i) It is also common to use the following types of end conditions for spline functions

a) $f''(t_0) = f''(t_n) = 0$ (natural end conditions)

b) $f^{(\nu)}(t_0) = f^{(\nu)}(t_n) \quad \nu = 0,1,2$ (periodic end conditions)

ii) Piecewise polynomial functions of higher degree also have certain minimum properties. In the set of all interpolating C^{2n} functions the spline of degree 2 n + 1 minimizes the integral

$$\int_a^b (f^{(n+1)}(t))^2 \, dt$$

(see [Späth '83]).

Fig. 1: natural cubic spline with chordal parametrization

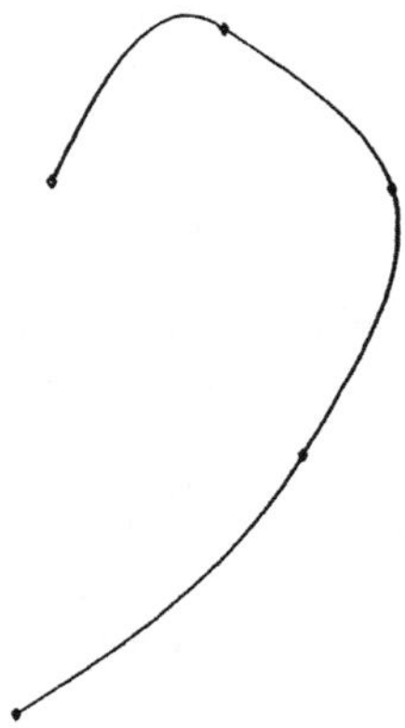

Fig. 2: weighted spline with weights $\omega_1 = 1, \omega_2 = 50, \omega_3 = 50, \omega_4 = 1$ and chordal parametrization

2. Splines in tension

Unwanted inflection points in an interpolating curve can often be removed by shortening the arc length of the curve. This fact inspired D. Schweikert to take the extremals of the functional

$$(2.1) \quad \int_a^b [f''(t)]^2 \, dt + \sigma^2 \int_a^b [f'(t)]^2 \, dt$$

into account (see [Schweikert, '66]). The Schwarz inequaltiy gives the following upper bound for the arc length L of a parametrized curve $x(t) = (x^1(t), x^2(t))$

$$L \le [(b-a) \int_a^b (x^{1\prime}(t))^2 + (x^{2\prime}(t))^2 \, dt]^{1/2}$$

Obviously the second part of the functional (2.1) has an effect on the arc length of the curve.

The Euler equation of the variational problem defined by (2.1) yields for each subinterval

$$f_i^{(4)}(t) - \sigma^2 f_i''(t) = 0 \qquad t \in [t_{i-1}, t_i]$$

and has the solution

$$(2.2) \quad f_i(t) = A_i + B_i t + C_i e^{-\sigma t} + D_i e^{\sigma t}.$$

Therefore the extremals of (2.1) are also called exponential splines. Among all functions of F_2 the spline under tension σ with the spline segments given by (2.2) minimizes the functional (2.1) (see [Pilcher, '74]).

The idea of exponential splines has been generalized by H. Späth and others by introducing different tension parameters σ_i $(i=1,\cdots,n)$ for each subinterval. This approach allows a local smoothing of the curve by varying the tension parameters.

Using the following representation of the exponential spline segment f_i $(i = 1,\ldots,n)$

$$f_i(t) = \alpha_i(t-t_{i-1}) + \beta_i(t_i - t) + \gamma_i \Psi_i(t - t_{i-1}) + \delta_i \Psi_i(t_i - t)$$

where

$$\Psi_i(t) = \frac{\sinh(\sigma_i t) - t\sinh(\sigma_i h_i)/h_i}{(\sinh(\sigma_i h_i)/h_i) - \sigma_i} \qquad i = 1,\ldots,n$$

and $h_i = t_i - t_{i-1}$, one gets from the interpolation conditions

$$\alpha_i = y_i / h_i \quad \text{and} \quad \beta_i = y_{i-1} / h_i \quad (i = 1,\ldots,n).$$

Introducing the quantities

$$\upsilon_i := \Psi_i'(h_i), \quad y_{i-1}' := f_i'(t_{i-1}^+), \quad y_i' := f_i'(t_i^-) \quad (i = 1,\ldots,n),$$

yields

$$\gamma_i = \frac{y_{i-1}' + \upsilon_i y_i' - (\upsilon_i + 1)(y_i - y_{i-1}) / h_i}{\upsilon_i^2 - 1}$$

$$\delta_i = \frac{y'_i + \upsilon_i y'_{i-1} - (\upsilon_i + 1)(y_i - y_{i-1})/h_i}{1 - \upsilon_i^2}$$

for $(i = 1,\dots,n)$. The unknowns $y'_2,\dots, y'_{n-1}$ can be determined from the tridiagonal system of equations

$$c_i\, y'_{i-1} + (c_i\, \upsilon_i + c_{i+1}\, \upsilon_{i+1})\, y'_i + c_{i+1}\, y'_{i+1}$$

$$= c_i(\upsilon_i + 1)\,(y_i - y_{i-1}) \,/\, h_i + c_{i+1}\,(\upsilon_{i+1} + 1)\,(y_{i+1} - y_i) \,/\, h_{i+1}$$

for $i = 1,\cdots,n-2$, where

$$c_i = \frac{\sigma i^2 \; \sinh(\sigma_i h_i)}{\upsilon_i^2 - 1} \quad .$$

The matrix of coefficients of this system is strictly diagonal dominant if $\sigma_i \geq 0$ for $i = 1,\dots,n$. For vanishing tension parameters σ_i the described algorithm computes the common cubic spline (see [Späth '69]).

G. Nielson introduced piecewise polynomials which combine the advantages of polynomial splines and exponential splines. By discretising the functional (2.1) he obtained a variational problem which is spline-in-tension-like but has polynomial extremals. These polynomial alternatives to exponential splines minimize the functional

$$(2.3)\quad \int_a^b [f''(t)]^2\, dt + \sum_{i=0}^{n} \upsilon_i\, [f'(t_i)]^2 \;, \quad \upsilon_i \geq 0,$$

in the set

$$H = \{f \in C^1\,[a,b] \mid f' \text{ is absolutely continuous } f'' \in L^2\,[a,b] \text{ and } f(t_i) = y_i \text{ for } i = 0,\dots,n\,,\; f'(t_0) = y'_0 \text{ and } f'(t_n) = y'_n\}$$

and are often called υ-splines.

A parametrized curve with υ-splines as component functions is of the class G^2 (especially curvature continuous at the knots) and can be influenced by modifying the point tension values υ_i. As the weight υ_i increases, the υ-spline curve becomes "tighter" at the i-th interpolation point because the magnitude of the tangent vector approaches zero.

Weighted υ-splines have been introduced by T. Foley to have additional weights ω_i which influence the curve shape on each segment. For given point tension factors $\upsilon_i \geq 0$ $(i= 1,\dots,n)$ and interval weights $\omega_i > 0$ $(i = 1,\dots,n-1)$ the weighted ν-spline interpolant is the unique C^1 piecewise cubic function s that minimizes the functional

$$(2.4) \quad \sum_{i=0}^{n-1} \left(\omega_{i+1} \int_{t_i}^{t_{i+1}} (f''(t))^2 \, dt\right) + \sum_{i=0}^{n} \upsilon_i \, (f'(t_i))^2$$

over all piecewise cubics in F_1 (see [Foley,'87]).

For vanishing ν_i the weighted ν-spline s is a common weighted spline and for $\omega_i = c$ $(i= 1,\dots,n)$ s is a common ν-spline with tension parameters ν_i / c. For the computation of these splines one represents each spline segment s_i $(i = 1,\dots,n)$ in the form

$$s_i(\tau) = y_{i-1} F_0(\tau) + y_i F_1(\tau) + y_{i-1}' G_0(\tau) + y_i' G_1(\tau) \quad \tau \in [0,1]$$

where the blending functions $F_0, \dots, G_1$ are the cubic Hermite basis functions. The unknowns $y_1', \cdots, y_{n-1}'$ can be computed by solving the following system of equations:

$$\begin{aligned} &2 c_i y_{i-1}' + (\upsilon_i + 4 c_i + 4 c_{i+1}) y_i' + 2 c_{i+1} y_{i+1}' \\ &= 6 c_{i+1} (y_{i+1} - y_i) / h_{i+1} + 6 c_i (y_i - y_{i-1}) / h_i \end{aligned}$$

for $i = 1,\cdots,n-1$ with $h_i = t_i - t_{i-1}$ and $c_i = \omega_i / h_i$.

The concept of ν-spline curves has been generalized to space curves with curvature and torsion continuity, the piecewise quintic τ-splines (see [Hagen, '85]). B-spline representations of ν- and τ-splines can be found in [Lasser, '88].

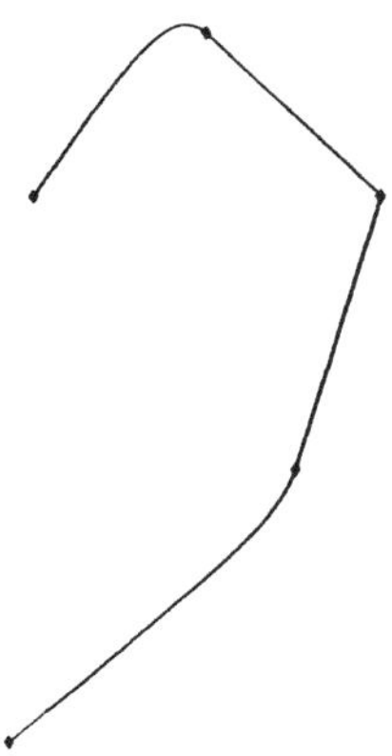

Fig. 3: exponential spline with
$\sigma_1 = 1$, $\sigma_2 = 50$, $\sigma_3 = 50$, $\sigma_4 = 1$

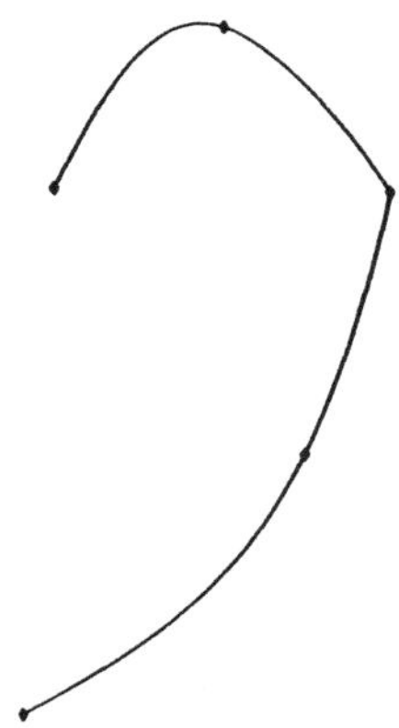

Fig. 4: ν-Spline with $\nu_0 = 0$,
$\nu_1 = 0$, $\nu_2 = 100$, $\nu_3 = 0$, $\nu_4 = 0$

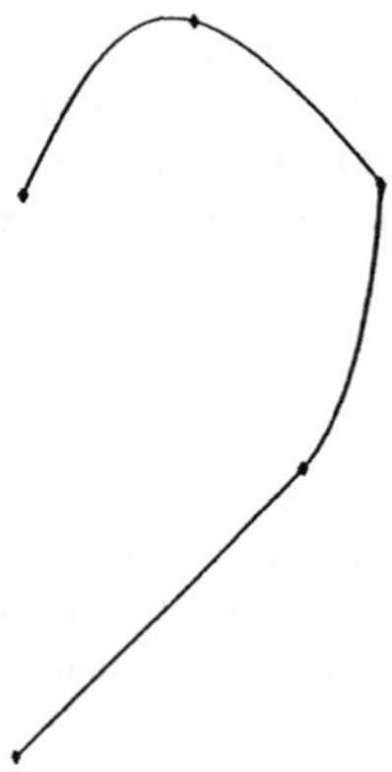

Fig. 5: weighted ν-spline with weights $\nu_0 = 0$, $\nu_1 = 0$, $\nu_2 = 100$, $\nu_3 = 0$, $\nu_4 = 0$ and $\omega_1 = 50$, $\omega_2 = 1$, $\omega_3 = 1$, $\omega_4 = 1$

3. Minimal energy splines

We now look at the extremals of the energy integral

$$(3.1) \quad E = \int_a^b \frac{(f''(t))^2}{[1+(f'(t))^2]^{5/2}}\, dt \;,$$

without taking the approximation (1.3) into account. The Euler equation of this problem is

$$(3.2) \quad f^{(4)}\,(1 + (f')^2)^2 - 10\, f'''\, f''\, f'\,(1 + (f')^2) - 5/2\,(f'')^3 + 15\,(f'')^3\,(f')^2 = 0$$

(see [Glass, '66]) which is a nonlinear differential equation while the equation $f^{(4)} = 0$ of a cubic polynomial is obviously linear. For this reason the extremals of the energy integral are sometimes called **nonlinear splines**. Equation (3.2) is equivalent to the differential equation

$$(3.3) \quad \frac{(f'')^2}{(1+(f')^2)^{5/2}} = A\, f' + B \qquad \text{with } A,B \in \mathbb{R}$$

(see [Woodford, '69], [Horn, 83]) which can be reduced to a differential equation of first order by introducing a special coordinate system (see [Schulze, '90]).

Algorithms for the interpolation of given data with nonlinear splines have been developed by Glass (1966), Woodford (1969) and Malcolm (1973) (see [Malcolm, '73]). The basic ideas used can be summarized as follows

— solve the differential equation (3.2) for each spline segment and certain derivatives at the knots, perform an optimisation process over all possible derivatives $y'_1, \cdots, y'_{n-1}$ at the knots $t_1, \cdots, t_{n-1}$ ([Glass, '66])

- perform an iterative procedure where the first derivatives at the knots are guessed and higher derivatives (up to the third derivative) are replaced according to the Taylor expansion ([Woodford, '69])

- discretise the energy integral and compute a discrete approximation for the minimal energy spline ([Malcolm, '73])

These methods however have not been accepted for practical use because they share numerical instabilities and they cannot be extended to the more general case of nonlinear spline curves.

Minimal energy spline curves are the extremals of the energy integral in the general setting

(3.4) $E = \int_a^b \varkappa^2(t) \mid x'(t) \mid \ dt$.

They are characterized by the differential equation

(3.5) $2\varkappa'' - \frac{2\omega'\varkappa'}{\omega} + \omega^2 \varkappa^3 = 0$

for the curvature function $\varkappa$ and the function $\omega(t) = \mid x'(t) \mid$ (see [Schulze, '90a]). For arc length parametrisized curves (3.5) reduces to the differential equation of J. Radon:

(3.6) $\varkappa''(s) + \frac{1}{2} \varkappa^3(s) = 0$

(see [Blaschke, '30], [Birkhoff / deBoor, 64]), which can be integrated by the use of the lemniscate function:

(3.7) $\varkappa(s) = \varkappa_m \sin \operatorname{lemn} (\varkappa_m(s - a) / 2), \ s \in [0,L]$.

Using the properies of the leminiscate function some interesting results about the segments of a minimal energy spline can be derived (see [Schulze, '90b]). The curvature function of each spline segment (also called free elastic curve) can be continued to a periodic function defined on the whole real line where the period T is given by

(3.8) $T = \frac{2\lambda}{\varkappa_m}$ with $\lambda = \frac{\Gamma^2(1/4)}{\sqrt{2\pi}}$

Furthermore the curvature function of a free elastic curve satisfies the relation

(3.9) $\varkappa(\varkappa_m / d,\ d \cdot a,\ d \cdot s) = (1/d)\ \varkappa(\varkappa_m,\ a,\ s)$

which causes the invariance of the curve under uniform scaling.

These and other properties of elastic curves are used in [Schulze, '90a] to derive an algorithm for the computation of minimal energy spline curves. From formula (3.7) it is obvious that the problem of determinating an interpolating free elastic curve for a given 2 point - 2 vector configuration can be reduced to the computation of the three relevant parameters $\varkappa_m$, L and a. Because of the periodicity of the lemniscate function it is possible to determine a two-parametric set of low degree polynomials $p = p(\varkappa_m,\ a,\ s)$ which gives a good approximation for the set of possible curvature functions. The intrinsic equation of the elastic curves is now replaced by

$$\varkappa(s) = P(\varkappa_m, a, s) \qquad s \in [0,L]$$

and the unknown parameters $\varkappa_m$, L and a can be determined by means of nonlinear optimization where the objective function is gained by integrating the intrinsic equation. If the choosen polynomials satisfy the relation (3.9) the generated curves share the scale invariance property which is useful for the efficient computation of the curves.

A high accurate approximation of an interpolating nonlinear spline can be obtained by applying an algorithm that proceeds in four steps.

First a common cubic spline curve is generated which interpolates the data (n points, 2 tangent vectors). The criterion for choosing the parametrization of this curve is to fulfil approximatively the relation (3.5). If one evaluates the left side of (3.5) in discrete points and takes the sum of squares of these values as a criterion of comparison, the most satisfactory results are obtained for the **chordal** parametrization (compared to other standard parametrizations). The tangents of the chordal spline are used as initial tangents of the nonlinear spline.

After determining initial tangents a sequence of free elastic curves is generated that interpolates the data. This sequence of curves is C^1 but has often drastic discontinuities in curvature.

In the third step the position of the initial tangents is corrected in order to decrease the curvature discontinuities. For this a nonlinear relation between the curvatures of an elastic segment in the starting point and the end point is used. After this procedure the elastic spline curve gained is nearly curvature continuous.

Finally another cubic spline curve is generated that interpolates not only the given data but also the midpoints of the free elastic segments of step 3. The parametrisatio of the splines is chosen according to the true arc lengths of the nonlinear spline segments. By this a curvature continuous interpolating curve is obtained with drastically reduced energy compared to the initial spline curve.

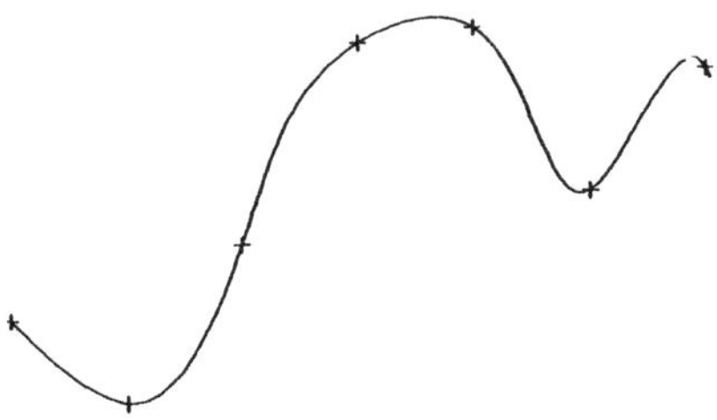

Fig. 6: chordal cubic spline with first derivative end condition

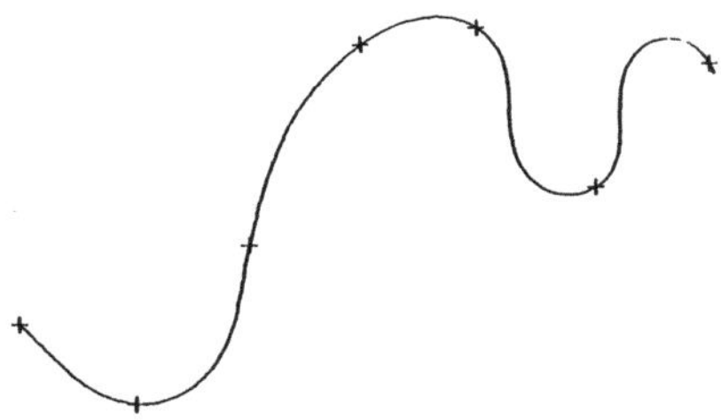

Fig. 7: interpolating C^1-sequence of free elastic curves

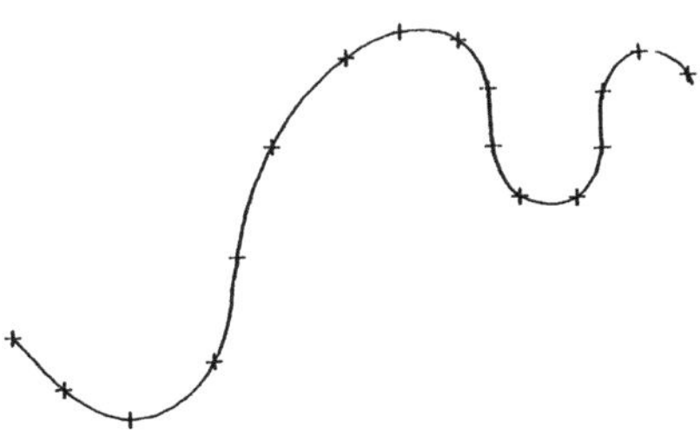

Fig. 8: final cubic spline according to step 4

4. Nonlinear splines in tension

The extension of the energy integral by a weighted arc length part gives the functional

$$(4.1)\quad \int_a^b \kappa^2(t)\,|\,x'(t)\,|\,dt + \sigma^2 \int_a^b |\,x'(t)\,|\,dt\ .$$

In analogy to section 2 the extremals of (4.1) are called **minimal energy splines in tension**. The segments of such a spline are called (general) elastic curves. The Euler equation of the problem is for arc length parametrized curves given by

$$\kappa'' + \frac{1}{2}\kappa^3 - \frac{1}{2}\sigma^2\kappa = 0$$

(see [Schulze, '90a]). This differential equation is equivalent to the differential equation of first order

$$(4.2)\quad (\kappa')^2 + \frac{1}{4}\kappa^4 - \frac{1}{2}\sigma^2\kappa^2 = C\ ,\ C \in \mathbb{R}$$

which is the equation of the elliptic p-function of Weierstraß:

$$\kappa^2(s) = \frac{2}{3}\sigma^2 - 4\ p(s+a,\ g_2,\ g_3)$$

$$g_2 = \frac{1}{3}\sigma^4 + C$$

$$g_3 = -\frac{1}{27}\sigma^6 - \frac{1}{6}\sigma^2 C$$

(see [Mehlum, '74], [Schulze, '90a]).

E. Mehlum developed two algorithms for the approximation of nonlinear splines in tension which have become known by the names KURGLA I and KURGLA II. The algorithm KURGLA I uses the following property of elastic curves:

$$(4.3)\quad \kappa(s) = -\frac{\lambda}{2}(x^2(s) - x^2(0)) + \frac{\mu}{2}(x^1(s) - x^1(0) + \kappa(0)\ .$$

Equation (4.3) can be interpreted as follows: the curvature varies linearly along some fixed direction in the plane (see [Mehlum, '74]). Using the interpolation conditions and additional assumptions Mehlum computes a sequence of linear functions which defines the wanted curve. Each of the linear functions is then approximated by a "staircase" function what results in an approximation of the spline curve by a sequence of circular arcs.

The algorithm KURGLA II is based on the differential equation (4.2) which can be written as

$$(4.4)\quad (x')^2 + (\frac{1}{2} x^2 - \frac{1}{2} \sigma^2)^2 = \Psi^2$$

Taking the assumption of high tension

$$x^2 \ll \sigma^2 ,$$

(4.4) gives approximately the equation

$$(x')^2 = \Psi^2 + \sigma^4/4 ,$$

which can be solved by a linear function:

$$x = \alpha s + \beta .$$

The method KURGLA II therefore uses Cornu spirals as spline segments, because these curves are defined by a linear dependence of curvature and arc length.

Another algorithm for the computation of an interpolating curve with Cornu spirals as segments can be found in [Hoschek/Lasser, '89, p. 116]. Other methods to approximate nonlinear splines in tension are described in [Kallay, '87] and [Jou, '89].

5. Further extremal problems

Using higher derivatives further smoothness criteria can be defined which also have meaning from the view of physics. In dynamics the quantity

$$(5.1)\quad R = \frac{d^3 x}{ds^3} = (x \cdot N)'$$

is called **jerk** and the quantity

$$(5.2)\quad SL = \frac{d^4 x}{ds^4}$$

denotes the **load** of the beam x.

In [Meier, '87] polynomial curves are derived from the principle of jerk minimisation

$$(5.3) \quad \int_0^L | R(s) |^2 \; ds \longrightarrow \min ,$$

as well as from the principle of minimizing the load

$$(5.4) \quad \int_0^L | SL(s) |^2 \; ds \longrightarrow \min .$$

Following Meier the most satisfactory results can be obtained using the functional

$$(5.5) \quad M = \int_0^L \alpha \, | x''(s) |^2 + \beta \, | x'''(s) |^2 \; ds$$

with $\alpha + \beta = 1$. This means performing a blended optimisation of energy and jerk.

6. Twist Estimation for Smooth Surface Design

The purpose of this section is to present a new twist estimation for smooth surface design based upon a calculus of variation approach.

The two main techniques for surface design are the Bezier- and B-Spline-methods and the Gordon-Coons-type surfaces. The fundamental idea of the Bezier- and B-Spline-methods is to evaluate and manipulate the curves and surfaces by a (small) number of control points.

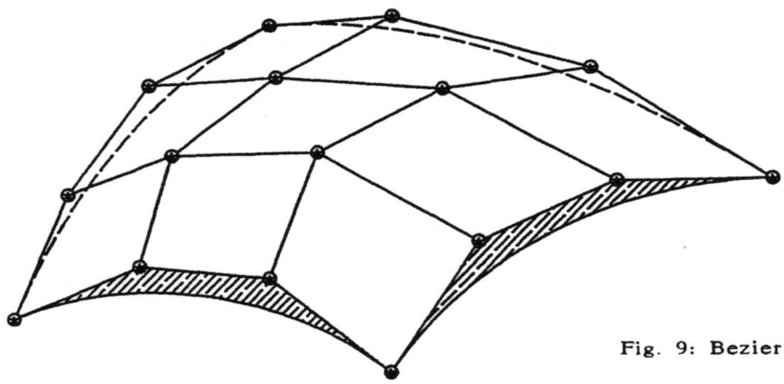

Fig. 9: Bezier-patch

The boundary control points of these patch-types can be supplied by a smooth network of curves. In this section we present a method to specify the inner control points of each patch in an efficient way to generate smooth surfaces.

Gordon-Coons-type surfaces can be written as a boolean sum of projectors, which are themselve interpolants to lower dimensional information. If the projectors involve derivative information, then the boolean sum involves mixed partial derivatives, the so-called twist vectors. The normal components of these twist vectors, which are responsible for the smoothness of the surface, can be supplied by our technique described in this paper.

(1) **Generalized Coons Patches:** This surface-patch technique was introduced by S. Coons (see [Coons, '64]) and W. Gordon pointed out (see [Gordon, '69]), that such a patch can be written as a boolean sum of projectors (for more detail see [Barnhill, '82]).

Using this technique, there were two major problems: appropriate input for the twist vectors and twist incompatibilities. To remove the incompatibility of the cross partial derivatives we can use either Gregory's square (see [Barnhill, '74] and [Barnhill-Gregory, '75]) or Nielson's convex combination extension of the boolean sum scheme (see [Nielson, '79]):

$$(6.1)\quad X(u,w) := P_1 + P_2 - \alpha P_1 P_2 - \beta P_2 P_1$$

with the cubic or quintic Hermite-projector P_1 and P_2 and appropriate convex combination functions $\alpha(u,w)$ and $\beta(u,w)$. Since we are interested in curvature-input, but not in partial derivatives of curvature functions, we set $X_{uuw} = 0$; $X_{uww} = 0$; $X_{uuww} = 0$. If we use cubic or quintic Hermite interpolants as representation of boundary curves, we get the following matrix representation of this patch:

$$(6.2)\quad \begin{pmatrix} H_0(u) \\ H_1(u) \\ \bar{H}_0(u) \\ \bar{H}_1(u) \\ \bar{\bar{H}}_0(u) \\ \bar{\bar{H}}_0(u) \end{pmatrix}^T \begin{pmatrix} X(0,0) & X(0,1) & X_w(0,0) & X_w(0,1) & X_{ww}(0,0) & X_{ww}(0,1) \\ X(1,0) & X(1,1) & X_w(1,0) & X_w(1,1) & X_{ww}(1.0) & X_{ww}(1,1) \\ X_u(0,0) & X_u(0,1) & \tilde{X}_{uw}(0,0) & \tilde{X}_{uw}(0,1) & 0 & 0 \\ X_u(1,0) & X_u(1,1) & \tilde{X}_{uw}(1,0) & \tilde{X}_{uw}(1,1) & 0 & 0 \\ X_{uu}(0.0) & X_{uu}(0,1) & 0 & 0 & 0 & 0 \\ X_{uu}(1.0) & X_{uu}(1,1) & 0 & 0 & 0 & 0 \end{pmatrix} \begin{pmatrix} H_0(w) \\ H_1(w) \\ \bar{H}_0(w) \\ \bar{H}_1(w) \\ \bar{\bar{H}}_0(w) \\ \bar{\bar{H}}_0(w) \end{pmatrix}$$

In this scheme there are no twist incompatibilities. One of the authors presented recently a solution for the twist input problem (see [Hagen, '88]). The twist vectors $\tilde{X}_{uw} = \alpha X_{uw} + \beta X_{wu}$ are replaced using the Gauß-frame-technique

$$\tilde{X}_{uw} = \langle \tilde{X}_{uw}, N \rangle \cdot N + \langle \tilde{X}_{uw}, X_u \rangle \cdot X_u + \langle \tilde{X}_{uw}, X_w \rangle \cdot X_w$$

N is the normal vector, $\langle , \rangle$ is the scalarproduct, and the scalar function $\langle \tilde{X}_{uw}, N \rangle$ can be supplied by the twist estimation method described in (3). The tangent components of the twist vector may be computed as in Selesnick's method (see [Selesnick, '81]).

(2) **Bezier- and B-Spline Surfaces:** The curves and surfaces now known as Bezier curves and surfaces were independently developed by P. de Casteljau and by P. Bezier. The underlying mathematical theory, based on the concept of Bernstein polynomials, was first introduced by R. Forrest (see [Forrest, '72]). The fundamental idea of this approach is to evaluate and manipulate the curves and surfaces by a (small) number of "control points". Since we build up Bezier-surfaces from curve-networks, we first consider Bezier-curves.

A Bezier-curve is a segmented curve. The segments $X_l(u)$, $l = 0,\dots,k$ of a Bezier-curve of degree m over the parameter interval $u_l \le u \le u_{l+1}$ are:

$$(6.3) \qquad X_l(u) := \sum_{i=0}^{m} b_{l \cdot m+i} \cdot B_i^m\left(\frac{u - u_l}{u_{l+1} - u_l} \right)$$

The Bernstein polynomials $B_i^m(t) := \binom{m}{i} (1-t)^{m-i} t^i; \quad 0 \le t \le 1$ are used as blending functions.

Bernstein polynomials are special degenerated B-Splines (see [Boehm, Farin, Kahmann, '84]). If we use B-Splines as blending functions, we can generalize the whole concept to so-called B-Spline-curves and surfaces (see [Gordon-Riesenfeld, '74]).

B-Spline curves are similar to Bezier-curves in that a set of blending functions combines the effect of n+1 control points:

$$(6.4) \qquad Y(u) := \sum_{j=0}^{n} d_j N_{j,N}(u)$$

The most important difference is the local support property of the B-Spline blending functions $N_{j,k}(u)$. Both curvetypes have the convex hull and variation diminishing property (for more details see [Boehm, Farin, Kahmann, '84]).

A Bezier-surface is a segmented surface. The segments $X_{pq}(u,w)$; $p = 0,\dots,k$; $q = 0,\dots,r$ of a Bezier surface of degree m,n over the rectangular parameter domain $u_p \le u \le u_{p+1}$; $w_q \le w \le w_{q+1}$ are:

$$(6.5) \qquad X_{pq}(u,w) := \sum_{i=0}^{m} \sum_{j=0}^{n} b_{p\cdot m+i/q\cdot n+j} \cdot B_i^m\left(\frac{u-u_p}{u_{p+1}-u_p}\right) B_j^n\left(\frac{w-w_q}{w_{q+1}-w_q}\right)$$

Instead of a control polygon a Bezier-surface(-segment) has a control polyhedron.

The definition of a B-Spline-surface over a rectangular parameter domain follows directly the same pattern:

$$(6.6) \qquad Y(u,w) := \sum_{i=0}^{m} \sum_{j=0}^{n} d_{ij} N_{i,M}(u) N_{j,N}(w)$$

A triangular Bezier patch is defined by

$$(6.7) \qquad X(u,w) := \sum_{I} b_{i,j,k} B_{i,j,k}^n\big(r(u,w),s(u,w),t(u,w)\big)$$

where r, s, t are local barycentric coordinates of the triangular parameter domain and "I" denotes summation over all $i,j,k \ge 0$, $i+j+k = n$. The $B_{i,j,k}^n$ are generalized Bernstein polynomials of degree n given by

$$B_{i,j,k}(r,s,t) := \frac{n!}{i!\ j!\ k!}\ r^i s^j t^k$$

These Bernstein polynomials have properties very much like the univariate ones. Therefore they are appropriate blending functions (see [Farin, '79]).

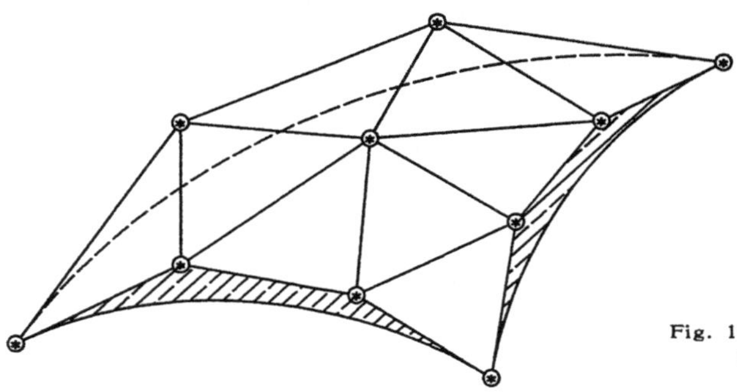

Fig. 10: triangular Bezier-patch

The boundary control points of these patch-types can be supplied by a smooth network of curves. In part (3) we present a method how to specify the "inner control points" of each patch in an efficient way to generate smooth surfaces.

(3) **Variational twist estimation**: The functional $\int_S (k_1^2 + k_2^2)\, dS$ (k_1 and k_2 are the principal curvatures of the surface S) is a standard fairness criterion for surfaces in engineering (see [Nowacki-Reese, '83]), because it is equivalent to the strain energy of flexure and torsion in a thin rectangular elastic plate of small deflection.

Hagen and Schulze used this functional for a variational formulation and solution of the surface fairing problem (see [Hagen-Schulze, '87]) under the assumption of orthogonal parameter lines. We present here a general solution without any regularity constraints:

The strain energy of flexure and torsion in a thin rectangular elastic plate of small deflection is equivalent to the functional

(6.8) $$G := \int_S (k_1^2 + k_2^2)\, dS \,.$$

Calculating the principal curvature of the surface we get

(6.9) $$G = \int_A \frac{(g_{11}h_{22} - 2g_{12}h_{12} + g_{22}h_{22})^2 - 2gh}{g^2} \sqrt{g} \;\; du\, dw$$

$$= \int_A (h_{12}^2 (2g + 4g_{12}^2) - h_{12} (4g_{12} (g_{11}\, h_{22} + g_{22}\, h_{11})$$

$$+ (g_{11}\, h_{22} + g_{22}\, h_{11})^2 - 2\, g\, h_{11}\, h_{22})\, g^{-3/2} \;\; du\, dw$$

$X : A\ E^3$ is a parametrization of the surface S and g_{11}, g_{22} and g_{12} are the components of the first fundamental form; h_{11}, h_{22} and h_{12} are the components of the second fundamental form and $h := h_{11}\, h_{22} - h_{12}^2$ and $g := g_{11}\, g_{22} - g_{12}^2$.

We have a functional of the form $G := \int\int F(u,w,h_{12}(u,w))\, du\, dw$, the Euler equation:

(6.10) $$\frac{\partial F}{\partial h_{12}} = 2\, h_{12} (2\, g + 4\, g_{12}^2) - 4\, g_{12} (g_{11}\, h_{22} + g_{22}\, h_{11}) = 0$$

gives a necessary condition for the energy minimum

(6.11) $$h_{12} = \frac{g_{12} (g_{11}\, h_{22} + g_{22}\, h_{11})}{g + 2\, g_{12}^2} \,.$$

The mean curvature $H := \frac{1}{2g} (h_{11}\, g_{22} - 2\, h_{12}\, g_{12} + h_{22}\, g_{11})$ measures the deviation of a surface from a minimal surface ($H \equiv 0$). Minimal surfaces correspond to the surface formed by a soap bubble between the boundary curves.

Because of the relation

$$h_{12} = 2\, g_{12} \cdot H$$

we consider a surface $X : A \rightarrow E^3$ with

$$h_{12} = \frac{g_{12}\,(g_{11}h_{22} + g_{22}h_{11})}{g + 2\, g_{12}^2} = 2\, g_{12} \cdot H$$

as smooth in the sense of $\int_S (k_1^2 + k_2^2)\; dS \rightarrow \min.$

This result gives "smooth normal components" of twist vectors. This information can also be used to specify the inner control points of Bezier-patches since the twist vectors of such a patch are given by

$$X_{uw} = (m-1)\,(n-1) \sum_{i=0}^{m-1} \sum_{j=0}^{n-1} \Delta^{1,1}\, b_{ij}\, B_i^{m-1}(u)\, B_j^{n-1}(w)$$

where $\Delta^{1,1}\, b_{ij} = b_{i+1,j+1} - b_{i,j+1} - b_{i+1,j} + b_{ij}$.

In the case of B-Spline surfaces we have an analogous situation. For more details see [Farin-Hagen, '90].

Remarks:

1 - If the surface consists of a rectangular patchwork $(g_{12} = 0)$ the theorem yields $h_{12} = 0$. That means the orthogonal parameter lines are lines of curvature.

2 - If the surface contains two asymptotic directions (two families of straight lines for example) we may have asymptotic parameter lines $(h_{11} = 0;\ h_{22} = 0)$. On this hyperbolic curvature situation our method creates locally flat surfaces, since the theorem yields in the case of $h_{11} = 0$ and $h_{22} = 0$ that h_{12} is zero too. For modeling surfaces with asymptotic parameter lines, Adini's twist (see [Barnhill-Brown-Klucewicz, '78]) is more appropriate than our twist estimation.

Example: We have constructed a turbine blade using this method

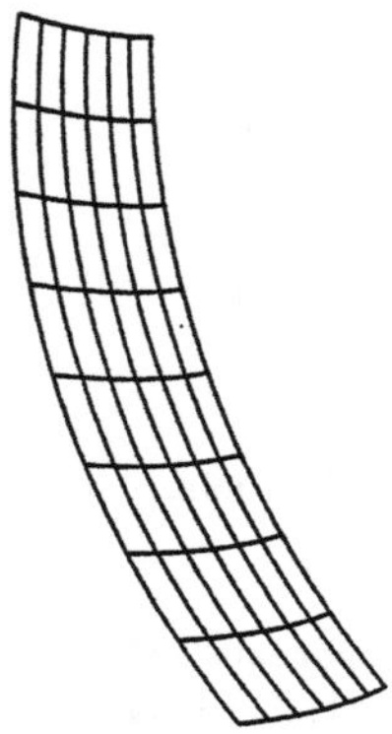

To visualize the "smoothness" and quality of this surface we use the reflectionline-analysis-method (for more details see [Barnhill-Farin-Fayard-Hagen, '88]).

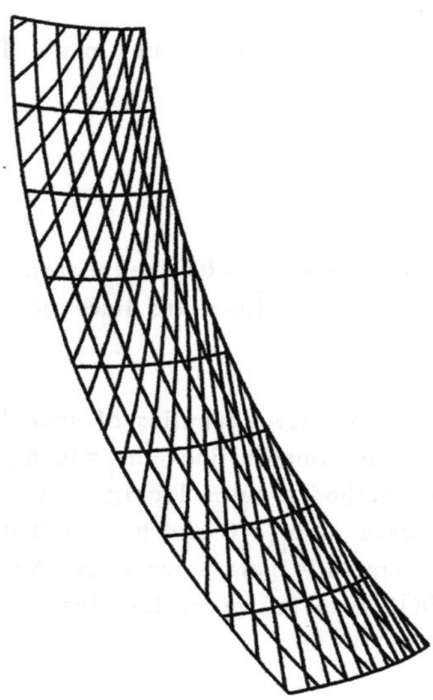

optimal twist input

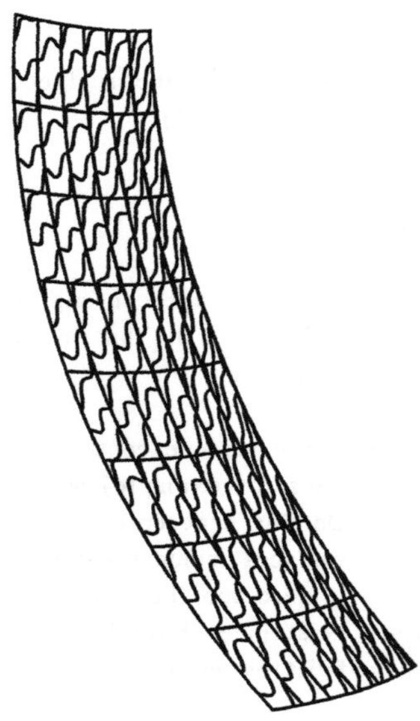

zero twist input

References:

Barnhill (1974): Smooth interpolation over triangles.
In: Barnhill-Riesenfeld (eds.): Computer Aided Geometric Design, Academic Press, New York, pp. 45-70

Barnhill (1982): Coons' patches.
Computer in Industry 3, 1982, pp. 37-43

Barnhill, Brown, Klucewicz (1978): A new twist in CAGD.
Computer Graphics and Image Processing 8, 1978, pp. 78-91

Barnhill, Gregory (1975): Compatible smooth interpolation in triangles.
J. Approx. Theory 15, 1975, pp. 214-225

Barnhill, Farin, Fayard, Hagen (1988): Twists, curvatures and surfaces interrogation.
CAD 20, 1988, pp. 342-346

Birkhoff, deBoor (1964): Piecewise polynomial interpolation and approximation.
In: H. L. Garabedian (ed.): Approximation of functions, Proc. of General Motors Symposium of 1964, Elsevier 1965

Blaschke (1930): Vorlesungen über Differentialgeometrie I.
Chelsea Publishing Company, New York 1967, reprint of 3rd edition of 1930

Boehm, Farin, Kahmann (1984): A survey of curve and surface methods in CAGD.
Computer Aided Geometric Design 1, 1984, pp. 1-60

Coons (1964): Surfaces for computer aided design.
Report MAC-TR-4; Project MAC, M.I.T.

Farin (1979): Subsplines über Dreiecken.
Diss. TU Braunschweig

Farin, Hagen (1990): A local twist estimator.
To be published in Hagen (ed.): Curve and Surface Design, SIAM, Philadelphia

Forrest (1972): Interactive interpolation and approximation by Bezier polynomials.
Computer J. 15, 1972, pp. 71-79

Foley (1987a): Local control of interval tension using weighted splines.
Computer Aided Geometric Design 4, 1987, pp. 281-294

Foley (1987b): Interpolation with interval and point tension controls using cubic weighted ν-splines.
ACM Transactions on Math. Software 13, 1987, pp. 68-96

Foley, Nielson (1989): Knot selection for parametric spline interpolation.
In: Lyche-Schumaker (eds.): Math. Methods in CAGD,
Academic Press, New York, pp. 261-273

Glass (1966): Smooth curve interpolation: a generalized spline-fit procedure.
BIT 6, 1966, pp. 277-293

Gordon (1969): Free-form surface interpolation through curve networks.
GMR-921, GM Research Labs.

Gordon, Riesenfeld (1974): B-spline curves and surfaces.
In: Barnhill-Riesenfeld (eds.): Computer Aided Geometric Design,
Academic Press, New York, pp. 95-126

Hagen (1988): Computer aided geometric design - methods and applications.
To be published in the proceedings of the conference on Engineering Graphics and Descriptive Geometry, Wien

Hagen, Schulze (1987): Automatic smoothing with geometric surface patches.
Computer Aided Geometric Design 4, 1987, pp. 231-235

Horn (1983): The curve of least energy.
ACM Transactions on Math. Software, Vol. 6, 1983, No. 4

Hoschek, Lasser (1989): Grundlagen der geometrischen Datenverarbeitung.
Teubner Verlag, Stuttgart 1989

Jou, Han (1989): Minimal energie splines I: plane curve segments.
Submitted to Math. Methods in the Applied Sciences

Kallay (1987): Method to approximate the space curve of least energy and prescribed length.
CAD, Vol. 19, 1987, No. 2

Lasser (1988): B-Spline-Bezier representation of Tau-Splines.
Technical Report NPS-53-80-006, Naval Postgraduate School, Monterey

Malcolm (1977): On the computation of nonlinear splines functions.
SIAM Numerical Analysis 14, 1977, No. 2, pp. 254-282

Mehlum (1974): Nonlinear splines.
In: Barnhill-Riesenfeld (eds.): Computer Aided Geometric Design, Academic Press, New York

Meier (1987): Der differentialgeometrische Entwurf und die analytische Darstellung krümmungsstetiger Schiffsoberflächen.
Fortschrittsberichte VDI-Reihe 20, No. 5, VDI-Verlag

Müller (1987): Approximationstheorie.
Akademische Verlagsgesellschaft Wiesbaden

Nielson (1974): Some piecewise polynomial alternatives to splines under tension.
In: Barnhill-Riesenfeld (eds.): Computer Aided Geometric Design, Academic Press, New York

Nielson (1979): The side vertex method for interpolating in triangles.
J. Approx. Theory 25, 1979, pp. 318-336

Nowacki, Reese (1983): Design and fairing of ship surfaces.
In: Barnhill-Boehm (eds.): Surfaces in CAGD, North Holland, Amsterdam, pp. 121-134

Pilcher (1974): Smooth parametric surfaces.
In: Barnhill-Riesenfeld (eds.): Computer Aided Geometric Design, Academic Press, New York

Schulze (1990a): Elastische Wege und Nichtlineare Splines im CAGD.
Dissertation Universität Kaiserslautern

Schulze (1990b): Properties of free elasic curves.
To be published in Hagen (ed.): Curve and Surface Design,
SIAM, Philadelphia

Schweikert: An interpolation curve using a spline in tension.
Journal of Math. and Physics, Vol. 45, pp. 312-317

Selesnick (1981): Local invariants and twist vectors in CAGD.
Computer Graphics and Image Processing 17, 1981, pp. 145-160

Späth (1969): Exponential spline interpolation.
Computing 4, 1969, pp. 225-233

Späth (1983): Splinealgorithmen zur Konstruktion glatter Kurven und Flächen.
3. Auflage, Oldenbourg

Woodford (1969): Smooth curve interpolation.
BIT 9, 1969, pp. 69-77

G^{n-1} – Functional Splines for Modeling

Josef Hoschek Erich Hartmann
Fachbereich Mathematik
Technische Hochschule Darmstadt

Abstract: Implicit curves and surfaces are used for interpolation, approximation, blending of curves , surfaces and solids and filling of surface holes. The introduced curves and surfaces can be interpreted as functional splines, which fulfil geometric continuity conditions.

1. Introduction

Most modeling systems are very limited in the complexity of the surfaces which they support. In general these systems use parametric representations of free formed surfaces like Gordon-Coons-Surfaces, Bézier-Surfaces, B-Spline-Surfaces. In such systems special problems occur if *splines with higher order continuity conditions* are required or *blending of surfaces* [11], [12], [14], [19], [20], [24] or *filling of surface holes* [2], [5], [7], [21] or *approximation of solids* are wanted.

Throughout this paper we are dealing with curves and surfaces which are described by **implicit** equations. We will identify the symbol of a curve (surface) with its equation. If f and g are two curves (surfaces), then we will consider composite curves (surfaces) of the form

$$F = (1-\mu)f - \mu g^n = 0 .$$

We call f the *base*-curve (-surface) and g the *transversal*-curve (-surface).

2. G^{n-1} - functional splines in the plane

In this section we want to extend the conic blending method in $\mathbb{R}^2$ introduced by [18] to sufficiently smooth implicit curves, further we give some statements about convexity and apply this method to the interpolation of polygons. In this paper we will only demonstrate results and applications. For detailed proofs compare [9], [10], [15], [17].

We start with two definitions.

Definition 1: For an implicit curve h we set

$$D^+(h) = \{ X \in \mathbb{R}^2 \mid h(X) \geq 0 \}.$$

Definition 2: Let f and g be two piecewise C^1 continuous curves in $\mathbb{R}^2$ with:

(i) $|f \cap g| \geq 1$,

(ii) $(f_x(P), f_y(P)) \neq (0,0)$ for all $P \in f \cap g$ (with f_x, f_y as partial derivatives).

Then the curve F defined by the equation

$$F = (1-\mu) f - \mu g^n = 0 , \quad 0 < \mu < 1 , n \geq 2 , \tag{3}$$

is called a *plane functional spline* related to the curves f and g . f is the *base curve* and g the *transversal curve* of F . n is the *exponent* and μ the *parameter* of F .

$$D(f,g) := D^+(f) \cap D^+(g) , D_0(f,g) := D(f,g) \setminus (f \cup g) .$$

$\Phi(f,g)$ denotes the family of the plane functional splines related to f and g (μ varies between 0 and 1).

The essential property of a functional spline gives

Theorem 1: Let F be a functional spline curve , f its base curve and g its transversal curve, n its exponent and $P \in f \cap g$ where f and g are C^n at P . Then at point P the functional spline curve F and the base curve f have contact of order (at least) n - 1 . If F has exponent $n \geq 3$, then F and f have the same curvature at P .

Remarks: 1. For n > 3 the curves F and f even have equal (n-3)-th derivatives of the curvature at P (s. [22]).

2. Similar results were developed independently in [6].

Definition 3: A functional spline curve F with exponent n and C^n-continuous base and transversal curves is called a *G^{n-1} functional spline curve.*

The application of Theorem 1 to the lines l_1, l_2, l_0 yields that the functional splines

$$F = (1-\mu) l_1 l_2 - \mu l_0^n = 0 \qquad (0 < \mu < 1 , n \geq 3) .$$

have curvature 0 at points P_1 and P_2 resp. (Fig. 1).

Remark: To generate the figures a marching algorithm is used (s.a. [8]):

(1) Determine in point X_i the tangent t_i (perhaps an approximation by the secant $\overline{X_{i-1}X_i}$) and take as the first approximation for X_{i+1} a point Q on the tangent t_i (Fig. 6).

(2) Iteration on a circle through Q with centre X_i yields X_{i+1} .

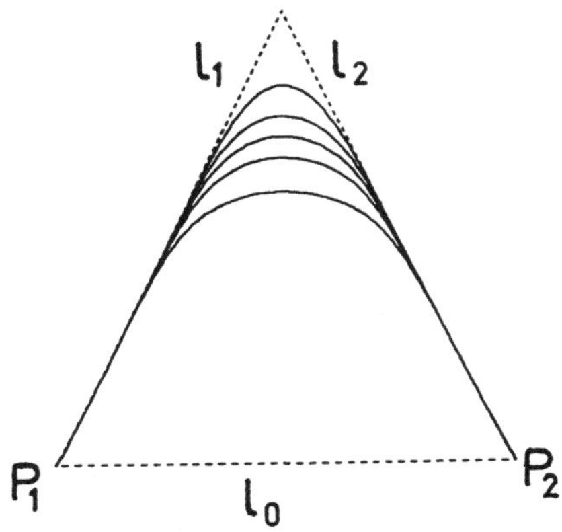

Fig. 1: Functional splines of degree n = 8 and different values of the parameter μ

Figure 2 shows an example with a non-algebraic base curve ($y = \cot|x|$) and a non-linear transversal.

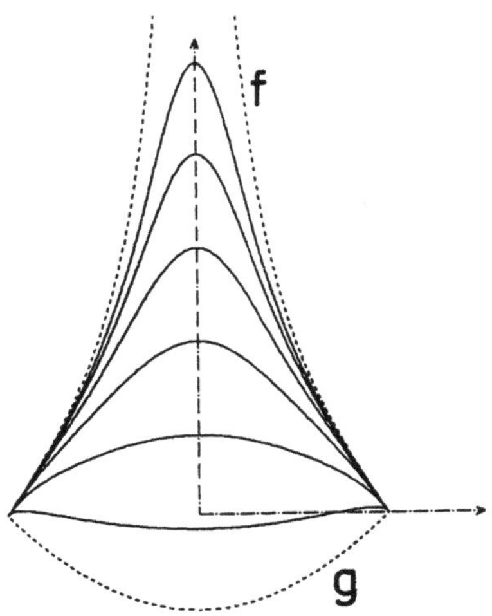

Fig. 2: Functional splines with base curve $f(x,y) = (\cos\frac{\pi}{2}x - y\sin\frac{\pi}{2}x)(\cos\frac{\pi}{2}x + y\sin\frac{\pi}{2}x)$ and transversal curve $g(x,y) = 2y + 1 - x^2$

We will now prove some global geometric properties of functional spline curves.

Proposition 1: A functional spline F related to two curves f and g has the property

$$F \cap f = F \cap g = f \cap g .$$

F lies totally within D(f,g) .

Proposition 2: Let $\Phi(f,g)$ be the family of functional splines related to the curves f and g and $P \in D_0(f,g)$. Then there is exactly one functional spline $F \in \Phi(f,g)$ passing through P .

Proposition 3: The following functional splines are convex

a) $F = (1-\mu)l_1 l_2 - \mu l_0^n = 0$ within the triangle l_1 , l_2 , l_0 (Fig.1) ,

b) $F = (1-\mu)\ l_1 l_2 l_3 - \mu\ l_0^n = 0$ within the *convex* quadrangle l_1, l_2 ,l_3, l_0.

c) $G(x,y) = (1-\mu)(\varphi(x)-y) - \mu y^n = 0$, if φ is a convex C function, within the "biangle" formed by $y = \varphi(x)$ and $y = 0$.

Application of Proposition 3 to a convex polygon yields

Theorem 2: For any plane convex polygon there exists an infinite number of convex G^2 interpolation functional splines containing the vertices. Such an interpolation curve consists of implicit segments of degree (at least) three.

Construction: Denote the points of the polygon by P_1 , P_2 , ... and the edges by l_{01}, l_{02}, ... Take to any point P_i an arbitrary line l_i such that all points $P_1, P_2, \ldots$ lie on one side of the line l_i . The set of functional splines

$$(1-\mu_i) l_i l_{i+1} - \mu_i l_{0i}^3 = 0 \quad (0 < \mu_i < 1),\quad i = 1,2,\ldots$$

are cubic curves which interpolate the given convex polygon.

The disadvantage of Theorem 2 is that at every vertex of the polygon the curvature is equal to zero . In order to avoid this defect one can for instance choose convenient quadratic curves c_i instead of the lines l_i . These curves c_i should be expressed in a local coordinate system by an explicit form in order to guarantee convexity by Proposition 3 . - Higher order of smoothness can be obtained by raising the exponents of the functional splines. – Figure 3 shows the interpolation of a square by functional splines with a) lines b) a circle as base curves.

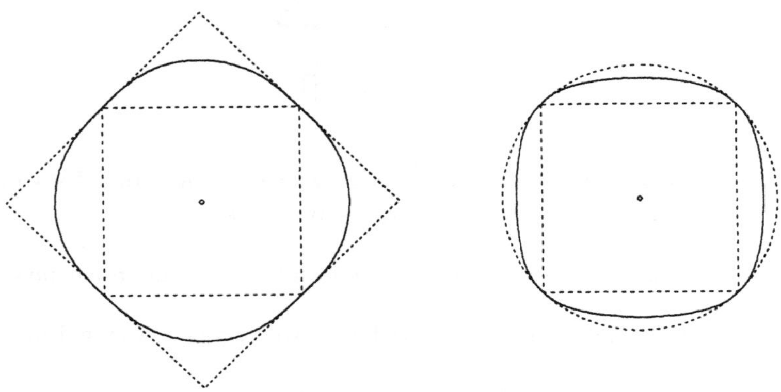

a) base curves: lines ($n = 4$, $\mu = \frac{2}{3}$) b) base curve: circle ($n = 4$, $\mu = 0{,}95$)

Fig. 3: Interpolation of a square

3. G^{n-1}-functional spline surfaces

In this section we extend the idea of functional spline curves to functional spline surfaces and show that in $\mathbb{R}^3$ we get analogous results.

First we extend Definitions 1 and 2 .

Definition 4: For an implicit surface h we set $D^+(h) = \{ X \in \mathbb{R}^3 \mid h(X) \geq 0 \}$.

Definition 5: Let f and g be two piecewise C^1-continuous surfaces in $\mathbb{R}^3$ where
(i) $f \cap g$ consists of a curve Γ ,
(ii) $(f_x(P), f_y(P), f_z(P)) \neq (0,0,0)$ for "nearly" all $P \in \Gamma$ (only a finite number of exceptions) .

Then the surface F defined by the equation

$$F = (1-\mu)f - \mu g^n = 0 , \qquad 0 < \mu < 1 , \qquad n \geq 2 ,$$

is called a *functional spline surface* related to the surfaces f and g . f is the *base surface* and g the *transversal surface* of F . n is the *exponent* and μ the *parameter* of F .

$$D(f,g) := D^+(f) \cap D^+(g) \ , \ D_0(f,g) := D(f,g) \setminus (f \cup g).$$

$\Phi(f,g)$ denotes the family of the functional splines related to f and g (μ varies between 0 and 1).

Theorem 3: Let F be a functional spline surface , f its base surface and g its transversal surface , n its exponent and $P \in f \cap g$ where f and g are C^n at P . Then at point P the functional spline F and its base surface f have contact of order (at least) n - 1 . If F has an exponent ≥ 3 , then F and f even have the same Dupinian indicatrix at P.

Remark: From Theorem 3 follows that for $n \geq 3$ F and f have (except a finite number of points) the same Gaussian curvature along Γ (s. [1], [3], [4], [16], [23]).

Definition 6: A functional spline surface F with exponent n and C^n-continuous base and transversal surfaces is called a *G^{n-1} functional spline surface* . .

3.1 Interpolation and approximation with G^{n-1} functional spline surfaces

First we will introduce convexity theorems. We start with a definition

Definition 7: If an implicit surface $f(x_1,\dots,x_m) = 0$ satisfies

$$\begin{vmatrix} f_{11} & f_{12} & \cdots & f_{1m} & f_1 \\ f_{21} & f_{22} & \cdots & f_{2m} & f_2 \\ \vdots & \vdots & & \vdots & \vdots \\ f_{m1} & f_{m2} & \cdots & f_{mm} & f_m \\ f_1 & f_2 & \cdots & f_m & 0 \end{vmatrix} \le 0 \qquad (*)$$

with

$$f_j = \frac{\partial f}{\partial x_j}, \qquad f_{ij} = \frac{\partial^2 f}{\partial x_i \partial x_j}$$

then we call the surface convex at the point $x = (x_1,\dots x_m)$.

Remark: For $m = 2$ and $m = 3$ this condition is necessary and sufficient for convexity [15].

Additionally we need the proposition

Proposition 4: Let $A = (a_{ij})$ be an $n \times n$ matrix. If A is non-negative definite, then for any n-vector $(b_1 \dots b_n) \in \mathbb{R}^n$ we have

$$D_{n+1} = \begin{vmatrix} a_{11} & a_{12} & \cdots & a_{1n} & b_1 \\ \vdots & \vdots & \vdots & \vdots & \vdots \\ a_{n1} & a_{n2} & \cdots & a_{nn} & b_n \\ b_1 & \cdots & \cdots & b_n & 0 \end{vmatrix} \le 0 \,. \qquad (*)$$

The proof follows by straightforward calculations. Thus we can prove

Theorem 4: The functional spline

$$F = (1-\mu)\, f - \mu g^n = 0$$

is convex in $\Omega(f,g)$, if $f,g \in C^2(\Omega)$ and M_f, M_g are non-negative definite matrices, here $\Omega(f,g) = \{x \in \mathbb{R}^m \mid f(x) \ge 0,\ g(x) \ge 0\}$ and

$$M_h = \begin{pmatrix} h_{11} & h_{12} & \cdots & h_{1m} \\ h_{21} & h_{22} & \cdots & h_{2m} \\ \vdots & \vdots & \cdots & \vdots \\ h_{m1} & h_{m2} & \cdots & h_{mm} \end{pmatrix}$$

with $h_{ij} = \dfrac{\partial^2 f}{\partial x_i \partial x_j}$.

From Theorem 4 we get

Proposition 5: The functional spline curve $F: (1-\mu) f(x,y) - \mu g^n(x,y) = 0$ or the functional spline surface $F = (1-\mu) f(x,y,z) - \mu g^n(x.y.z) = 0$ are convex in $\Omega(f,g) = \{(x,y,z) \in \mathbb{R}^3 : f(x,y,z) \geq 0, g(x,y,z) \geq 0\}$ if f and g fulfil condition (*).

Let us now consider some applications of the last theorem. Figure 4 contains the interpolation of a circle c with radius r by a functional spline: the base surface is the upper half sphere containing c.

$$(1-\mu)(x^2+y^2+z^2-r^2) - \mu z^n = 0 \qquad \text{and} \qquad n = 3, 8.$$

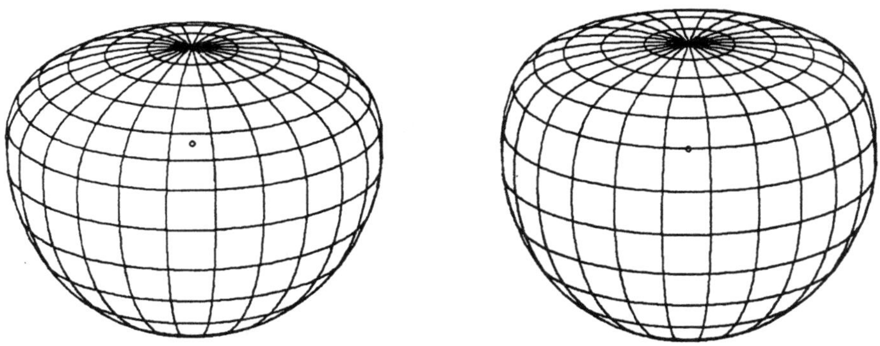

Fig. 4: G^2-Interpolation of a circle. Base surface is a sphere, transversal surface the plane through the midpoint of the sphere and the circle

Additionally we establish a G^2- interpolation of two spatial curves with a common segment:
Let Γ_1, Γ_2 be two sufficiently smooth curves with the common segment Γ (Fig. 5). Take (if possible) convenient surfaces f_1, f_2 containing Γ_1 resp. Γ_2 and a surface f containing Γ. Further we need transversal surfaces g_1, g_2 passing $\Gamma_1 \cup \Gamma$ resp. $\Gamma_2 \cup \Gamma$. Then the functional splines

$$F_1 = (1-\mu) f_1 f - \mu g_1^n = 0 \quad, 0 < \mu < 1 \;, \quad n \geq 3,$$
$$F_2 = (1-\lambda) f_2 f - \lambda g_2^m = 0 \quad, 0 < \lambda < 1 \;, \quad m \geq 3 \;,$$

interpolate $\Gamma_1 \cup \Gamma$ resp. $\Gamma_2 \cup \Gamma$ and have (because of Theorem 3) along the common curve Γ the same tangent planes and Dupinian indicatrices. Thus $F_1 \cup F_2$ is a G^2-interpolation surface containing $\Gamma_1 \cup \Gamma_2 \cup \Gamma$.

Fig. 5: Interpolation of neighboring curves

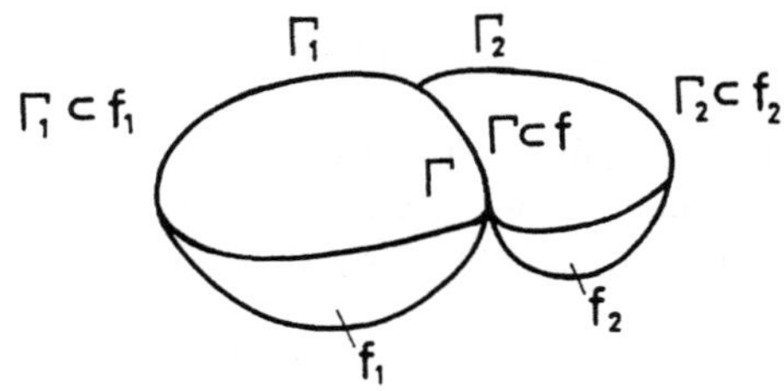

Figure 6 contains the following example to the last result: Γ_1 and Γ_2 are the halves of the circle c: $x^2 + y^2 = r^2$ with $x \geq 0$ resp. $x \leq 0$, Γ is a diameter of the circle. f_1 , f_2 are quarters of the sphere containing the circle c , f is the plane $x = 0$ and g_1 and g_2 are the plane $z = 0$. Thus we get the functional splines

$$F_1 = (1 - \mu)(r^2 - x^2 - y^2 - z^2)x - \mu z^3 = 0 \qquad \text{(left part)}$$

$$F_2 = (1 - \lambda)(r^2 - x^2 - y^2 - z^2)(-x) - \lambda(-z)^3 = 0 \qquad \text{(right part)}$$

Fig. 6: Interpolation of a circle and a line with help of the sphere

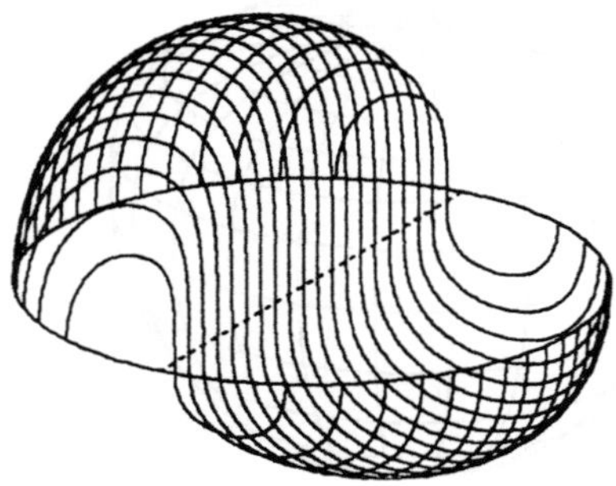

Now we will extend the functional spline method to interpolation and approximation of solids. First we have

Theorem 5: For any convex polyhedron in $\mathbb{R}^3$ there is an infinite number of convex interpolation surfaces which pass through all edges and are C^2 everywhere except at the vertices. To a face π_i of the polyhedron with N_i edges the corresponding part S_i of the interpolation surface has degree $\geq N_i$.

Construction: We construct through each edge of the polyhedron a new plane (different from all planes containing faces) such that the polyhedron lies on one side of the plane (s. Fig. 7). If p_0 is the plane that contains the face π_i and if $p_1, p_2, \ldots$ are the "new" planes corresponding to the face π_i , we interpolate this face by a functional spline

$$F = (1 - \mu)p_1 p_2 \ldots - \mu p_0^3 = 0 \qquad (0 < \mu < 1) \qquad (*)$$

The set of all functional splines (corresponding to the faces) make up a piecewise interpolation surface of the polyhedron.

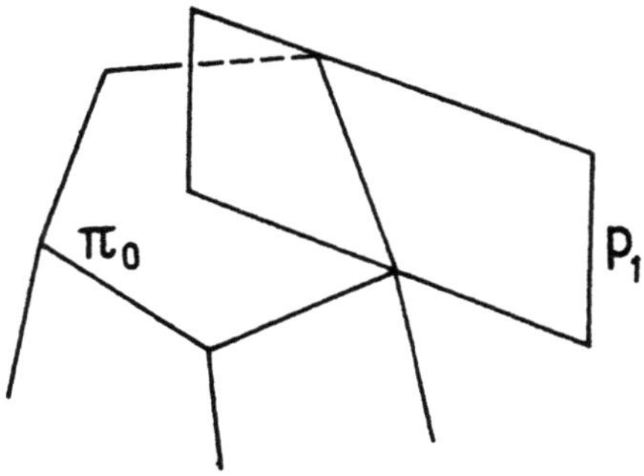

Fig. 7: Interpolation of the face of a polyhedron

Although the above surface is joining G^2-continuously, there are flat points along the edges. In order to avoid this defect, we can use other base surfaces along the edges, for example quadrics .

The following figures demonstrate the effectiveness of the introduced method: Figures 8 - 12 show the interpolation of a square, a cube and a dodecagon. For the representation of the figures the methods introduced in [HARTMANN '88] are used: Figure 8 describes the interpolation of a square with the function

$$(1-\mu)(z+x-a)(z-x-a)\,(z+y-a)(z-y-a) - \mu z^3 = 0 \ ,$$

where planes are used as base surfaces. In Figure 9a the surface out of Figure 8 is transformed to the faces of a cube. Along the edges the interpolation surface has flat-points (G^2-continuous to a plane!). This shortcoming is avoided in Figure 9b: we use cylinders as base surfaces

$$(1-\mu)\ (z^2+x^2-\frac{a^2}{2})\ (z^2+y^2-\frac{a^2}{2}) \ - \mu\,(z-\frac{a}{2})^3 \ = 0$$

The interpolation surfaces in Figure 9a,b have singularities at the vertices of the cube. The singularity at the vertex $V\,(\frac{a}{2}, \frac{a}{2}, \frac{a}{2})$ can be removed if we cut out a region Φ about V with the plane $\Pi: x+y+z-a=0$ and replace Φ by the surface with the equation

$$F(x,y,z) = (1-\lambda) f_1(x,y,z) f_2(x,y,z) f_3(x,y,z) - \lambda (x+y+z-a)^3 = 0 \ , \quad 0 < \lambda < 1 \ ,$$

with f_i as the corresponding face-surfaces of the interpolation surface in the vertex V (see Fig. 9c). This *approximating* functional spline surface is totally G^2- continuous.

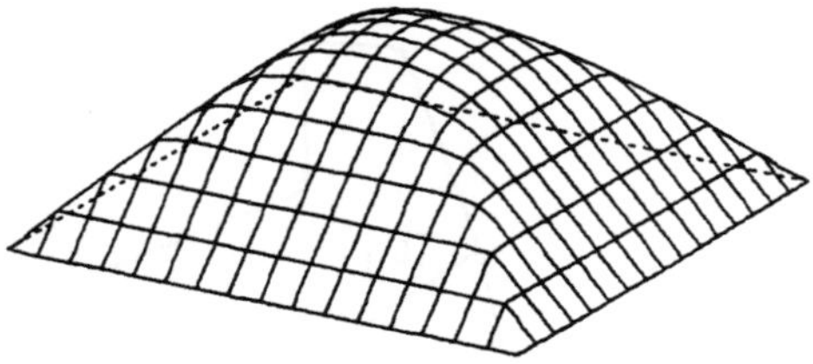

Fig. 8: Interpolation of a square. Besides the vertices the interpolating surface is G^2-continuous

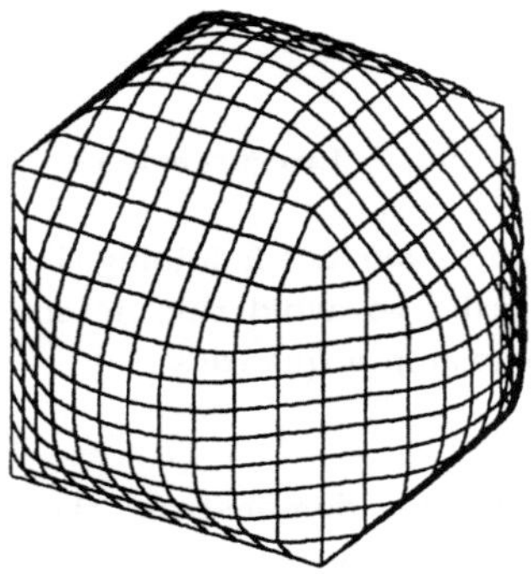

Fig. 9a: Interpolation of a cube with help of planes (observe the flat points at the edges)

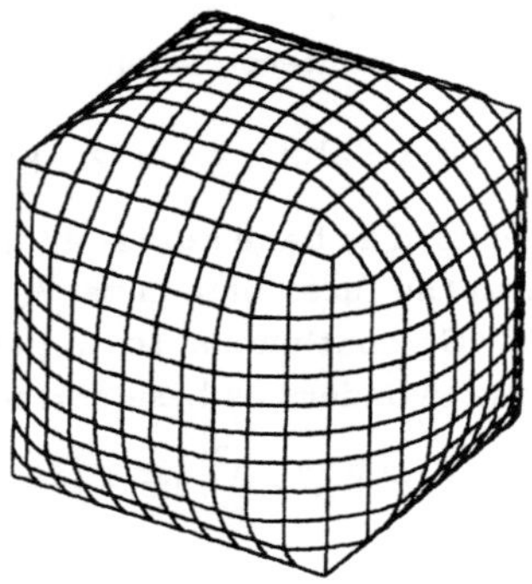

Fig. 9b: Interpolation of a cube with help of cylinders

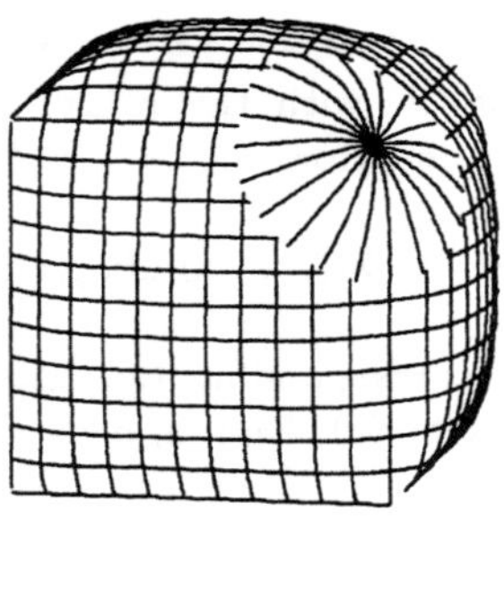

Fig. 9c: Removal of a singularity from the example in Figure 9a

Figure 10 contains the interpolation of a dodecagon by the functional spline

$$(1-\mu)\, l_1 \ldots l_{12} - \mu\, z^3 = 0. \qquad \text{(with } l_i \text{ as suitable chosen planes).}$$

The interpolation surface is smooth. The corners, which appear in the figures, follow from the projection of the surfaces curves.

Fig. 10: Interpolation of a dodecagon (besides the vertices the interpolation surface is regular)

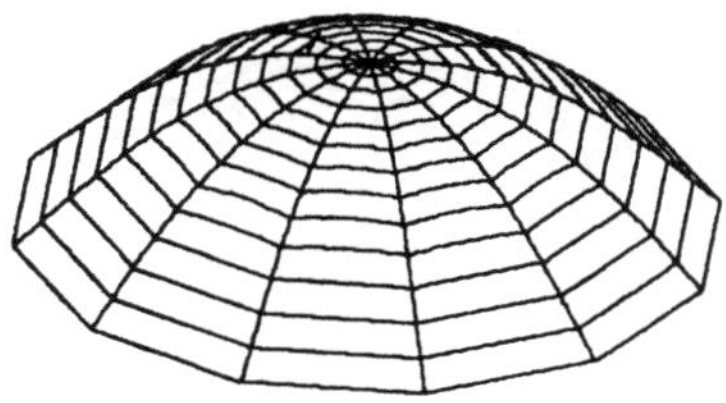

In Figures 8, 9 the interpolation of a cube is developed while the interpolation surface contains the edges of the cube. We now will describe a procedure for constructing a family of G-interpolation surfaces which pass **only** the corners of a cube. The method used can be generalized to arbitrary convex polyhedrons.

Step 1: For each corner P_i we choose a plan ε_i that contains P_i and supports the cube, i.e. all points of the cube are situated on one side of ε_i. Here we take the regular octahedron shown in Figure 11a.

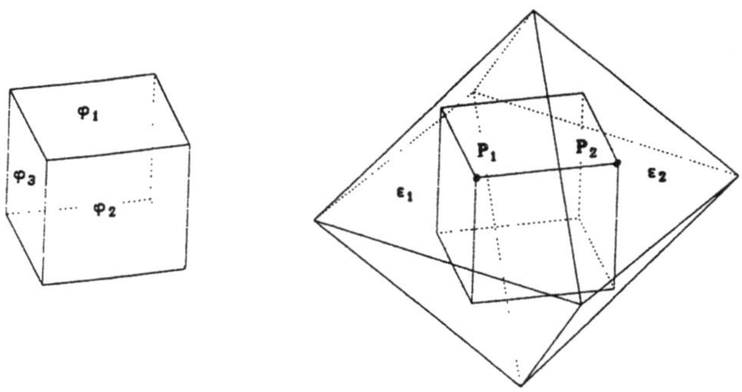

Fig. 11a: Base surfaces (octahedron) for the interpolation of the corners of a cube

Step 2: The plane φ_i, that contains the i-th face of the cube, cuts out of the octahedron a pyramid (Fig. 11b). With the method introduced by Theorem 5 we construct upon φ_i a G^2-interpolation surface Φ_i that contains the base quadrangle of the pyramid and the corners of the i-th face of the cube. If we replace the i-th pyramid of the octahedron by Φ_i we get a convex surface which is the boundary of a convex body B_i. In the same way we round all the corners of the octahedron. If we intersect all the convex bodies B_i, $i = 1,\ldots,6$, we get a convex body whose boundary Φ consists of parts of the surfaces Φ_i, $i=1,\ldots,6$, as boundary patches (Figure 11b shows 3 neighboring patches.). The surface Φ has singular points, namely the intersection curves of neighboring surfaces Φ_i.

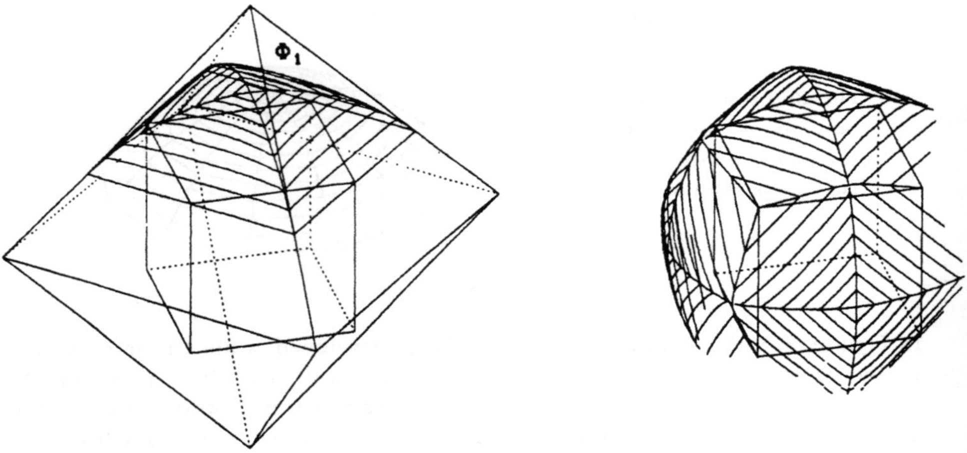

Fig. 11b: Interpolation surfaces of the corners of the cube

Step 3: We intersect every plane φ_i with the surface Φ and get a set of auxiliary curves on Φ (Fig. 11c).

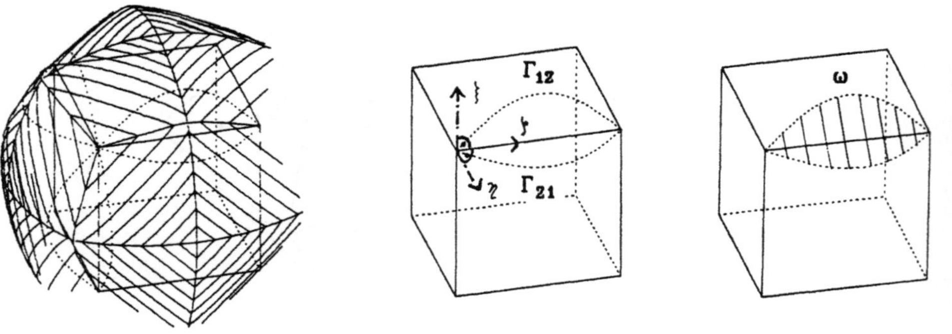

Fig. 11c: Auxiliary curves of $\varphi_i \cap \Phi_i$

Let Γ_{12}, Γ_{21} be the auxiliary curves belonging to the planes φ_1, φ_2 and let e_{12} be the edge between faces 1 and 2 of the cube. We introduce a local ξ-η-ζ-coordinate system such that e_{12} lies on the ζ-axis, the ξ-η-plane is orthogonal to e_{12} and φ_1 contains the ξ-axis, φ_2 contains the η-axis (Fig. 11c). Then Γ_{12} and Γ_{21} can be described by equations $\xi = \gamma_1(\zeta)$ and $\eta = \gamma_2(\zeta)$ resp. We take the ruled surface ω with equation

$$g(\xi,\eta,\zeta) = \xi\,\gamma_2(\zeta) + \eta\,\gamma_1(\zeta) - \gamma_1(\zeta)\,\gamma_2(\zeta) = 0$$

as the base surface of functional spline Ψ :

$$F = (1-\lambda)f_1 f_2 - \lambda g^n = 0\,, \quad 0 < \lambda < 1\,, \quad n > 2\,,$$

where $f_1(\xi,\eta.\zeta) = 0$ and $f_2(\xi.\eta.\zeta) = 0$ are the equations of Φ_1 and Φ_2 resp. The functions γ_1, γ_2 are convex, even in the general case. In the case of the cube we can choose the parameters of the functional splines f_1, f_2 such that $\gamma_1 = \gamma_2$. Figure 11d shows this G^2-interpolation of the vertices of a cube

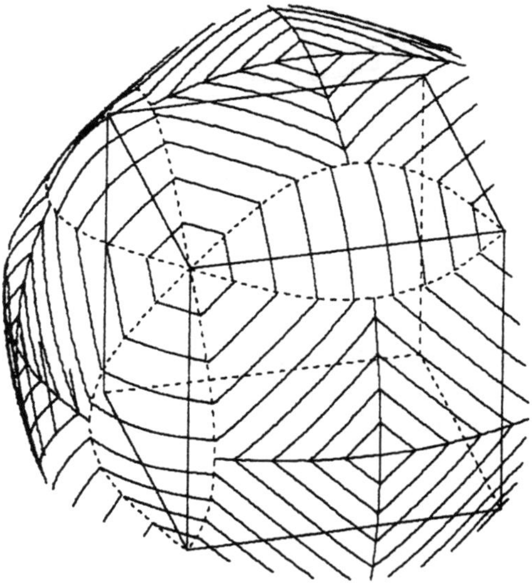

Fig. 11d: G^2-interpolation of vertices of a cube: at the vertices the surface has flat points

We also can construct pure approximation surfaces of a polyhedron with help of the product of the planes p_i of the faces of the polyhedron. By straightforward calculation with help of the convexity criterion in Definition 7 we can prove

Theorem 6: Let p_i be planes in $\mathbb{R}^3$, then the function

$$F = p_0\, p_1 \cdots p_n - C = 0$$

with $C \in \mathbb{R}, C \neq 0$ describes a convex approximation surface of the convex domain determined by the planes p_i.

Figure 12 demonstrates an application of this result: we approximate a cube with the function (planes contain the faces of the cube)

$$F = (a-x)(a+x)(a-y)(a+y)(a-z)(a+z)\cdot a^{-6} - C = 0$$

with $2a$ as lengths of the edge of the cube and $C = 0.1$.

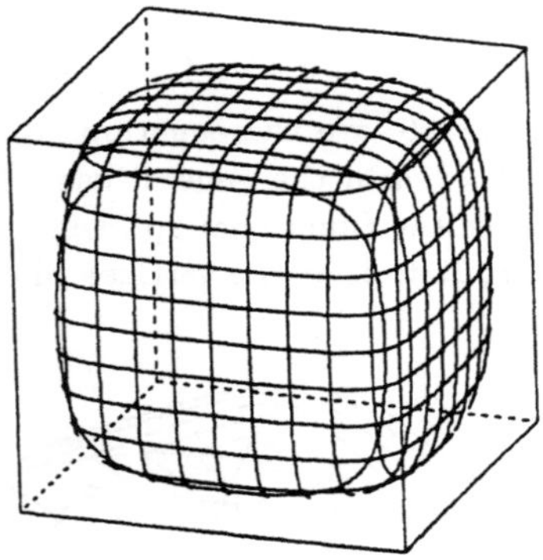

Fig. 12: Approximation of a cube

3.2 Rounding of solids

Now we show an application of the functional splines on the problem of rounding of the edges and the vertices of solids: We start with the cube (Fig. 12a) (0,0,0), (a,0,0), (0,a,0), (a,a,0), (0,0,a), (a,0,a), (0,a,a), (a,a,a).

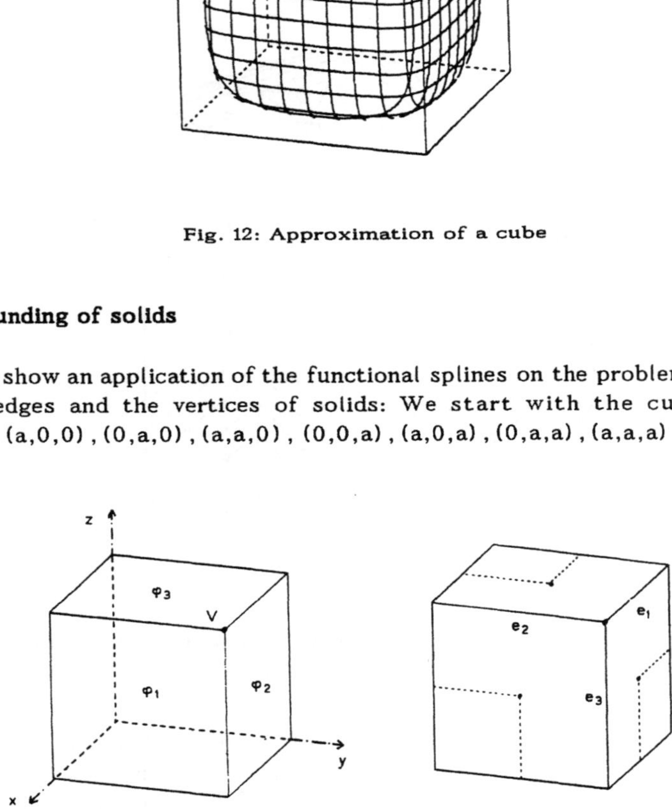

Fig. 12a: Corner V of a cube which should be rounded

and consider the edges e_1, e_2, e_3 of the faces φ_1, φ_2, φ_3 resp. and the corner V. we smooth the edges in the marked areas (dotted lines in Fig. 12a) by the functional spline surfaces:

$$f_1(x,y,z) = (1-\mu)(a-z)(a-y) - \mu(z+y-3a/2)^3 = 0, \quad a/2 \le y,z \le a,$$

$$f_2(x,y,z) = (1-\mu)(a-z)(a-x) - \mu(z+x-3a/2)^3 = 0, \quad a/2 \le x,z \le a,$$

$$f_3(x,y,z) = (1-\mu)(a-x)(a-y) - \mu(x+y-3a/2)^3 = 0, \quad a/2 \le x,y \le a,$$

$$0 < \mu < 1.$$

The cylindrical (but not quadric) surfaces f_1, f_2, f_3 intersect in curves c_1, c_2, c_3 which meet in the point W . f_1 has G^2-connection to the plane $\varepsilon_2 : y = a$ resp. $\varepsilon_3 : z = a$ that contains the face φ_2 resp. φ_3 of the cube. Analogous statements hold for f_2 and f_3.

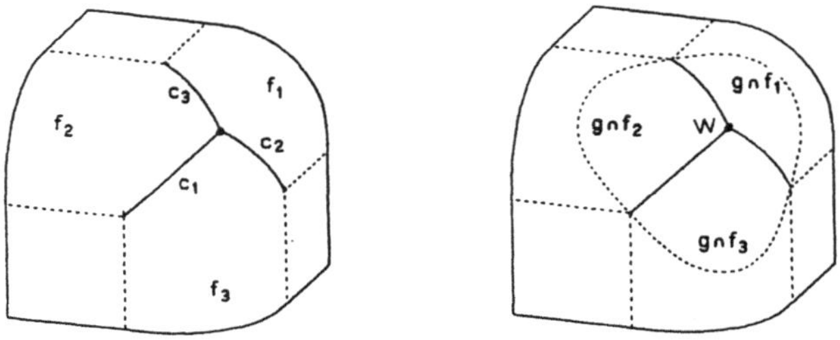

Fig. 12b: Intersection curves of the rounding cylinders

For smoothing the corner W we use the functional spline surface

$$F = (1-\lambda)f_1f_2f_3 - \lambda\, g^3 = 0 , \qquad 0 < \lambda < 1 ,$$

where g is the plane $g(x,y,z) = x + y + z - 2a = 0$ passing the points $(a,\frac{a}{2}, \frac{a}{2})$, $(\frac{a}{2}, a, \frac{a}{2})$, $(\frac{a}{2},\frac{a}{2}, a)$. F has G^2-connections to the surfaces f_1, f_2, f_3 . In order to make the surface F visible we draw curves which are sections with planes containing the diagonal of the cube. The pictures below show solutions for various μ and λ .

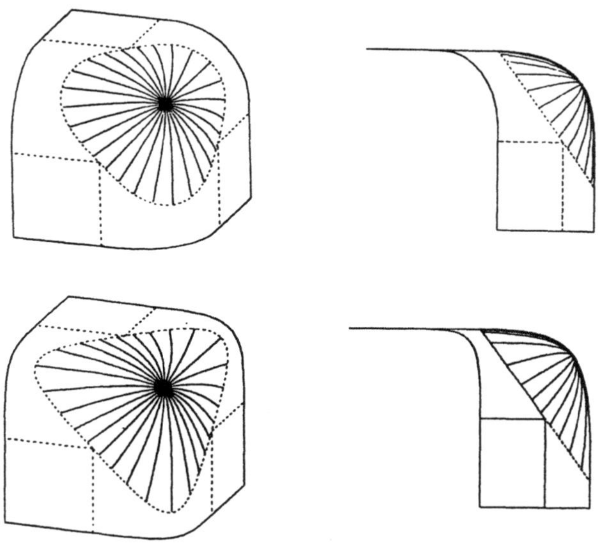

Fig. 12c: Rounding surfaces of a vertex of the cube

A different joining (Fig. 13) of the "cylinders" $f_1 = 0$, $f_2 = 0$, $f_3 = 0$ can be reached if we choose the transversal surface:

$$g(x,y,z) = (x - \tfrac{a}{2})(y - \tfrac{a}{2})(z - \tfrac{a}{2})(x + y + z - 2a) . \quad \text{(g is a product of 4 planes.)}$$

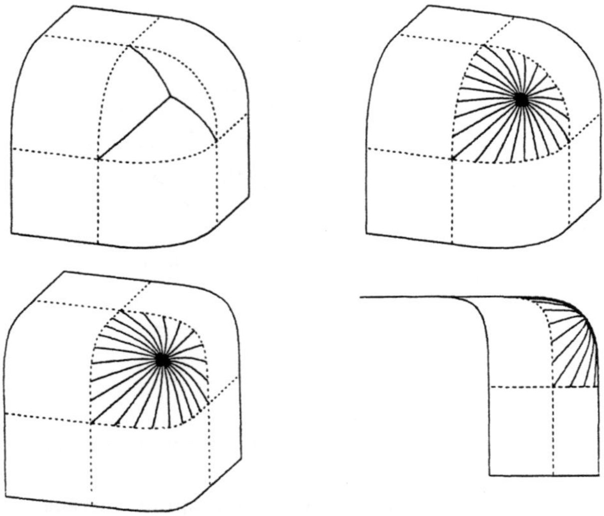

Fig. 13: Other solutions of the rounding problem

Until now we have constructed symmetric rounding surfaces. We can also introduce rounding surfaces with less symmetries: We consider two surfaces Φ_1 , Φ_2 which have a **plane intersection** curve, then it is convenient to take also plane auxiliary curves Γ_1 , Γ_2 . Plane intersection curves often occur caused by symmetries. For example if the two blended edges of a cube are taken (Fig. 14):

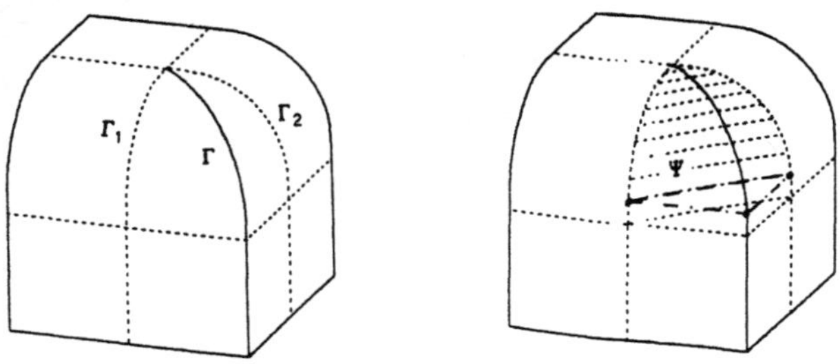

Fig. 14: Rounding of two edges of a cube and ruled surface Ψ

The intersection curve Γ obviously lies in a plane. In this case it is convenient to choose the dotted plane curves Γ_1 , Γ_2 as auxiliary curves.

There is a natural correspondence between points of Γ_1, Γ_2 and Γ with same z-coordinates. Corresponding points of Γ_1, Γ_2 are connected by lines and yield the ruled surface Ψ. With a convenient coordinate system (origin in the hidden corner of Fig. 14) the blending surfaces of the two horizontal edges of the suitcase corner (Fig. 14)) have equations

$$f_1(x,y,z) = (1-\mu)(a-x)(a-z) - \mu(z+x-\tfrac{3}{2}a)^n = 0$$

$$f_2(x,y,z) = (1-\mu)(a-y)(a-z) - \mu(z+y-\tfrac{3}{2}a)^n = 0 \qquad n > 2.$$

Let $x = \varphi(z)$ and $y = \varphi(z)$ be the explicit forms of the equations $f_1 = 0$ and $f_2 = 0$ resp. Then the ruled surface Ψ (Fig. 14) can be described by the equation

$$g(x,y,z) = x + y - \varphi(z) - \tfrac{a}{2}$$

and the blending surface Θ (Fig. 15) that rounds the suitcase corner has the equation

$$F(x,y,z) = (1-\lambda)(x-\varphi(z))(y-\varphi(z)) - \lambda(x+y-\varphi(z)-\tfrac{a}{2})^n = 0, \quad 0<\lambda<1,\ n\geq 3.$$

The surface $F = 0$ joins Φ_1 and Φ_2 G^2-continously along Γ_1 and Γ_2 For displaying the transition surface in Figure 15 sections of Θ with horizontal planes are used .

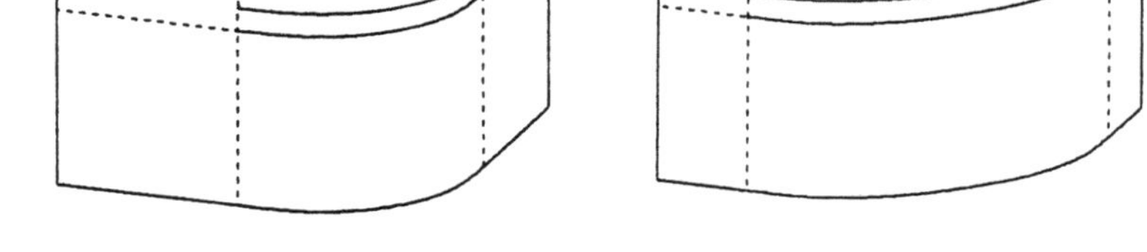

Fig. 15: G^2-blending of a suitcase corner

If auxiliary curves as shown in Figure 15b are chosen, one gets "toroidal" solutions.

The same idea (plane auxiliary curves) is used in the following examples (Figs. 16, 17). The equations of the ruled surfaces and the blending surfaces can be evaluated in the same manner shown above.

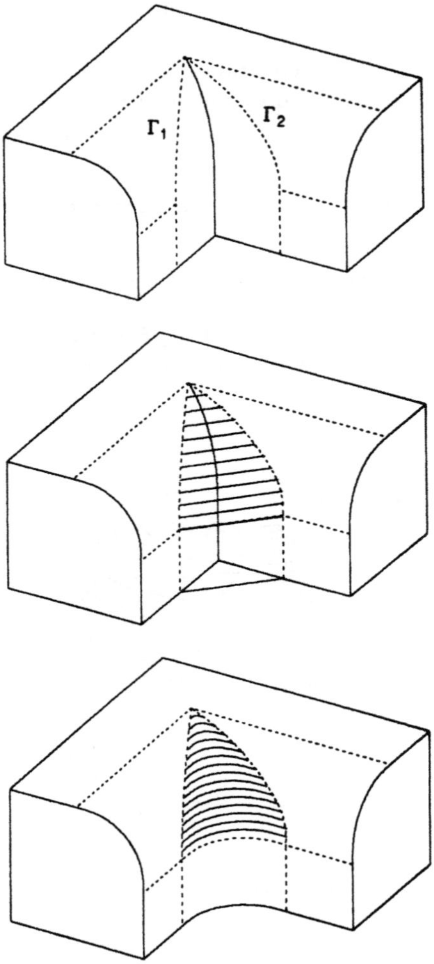

Fig. 16: G^2-blending of a "2-beam" corner

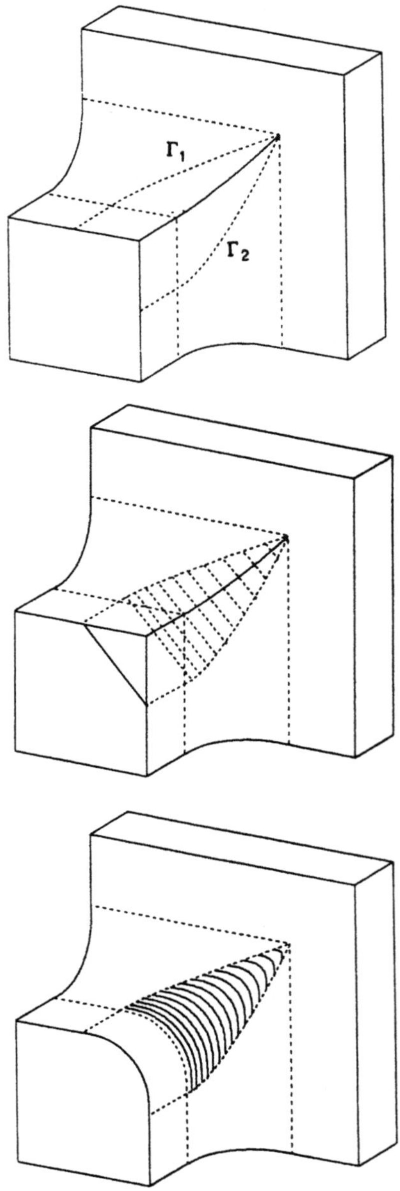

Fig. 17: G^2-blending of a "beam-plane" corner

Further reasonable solutions of the above rounding problems can be achieved, if other plane auxiliary curves Γ_1 , Γ_2 are chosen (Figs. 18, 19):

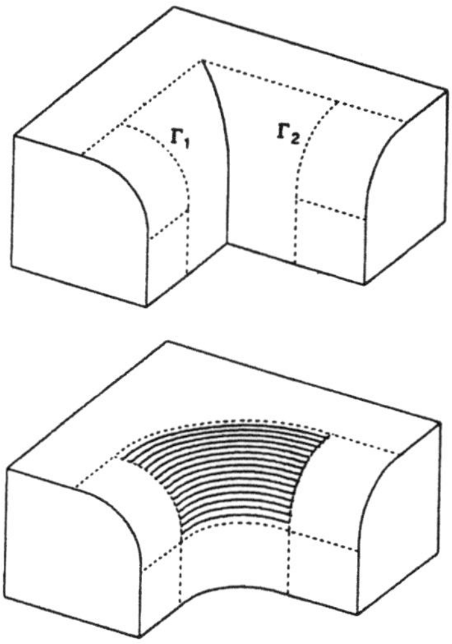

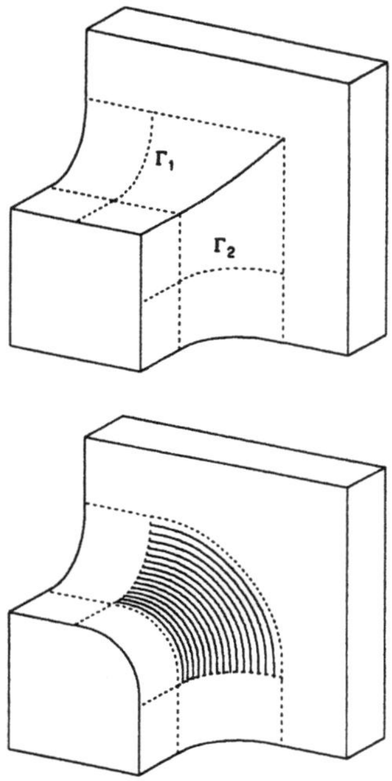

Fig. 18: toroidal solution for the 2-beam-corner

Fig. 19: toroidal solution for the beam-plane-corner

The "toroidal" solutions of the "suitcase" problem (Fig. 15) and the "2-beam" problem (Fig. 18) lead to a G^2-blending of two intersecting quadric tubes (Fig. 20):

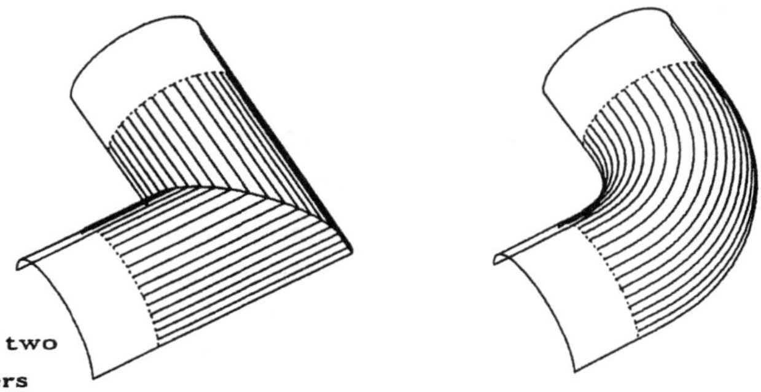

Fig. 20: G^2-blending of two quadric cylinders

We can also find a rounding surface for a 3-beam corner, a problem which sometimes is called "filling of holes".

First we introduce a coordinate system and give some assignments: We choose the coordinate system such that the corner $V_0 = (0,0,0)$ and the planes through V_0 are described by the equations $x = 0$, $y = 0$, $z = 0$. We denote the edges which are concurrent in V_0 by e_1, e_2, e_3, e_4, e_5, e_6. The area of interpolation is marked by dotted lines :

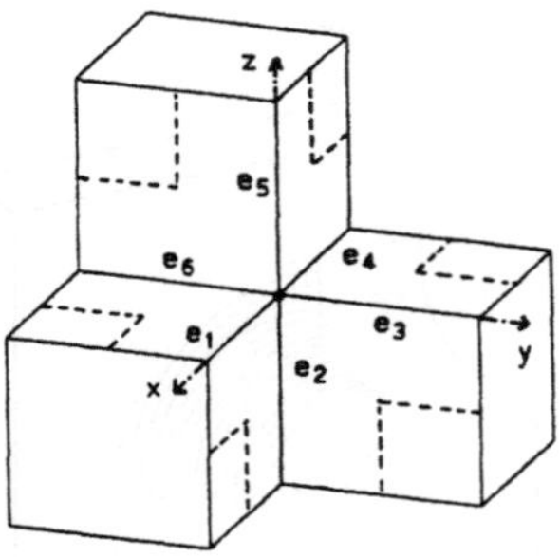

Fig. 21a: 3-beam corner

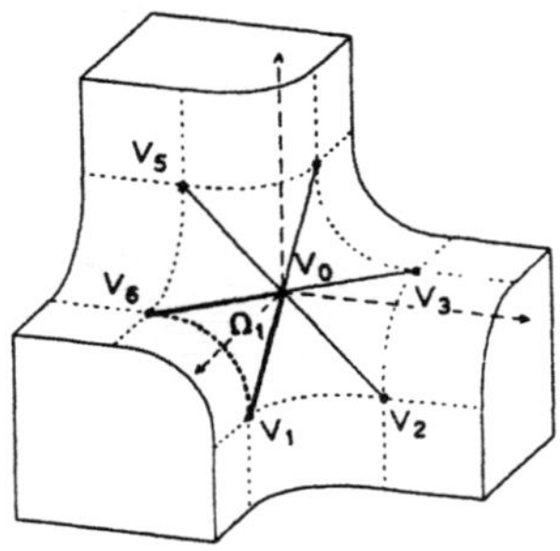

Fig. 21b: Rounding of the edges of the 3-beam corner

We notice that the points V_1, V_2, V_3, V_4, V_5, V_6 are contained in the plane Π: $x + y + z = 0$. This plane will play an important role. We will construct an interpolation surface Φ that passes through the diameters of the hexagon $V_1,\dots,V_6$ and which has Π as tangent plane in the points of the diameters. Thus the interpolation surface Φ will be determined if we define it in the marked area Ω_1 (Fig. 21b).

For smoothing the **edge** e_1 in the area $x \geq r_C$ we use the method given above (Sect. 3.2) for the edges of a cube. Here we get the functional spline:

$$f_1(x,y,z) = (1-\mu)yz - \mu(y + z + r_C)^3 = 0 \quad , \; 0 < \mu < 1 .$$

We call this part of the interpolation-surface "cylinder" Σ.

For the interpolation surface Φ in the area $0 \leq x \leq r_C$ of Ω_1 we construct a functional spline: Let Π be the plane $f_2(x,y,z) = x + y + z = 0$ and Γ be the "cone" (ruled surface) with vertex V_0 and generating curve $f_1 = 0$, $x = r_C$. Γ contains the diameters d_1, d_6 and can be described by the equation

$$g(x,y,z) = (1-\mu)\, x\, y\, z / r_C - \mu(x + y + z)^3 = 0$$

where μ is the parameter of f_1.

The functional spline surface

$$F = (1-\lambda) f_1 f_2 - \lambda g^3 = 0 \quad , \quad 0 < \lambda < 1 .$$

has the demanded properties (contact of order two with Σ and Π) .

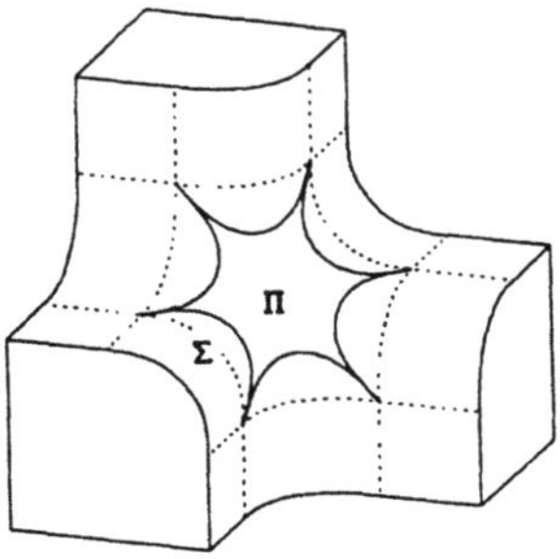

Fig. 22a: "cylinder", plane Π

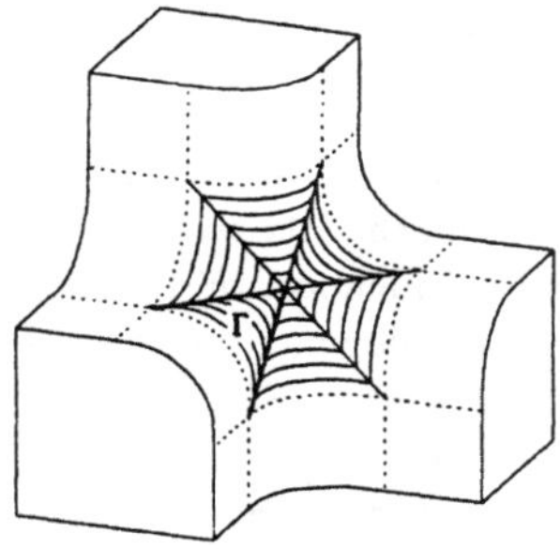

Fig. 22b: "cone"

The following mappings give us the solution in the other areas Ω_2,..

$$\begin{aligned}(x,y,z) &\longrightarrow (-x, -y, -z)\\ &\longrightarrow (\pm y, \pm x, \pm z)\\ &\longrightarrow (\pm y, \pm z, \pm x)\end{aligned}$$

The functional splines in Figure 23a have the same parameter μ as in the figures above. Figure 23b shows solutions for different μ and λ . The curves to be seen are plane sections of the interpolation surfaces.

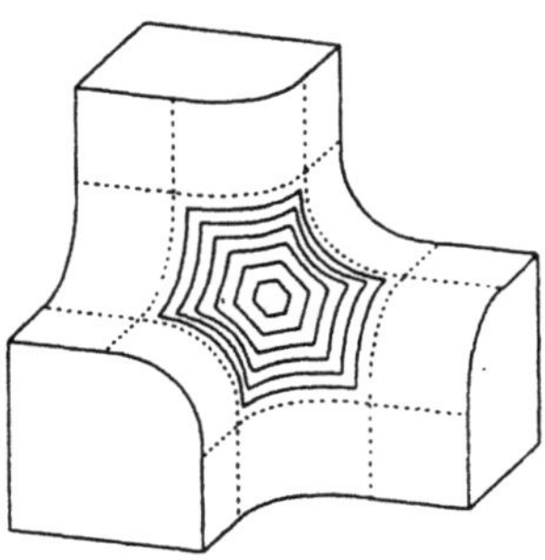

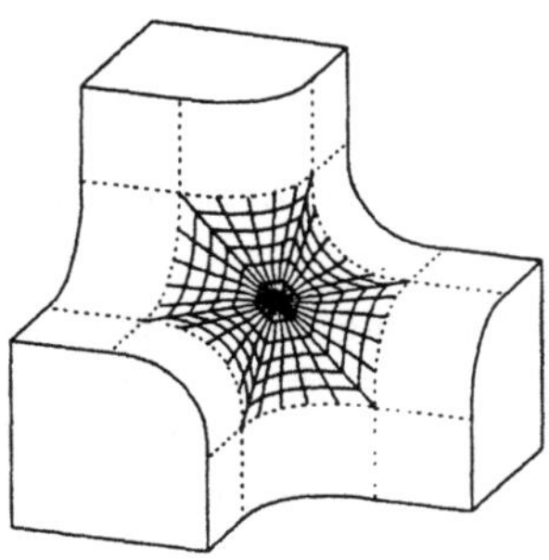

Fig. 23a,b: Different solutions for the rounding of a 3-beam corner

3.3 Blending with G^2-functional splines

First a general algorithm is given for blending two implicit surfaces using plane G^2-functional splines which join G^2-continously the given base surfaces, then it is shown that for some special cases the equations of the transition surfaces can be determined and the G^2-continuity is guaranteed.

Let Φ_1 and Φ_2 be two intersecting surfaces with intersection curve Γ . The first step of the method is to determine two **auxiliary curves** Γ_1 and Γ_2 on Φ_1 resp. Φ_2 in the neighbourhood of Γ (Fig. 24) . The local distance between Γ_1 and Γ or Γ_2 and Γ should depend on the angle between the surfaces in the corresponding intersection curve (Fig. 24). (In classical descriptive geometry Γ_1 and Γ_2 are determined by a "rolling sphere" which contacts both surfaces [3].) Now a convenient **ruled surface** Ψ containing the curves Γ_1 and Γ_2 is taken for constructing an implicit **interpolation surface** Θ that will substitute the surfaces Φ_1 and Φ_2 in the neighborhood of the intersection curve Γ (Fig.24).

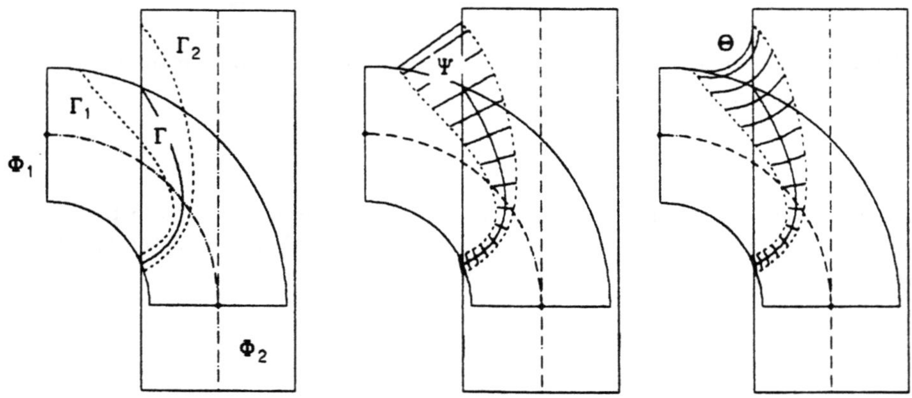

Fig. 24: Auxiliary curves Γ_1, Γ_2 ruled surface Ψ transition surface Θ

The blending of two planes Π_1 , Π_2 is a simple task. If a "rolling sphere " with radius r_0 is chosen then Γ_1 and Γ_2 are lines parallel to the intersection line $\Gamma = \Pi_1 \cap \Pi_2$. The distance ρ between Γ_i and Γ depends on the angle φ enclosed by the planes, which is $\rho = r_0 \tan(\varphi/2)$ (Fig 25).

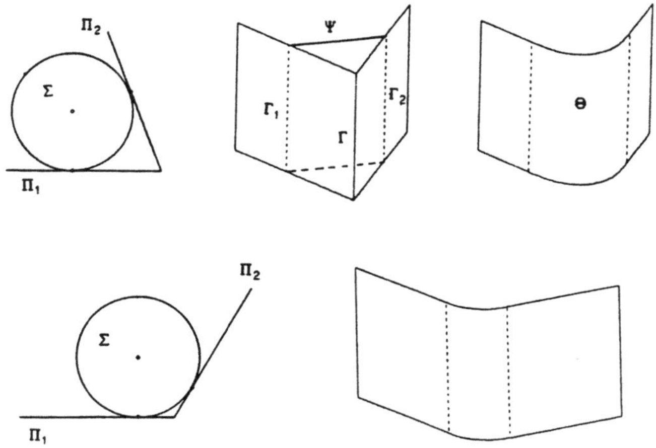

Fig. 25: Blending two planes

The ruled surface Ψ (mentioned above) is here the plane containing the lines Γ_1 , Γ_2 . Let $f_i = 0$ be the equation of the plane Π_i , $i = 1,2$, and $g = 0$ the equation of Ψ. Then , if the signs of f_1, f_2, g are suitably chosen, the equation

$$F = (1-\mu) f_1 f_2 - \mu g = 0 , \qquad 0 < \mu < 1 .$$

describes a G^2-transition surface Θ .

Now more general surfaces are to be blended . Let Φ_1 and Φ_2 be two intersecting surfaces with equations $f_1 = 0$, $f_2 = 0$ resp. and intersection curve Γ . In order to avoid essential difficulties while introducing the method some restrictions are imposed :

(i) f_1 and f_2 are differentiable in the area of consideration.

(ii) There is no point in $\Phi_1 \cap \Phi_2$ where the tangent planes coincide.

The idea of the current blending method is: Describe the transition surface Θ by plane sections with planes normal to the intersection curve Γ .

Step 1: Determination of points of the auxiliary curves Γ_1 and Γ_2. With the aid of a "rolling sphere" Σ (radius r_0) to each point X_i two points $P_1 \in \Gamma_1$ and $P_2 \in \Gamma_2$ are determined: Let Π_1 , Π_2 be the tangent planes of the surfaces Φ_1 , Φ_2 and Ω the plane containing X_i which is orthogonal to the intersection curve Γ . The "rolling sphere" has its midpoint in the plane Ω and contacts the planes Π_1 , Π_2 in the points Q_1 , Q_2 (Fig. 26a). Let γ be the circle through the point Q_1 (and Q_2) with midpoint X_i . Then P_1 and P_2 are the intersection points $\gamma \cap \Phi_1$, $\gamma \cap \Phi_2$ resp. (Fig. 26).

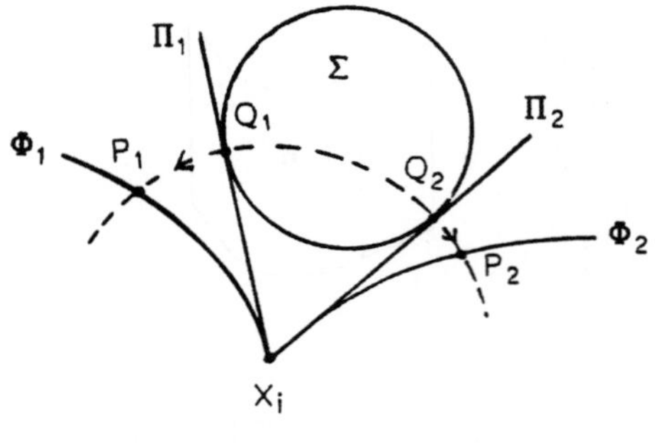

Fig. 26a: Determination of P_1, P_2

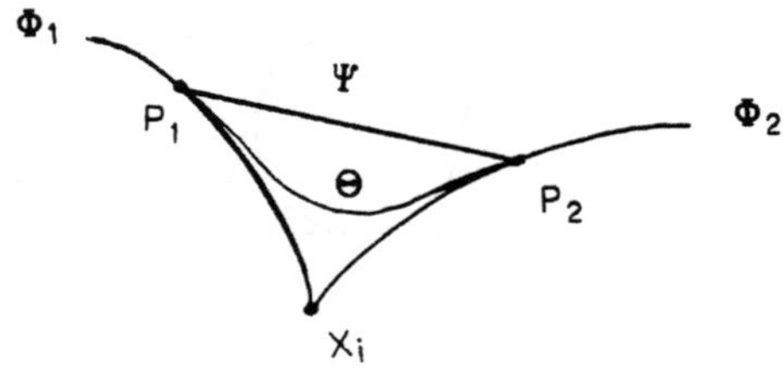

Fig. 26b: Ruled surface Ψ functional spline Θ

Step 2: Determination of the transition surface Θ by its plane sections with planes orthogonal to the intersection curve Γ . We use the denotations of Step 1: If the ruled surface Ψ is described in plane sections with planes orthogonal to Γ by the equation $g(x,y,z) = 0$ where $g(x,y,z)$ is the (oriented, normal) distance of a point (x,y,z) (between Ψ , Φ_1 and Φ_2) to Ψ (Fig. 26b) then the equation

$$F = (1-\mu)f_1 f_2 - \mu g^3 = 0 , \qquad (0 < \mu < 1)$$

represents a **transition surface** Θ . (Pay attention to the signs !) With the marching method from the introduction we can determine for each point X_i of the intersection curve the plane section $\Omega(X_i) \cap \Theta$ (Fig. 26b).

The pictures below show applications of this blending method:

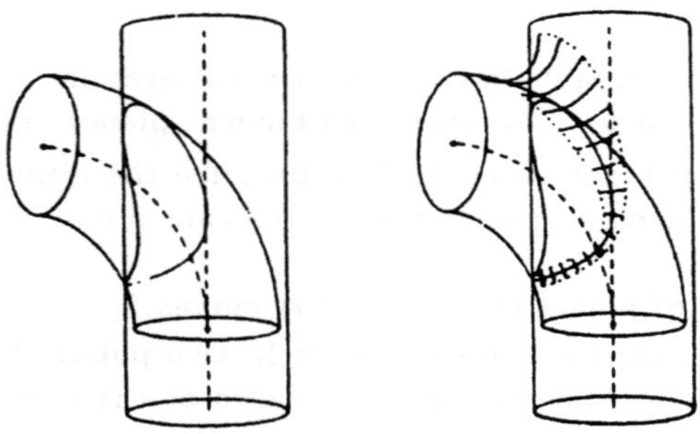

Fig. 27b: Blending a torus and a cylinder

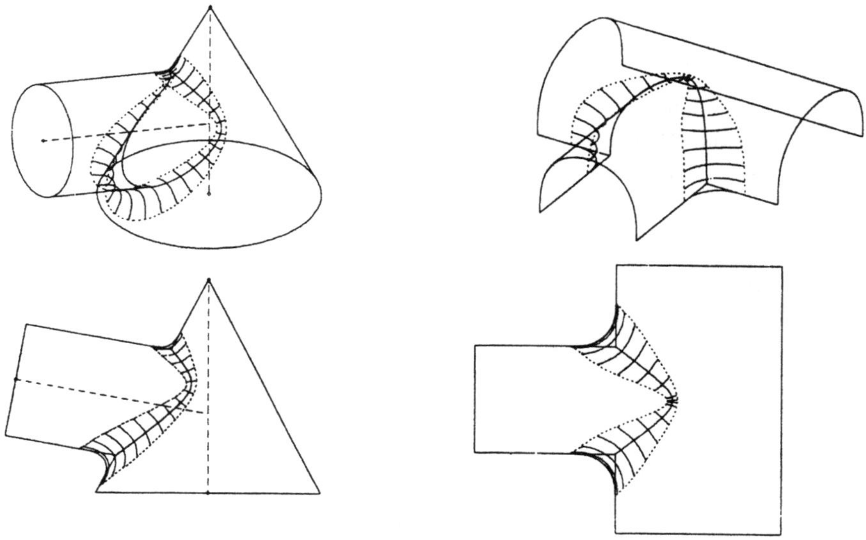

Fig. 28: Blending cones and cylinders

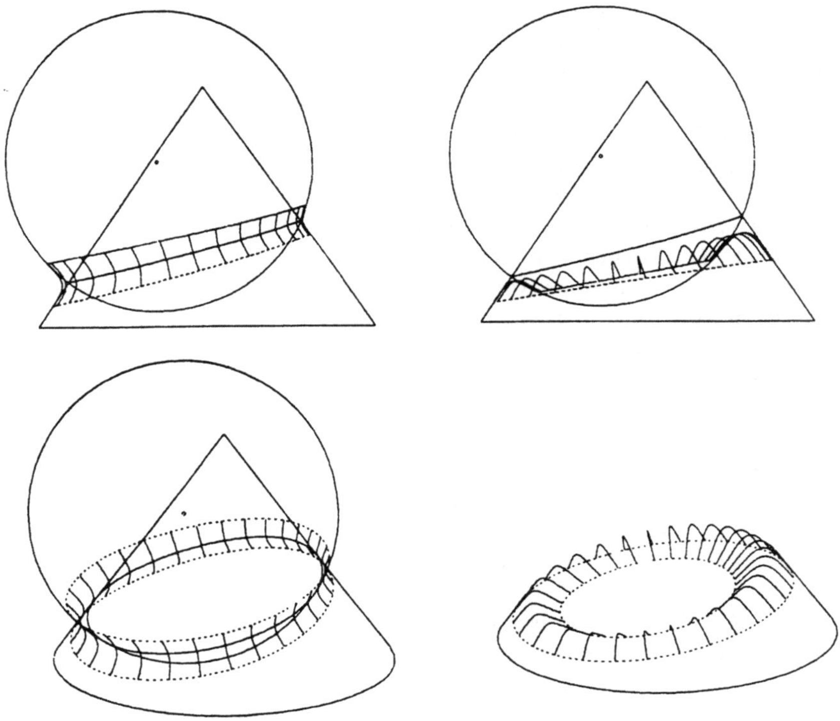

Fig. 29: Additve and subtractive blending
of a sphere and a cone

REFERENCES

1 Cohen, S.: Beitrag zur steuerbaren Interpolation von Kurven und Flächen. Dissertation TU Dresden 1982.

2 De Rose, T.D., Loop, Ch.T.: S-patches: A Class of Representations for Multi-Sided Surface Patches. Technical Report 88-05-02 University of Washington, 1988.

3 Do Carmo, M.. P.: Differential Geometry of Curves and Surfaces. Prentice Hall 1976.

4 Favard, J.: Cours de Géométrie différentielle locale. Gauthier 1957.

5 Gregory, J.A., Hahn, J.M.: Geometric continuity and convex combination patches. Computer Aided Geometric Design **4**, 79-89 (1987).

6 Garrity, Th, Warren, J.: Geometric Continuity. Computer Aided Geometric-Design 1990.

7 Hahn, J.M.: Filling Polygonial Holes with Rectangular Patches. in Strasser, W.(ed.): Theory and Practice of Geometric Modeling, Springer 1989.

8 Hartmann, E.: Computerunterstützte Darstellende Geometrie. Teubner 1988.

9 Hartmann, E., Li, J.: Smoothing of Corners with Functional Splines. Preprint Fachbereich Mathematik, Technische Hochschule Darmstadt (1989).

10 Hartmann, E.: Blending of Implicit Surfaces with Functional Splines. Submitted to Computer-aided design 1990.

11 Hoffmann, Ch., Hopcroft, J.: Automatic surface generation in Computer-aided design. Visual Computer **1**, 92-100 (1985).

12 Hoffmann, Ch., Hopcroft, J.: The Potential Method for Blending Surfaces and Corners. In: Farin, G. E. (ed): Geometric Modeling: Algorithms and new Trends, 347 - 366, SIAM 1987.

13 Hohenberg, F.: Konstruktive Geometrie in der Technik. Springer 1966.

14 Holmström, I.: Piecewise quadric blending of implicitly defined surfaces. Computer Aided Geometric Design **4**, 171 - 190 (1987).

15 Hoschek, J.; Hartmann, E.; Li, J.; Feng, Y.Y.: G^{n-1}-Functional Splines for Interpolation and Approximation of Surfaces and Solids. International Series in Numerical Mathematics, Birkhäuser 1990.

16 Hoschek, J. - Lasser, D: Grundlagen der geometrischen Datenverarbeitung. Teubner 1989.

17 Li, J.; Hoschek, J.; Hartmann, E.: G^{n-1} functional splines for interpolation and approximation of curves and surfaces and solids. Computer Aided Geometric Design 1990.

18 Liming, R.A.: Practical Analytical Geometry with Applications to Aircraft. Macmillan 1944.

19 Pratt, V.: Direct Least-Squares Fitting of Algebraic Surfaces. Computer Graphics **21**, 145-151 (1987).

20 Ricci, A.: A constructive geometry for computer graphics. Computer Journal **16**, 157-160 (1973).

21 Sabin, M.A.: Non-rectangular patches suitable for inclusion in a B-Spline surface. In: ten Hagen (ed.): Proceeding Eurographics, North-Holland 1983, 57-69.

22 Scheffers, G.: Anwendungen der Differential- und Integralrechnung auf Geometrie. Vol. I, Springer 1922.

23 Scheffers, G.: Anwendungen der Differential- und Integralrechnung auf Geometrie. Vol. II, 3. ed., Springer 1922.

24 Woodwark , J. R.: Blends in Geometric Modelling. In: Martin, R. R. (ed.): The mathematics of surfaces II, 255 - 297. Clarendon Press 1987.

Numerically-Controlled Milling of CAD Surface Data

Reinhold Klass and Peter Schramm

Mercedes Benz AG, Abteilung EP/ADTK, D-7032 Sindelfingen, Germany

Abstract

The article reviews CAM methods applied in the milling of freeform surfaces. Based on a brief presentation of the technological constraints, the main steps in the extraction of machine-understandable information from surface data are discussed. Special emphasis is laid on the problem of collisions between tool and surface model.

1. Introduction

In recent years the use of computers in the processing of geometric surface models has attracted considerable scientific and even more industrial interest. In fields like car or airplane body design almost all steps from styling to the production of models or tools are nowadays to some extend aided by computer programs allowing for more versatility and control of the process. Along the way, numerous 3-letter-abbreviations for different parts of the field have been created, the most popular of which are CAD (=Computer Aided Design) and CAM (=Computer Aided Manufacturing).

Most publications have focussed on various aspects of geometric curve and surface design [1,2] The CAM part of the problem may well have attracted an equal amount of effort among software groups but very little of that is available in print [3,4]. One reason for this relative lack of publications is that as soon as one comes close to manufacturing, the general principles have to be supplemented by so many technological constraints and specialities that it is hard to still extract something vaguely interesting for anybody else.

In the present article, we will attempt to describe the main problems involved in the numerically-controlled (NC) milling of surfaces using a CAD/CAM system. Our considerations are based on the methods used in our CAD/CAM system SYRKO which has been developed and applied by Mercedes Benz over the last years. Hence, our examples stem from car body design, but the problems and solutions are quite the same as in other CAD applications.

It is beyond our scope to describe any of the various techniques involved in the design of something as complicated as a car body. Suffice to say that in CAD a surface is usually described as a piecewise polynomial function of two parameters. The most common descriptions are the Bezier and B-Spline representations [1,2] . On a graphic terminal, a surface is then depicted as a net of isoparametric curves (Fig.1). In practise, most surfaces are trimmed, i.e. the range of the spline parameters is restricted by boundary curves.

A real part of a car is usually too complicated to be described by a single surface (Fig.2). Hence, the process of manufacturing is complicated by the fact that when working on a part we always have to take into account several trimmed surfaces. Naturally, two adjacent surfaces cannot have the exact same boundary curves. By means of special approximation methods we can guarantee, however, certain continuity and smoothness conditions with small tolerances where surfaces contact each other.

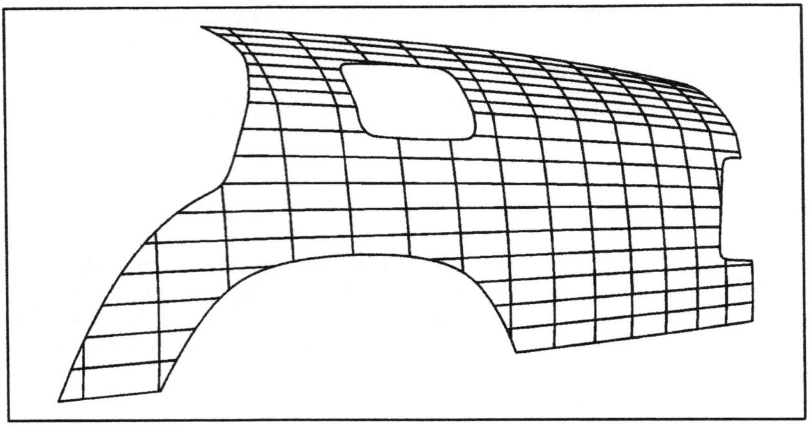

Fig. 1. : Graphic representation of a trimmed surface

Assembling parts like that in Fig.2 we get the surface model of an entire car (Fig.3). Generally, the whole outer surface is available as CAD data whereas the percentage of inner parts which are numerically designed varies from company to company.

Parts of a car body like the fender depicted in Fig.2 are manufactured out of planar metal sheets by use of dies having the same shape as the desired surface. To build these dies, the data of a part usually have to be extended by the so-called geometry of redesign (Fig.4) which serves to position the sheet correctly before it can be formed.

The construction of tools such as dies which are used to produce parts of the car body is only one problem where NC-milling is involved. Another important application is the making of 3-D models which despite all progress in computer graphics are still needed to control a designed shape. Further, gauges which are used to control whether the finished product is close enough to the desired measure are also made starting from CAD data.

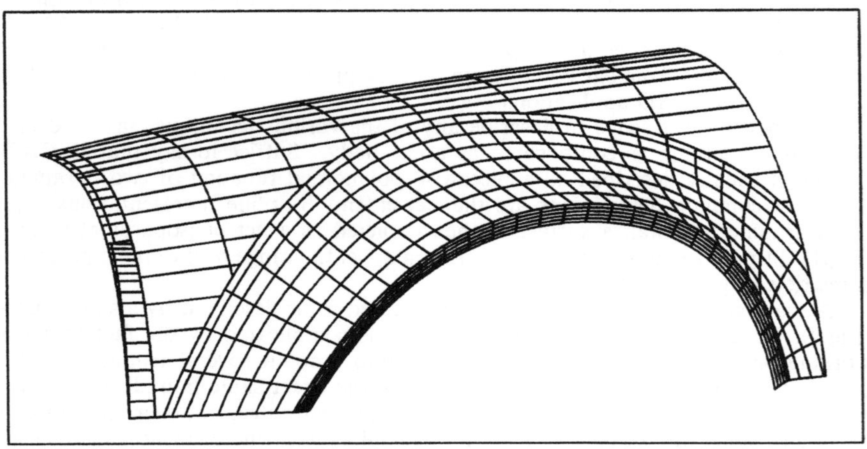

Fig. 2. : Surface model of a fender

Fig. 3. : Shaded picture of the CAD model of a car

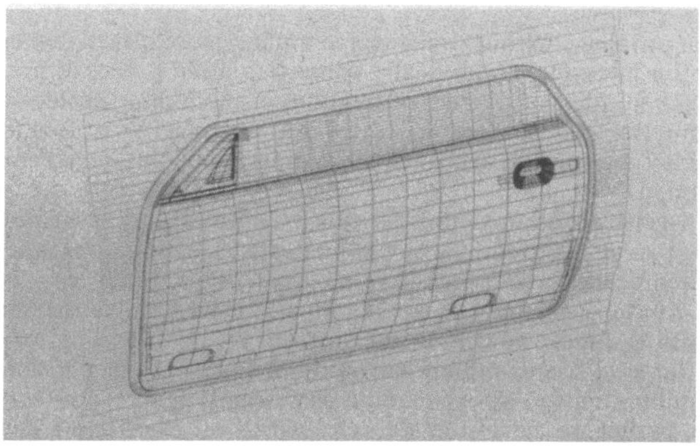

Fig. 4. : The geometry of redesign is an extension of the surface model required for tools in order to fabricate the actual car part

Let us now outline how this article is organised. In the following section we will briefly discuss several technological aspects of NC-milling such as typical shapes of milling cutters and a classification of the tools and methods. In section 3 we describe how the relevant information for a milling tool, i.e. the milling courses, can be obtained from numerical surface data. One of the most interesting steps in this process is presented in a little more detail in the subsequent section. The so-called collision check ensures that while the milling cutter is on its path across the surface under work, it does not collide with parts of the surface model anywhere else. The problem of cutter interference has been treated by several authors [5,6] but our approach is different inasmuch as we do not distinguish between collisions along and perpendicular to the milling course. Finally, we conclude with a few remarks on how the information is transferred from the CAM system to the NC-machine.

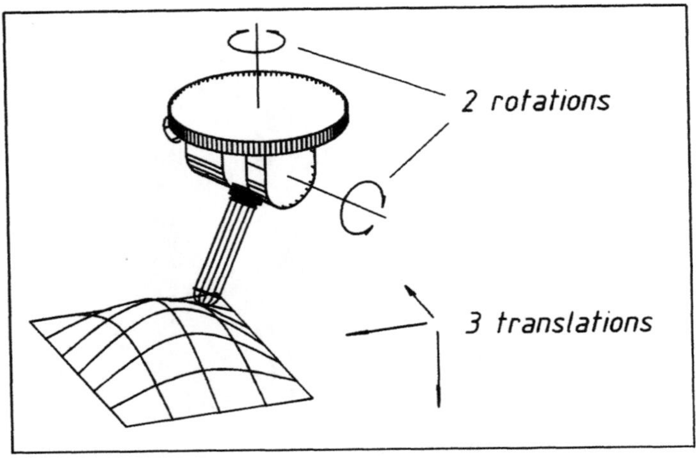

Fig. 5. : Schematic picture of a 5-axis-milling machine

2. A few Remarks about Technology

To describe the problems involved in surface milling more clearly, we have to discuss the most common methods. Numerically controlled milling devices are generally distinguished by the number of degrees of freedom of the milling cutter. The most basic apparatus to mill freeform surfaces has 3 translational degrees of freedom. Hence, the milling cutter may move quite arbitrarily in 3 dimensions but its axis remains pointed in a fixed direction.

A powerful generalization of this so-called 3-axis-milling is obtained by using a tool with 5 degrees of freedom. In addition to the 3 translations, the shaft of a 5-axis-milling machine can be oriented in (almost) any direction by means of 2 rotations (Fig.5). These rotations of the axis are commonly used to have the shaft pointing in a direction which is almost normal to the surface.

The advantages of 5-axis-milling are quite obvious. Let us consider a cylindrical shape of the milling cutter. As is readily seen from Fig.6, the effective width of a milling course, i.e. where the cutter actually works on the desired surface, may become very narrow for a 3-axis-machine as the angle between surface normal and tool axis increases. On the other hand, in 5-axis-milling the spacing between adjacent milling courses can remain close to the diameter of the cutter as long as the curvature of the surface is not too large.

As a result, we usually have a smaller number of milling courses, better surface quality, and less need for postprocessing in 5-axis-milling. This gain in processing time becomes particularly large once a part can be fabricated in one go on a 5-axis-machine whereas several reorientations would be required in 3-axis-milling. A further advantage is that a 5-axis-machine has more possibilities to avoid collisions. In addition to the standard procedure of lifting the cutter it can also be rotated away (cf. section 4). On the other hand, 5-axis-machines are more expensive and the programming is more involved. Therefore, the technique is often reserved for more "elaborate" problems.

Set between the techniques we have discussed is 4-axis-milling. Here, the orientation of the cutter can be varied around only one rotational axis. The method is usually applied to problems with planar milling courses such that the shaft may move in the same planes.

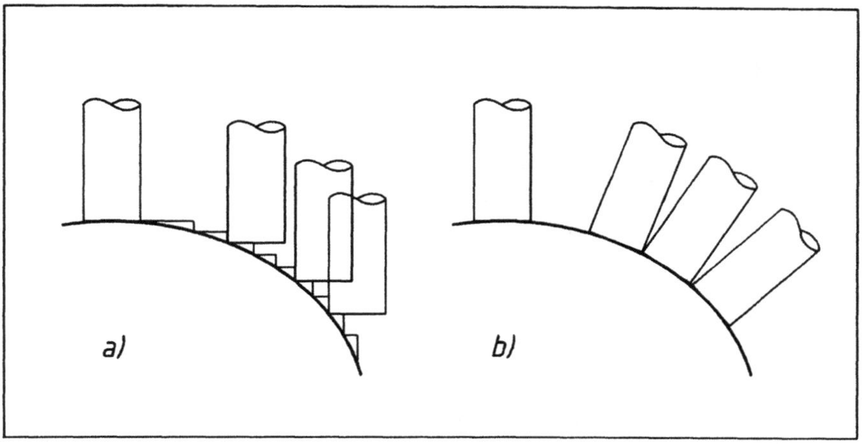

Fig. 6. : Width of milling courses for a) 3- and b) 5-axis-milling

As a final example of common techniques we mention milling in 2.5 dimensions. This may sound terribly complicated but the principle is straightforward. The motion of the milling cutter is totally free in 2 dimensions, but in the third direction there is only up or down. Hence, the method can be readily applied to problems like contour milling. Since we are mostly interested in the fabrication of freeform surfaces, the most important techniques are 3- and 5-axis-milling.

In addition to selecting the type of machine, one also has to decide which milling cutter is to be used when working on a part. Clearly, the diameter of the cutter shaft should be chosen with careful regard of the geometry. But furthermore, the shape of the cutter has substantial influence on both theory and practical results. The most common types of cutters are depicted in Fig.7 but many more special shapes are also in use.

A spherical milling cutter (Fig.7a) is the easiest shape as far as theory is concerned. In practise, an important property is that it automatically creates fillets if it properly moves along a convex edge between two surfaces. This feature is of particular advantage for fillets with small radii which are sometimes not included in a surface model. For milling large smooth surfaces, however, a spherical shape is not very effective, since a great number of milling courses is required (see also section 3).

Cylindrical milling cutters (Fig.7b) are both effective and easy to manufacture. The theory is not too hard to handle either. A disadvantage, however, is that cylindrical cutters are subject to high attrition since they mill mostly with a small area at the edge of the cylinder where the forces are very high.

Modifications of the cylindrical shape can improve the durability of the tool. In Fig.7c the cylinder shaft and the planar bottom of the cutter are connected by a radius. This shape is referred to as ball flat (BAFL). Another variation of the cylinder shape is the conic end shown in Fig.7d. Naturally, the theory for these more complex shapes is more involved, especially if one has to check collisions between tool and surface. We note that worn-off cylindrical cutters are sometimes used in one of the modified shapes after additional grinding.

3. From Surfaces to Milling Courses

Let us now turn to the central problem in surface milling, of how to get milling courses (i.e. information which can be processed by the machine) from surface data.

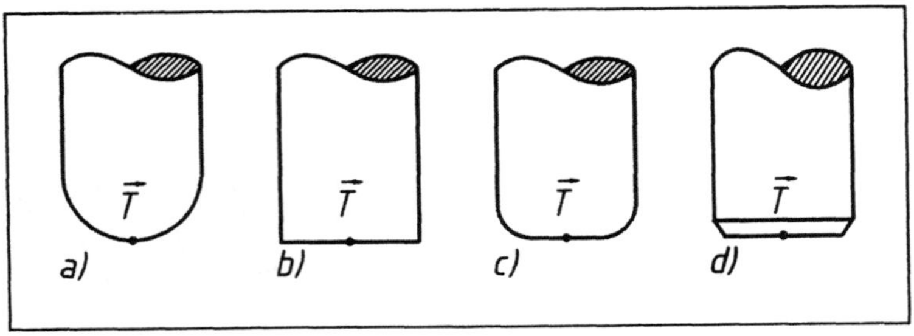

Fig. 7. : Important types of milling cutters: a) spherical, b) cylindrical, c) BAFL (ball flat), d) cylindrical with conic end. The point marked $\vec{T}$ is the so-called tool endpoint

Naturally, the first step is to find suitable curves. These curves mark the path along which the milling cutter actually works on the surface. Clearly, the point where cutter and surface contact each other is not fixed on the cutter but depends on the orientation and shape of both cutter and surface.

The contact curves are usually described by a series of points where each point also bears information about the orientation of the cutter shaft. We will denote such a sequence of points plus vectors as a master dimension identifier (MDI). Splines are not employed in this context since polygonal information can be forwarded more easily to the machine. By means of approximation methods we can guarantee that the polygonal path does not deviate from the curved surface by more than a given tolerance. Edges of the surface model are usually represented by double points in the MDI where the points have the surface normal of either side of the edge.

Now, there are several familiar methods to define contact-MDIs on a surface:

- One can simply use isoparametrics of the surface (Fig.8a). This has the disadvantage, however, that the spacing between neighboring paths has to be very narrow in parts of the geometry if the isoparametrics deviate strongly from parallels.
- This problem is not met for the milling along parallel planar intersections of the surface (Fig.8b), which is particularly suited for 3- or 4-axis milling.
- Furthermore, the contact MDIs can also be chosen parallel to a given curve, e.g. one serving as a "spine" for the surface (Fig.8c). Here, parallel describes two curves with constant spacing in between.
- More generally, one can also choose two curves on either side of the surface. The middle curves are then obtained as weighted averages of the outer two (Fig.8d).
- As a final method, we note that contact curves can also be obtained as the projection of a given set of curves onto the surface (Fig.8e).

It should be understood, however, that a milling course is by no means simply a surface curve. Rather, it is the path of the endpoint of the tool as it cuts its way across the surface. Since the tool endpoint (marked by $\vec{T}$ for the milling cutters in Fig.7) differs from the point of contact between cutter and surface, a milling course can only be calculated from a surface curve if the orientation and shape of the cutter are specified. For a BAFL cutter this is illustrated in Fig.9. It is readily seen that the tool endpoint $\vec{T}$ can be obtained from the contact point $\vec{C}$ by means of the relation

(1) $\vec{T} = \vec{C} + r_2(\vec{n} - \vec{m}) + (r_1 - r_2)(\vec{n} \times \vec{m}) \times \vec{m}$

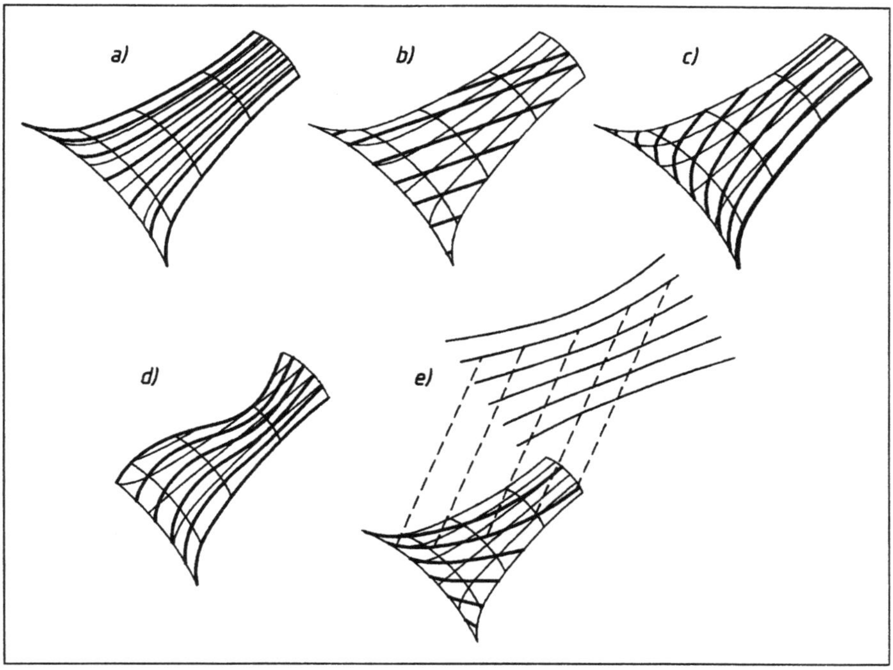

Fig. 8. : Ways of defining contact-MDIs on a surface: a) isoparametrics, b) planar intersections, c) parallel to one curve, d) average of two curves, e) projection curves

where the cross denotes the vector product. Here, $\vec{n}$ and $\vec{m}$ are normalized vectors in the direction of the surface normal and the tool axis, respectively. Further, r_1 is the radius of the cutter shaft, and r_2 is the fillet radius at the bottom of the tool.

The necessity to distinguish between surface normal and shaft orientation is quite obvious for machines with 3 or 4 degrees of freedom. Note, however, that this distinction remains valid also in 5-axis-milling. There the process is more effective and the surface quality improves if the two directions differ slightly from each other. Angles between 1 and a few degrees are quite common. The rotation is usually performed in the plane spanned by the two vectors $\vec{m}$ and $\vec{n}$.

By means of relations like (1) we can transform contact MDIs into paths of the tool endpoint. In general, however, the resulting curves are still not the desired milling courses. By now, we have only considered how the cutter is supposed to work on the surface. The cutter shaft, however, is an extended device, and we also have to make sure that there are no unwanted contacts between device and the entire surface model. This procedure is called collision check and will be discussed in more detail in the following section. The result is a modified path for the tool endpoint, i.e. our first milling course.

Now, one question we have not yet adressed is how to find the correct spacing between two adjacent paths. Clearly, the shaft diameter is a natural upper limit, but once either surface or cutter are curved the problem is slightly more complicated. Looking at the cross section of the surface depicted in Fig.10 we have to make sure that the scallop height between neighboring paths of the cutter is not higher than a certain tolerance. Further, cusps like in Fig.10 can necessitate postprocessing because they are seen even if the tolerance is not violated. Again, this problem depends strongly on the shape of the cutter.

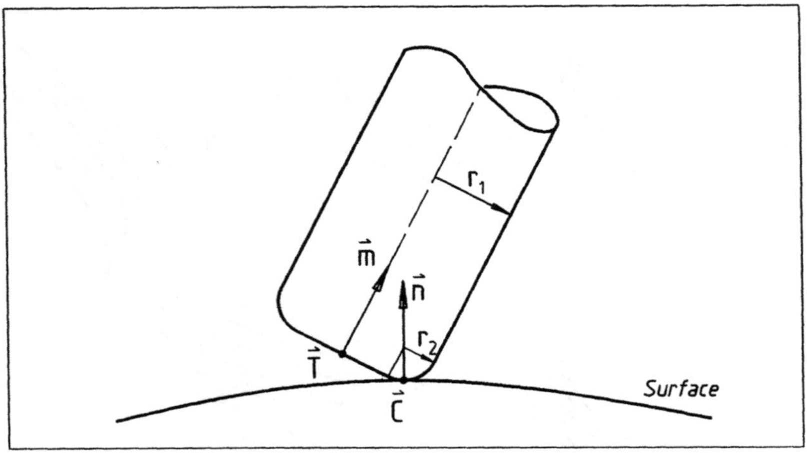

Fig. 9. : Tool endpoint $\vec{T}$ and point of surface contact $\vec{C}$ for a BAFL milling cutter

Once we have determined the necessary spacing to the next milling course, we can find the corresponding contact curve, such that this spacing is nowhere exceeded. Then the steps leading from contact curve to milling path are continued until the entire surface is covered.

4. The Collision Check

In the preceding section we have discussed how surface data can be transformed into paths of the milling cutter which can be used to manufacture the surface. However, the necessity for corrections of the MDI for the tool endpoint due to collisions with the surface model was only briefly mentioned. Let us now consider this problem in more detail.

The starting point is the original trajectory of the tool endpoint which follows from the contact curve by means of relations like (1). This path was obtained by requiring that the tool touches the desired surface at the point $\vec{C}$ (cf. Fig.9). On the other hand, the surface may easily be violated by the cutter shaft somewhere else

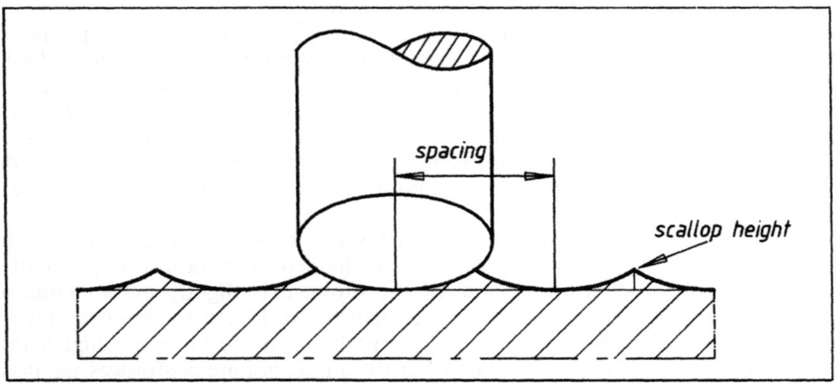

Fig. 10. : Allowed spacing between neighboring milling courses

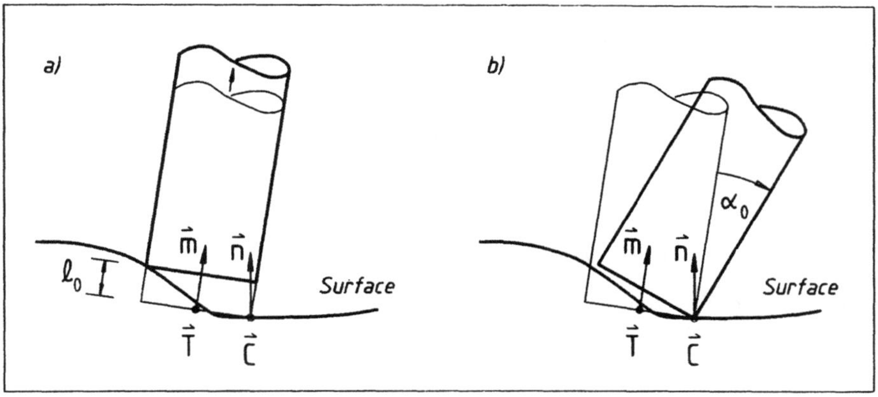

Fig. 11. : Collisions between milling cutter and surface can be avoided by a) lifting, and b) rotating the tool

(Fig.11). Such collisions have to be carefully examined and avoided in the process of manufacturing. Obviously, we have to check all surfaces within a certain neighborhood of $\vec{C}$. For complicated surface models this includes an enormous amount of data.

Let us now examine the possibility of a collision at a given point of the milling course. In principle, we have a standard intersection problem. However, simply intersecting the cutter surface with the entire model would be too time consuming. Fortunately, we can reduce the data by considering appropriate boxes first. Further, we shall see that even in critical areas we don't have to determine the full surface-surface intersection.

But let us first discuss how a collision can be avoided. Clearly, the MDI for the tool endpoint has to be modified, and there are two ways to do so: we can either change the points or the vectors.

Changing the point is an obvious choice which can always be applied. The tool is lifted in the direction of its axis $\vec{m}$ until the collision is avoided (Fig.11a). Note that this reaction includes that the tool cannot fabricate the desired surface in this area since it no longer reaches the point $\vec{C}$. Hence, we need further processing with another tool of different size and/or shape.

A second possible reaction to collisions is to change the axis of the cutter. It is rotated around an axis which goes through $\vec{C}$ and is rectangular on both surface normal $\vec{n}$ and tool axis $\vec{m}$ (Fig.11b). Clearly, this is only possible in 5-axis-milling or, in a more restricted way, in 4-axis milling where the axis cannot be chosen normal to $\vec{m}$, in general. The advantage of this second method is that the contact at $\vec{C}$ is retained which means that the surface may still be fabricated in the area under consideration.

Now, the problem we have to solve is to determine the length l_0 by which the tool has to be lifted, or the angle α_0 by which the shaft must be rotated to avoid the collision. Again, the solution depends on the shape of the tool. Here, we will only discuss the case of a cylindrical milling cutter (cf.Fig.7b).

Projection of a Point onto a Surface

An important underlying problem in order to determine the corrections for the milling cutter is to find the projection of a point onto a surface in an arbitrary direction. Let

(2) $$\vec{F}(u,v) = \sum_{i,k=0}^{n} N_i(u) N_k(v) \vec{B}_{ik}$$

be the B-spline representation of the surface. Here, $\vec{B}_{ik}$ are the de Boor points while $N_i(u)$ are the standard basis polynomials for B-splines of order n. Now, the projection $\vec{\mu}$ of a given point $\vec{X}$ onto the surface is obtained as a solution of

(3) $$\vec{\mu}(\vec{X}) := \vec{X} + l\vec{m} = \vec{F}(u_0,v_0)$$

where $\vec{m}$ is the direction of projection. Choosing two vectors $\vec{m}_1$ and $\vec{m}_2$ which are orthogonal to $\vec{m}$ we readily have

(4) $$\begin{aligned} f^1(u_0,v_0) &= <(\vec{F}(u_0,v_0) - \vec{X}), \vec{m}_1> = 0 \\ f^2(u_0,v_0) &= <(\vec{F}(u_0,v_0) - \vec{X}), \vec{m}_2> = 0 \end{aligned}$$

where the angular brackets denote the scalar product. Hence, we have two equations for the unknown parameters u_0 and v_0 For the numerical solution of (4) we can apply Newton's method where the iteration is given by

(5) $$\begin{pmatrix} u_{i+1} \\ v_{i+1} \end{pmatrix} = \begin{pmatrix} u_i \\ v_i \end{pmatrix} - \frac{1}{D} \begin{pmatrix} <\vec{F}_v(u_i, v_i), \vec{m}_2> & <-\vec{F}_v(u_i, v_i), \vec{m}_1> \\ <-\vec{F}_u(u_i, v_i), \vec{m}_2> & <\vec{F}_u(u_i, v_i), \vec{m}_1> \end{pmatrix} \begin{pmatrix} f^1(u_i, v_i) \\ f^2(u_i, v_i) \end{pmatrix}$$

with

(6) $$D := <\vec{F}_u \times \vec{F}_v, \vec{m}_1 \times \vec{m}_2>$$

In this expression $\vec{F}_u$ and $\vec{F}_v$ denote the partial derivatives of the polynomial $\vec{F}(u,v)$. Initial values (u_1,v_1) may be obtained by considering an approximate surface with planar patches.

Avoiding Collisions by Lifting the Cutter

The length l_0 by which we have to lift the tool in order to avoid a collision can readily be obtained in terms of the projection problem. The direction of projection is the tool axis $\vec{m}$ and we have

(7) $$l_0 = \max_{X \in K} <(\vec{\mu}(\vec{X}) - \vec{X}), \vec{m}>$$

In this expression, X are all points of the circular surface K at the bottom of the cutter. In most practical cases, the maximum is at the rim so that (7) simplifies to read

(8) $$l_0 = \max_{t \in [0, 2\pi]} <(\vec{\mu}(\vec{K}(t)) - \vec{K}(t)), \vec{m}>$$

where $\vec{K}(t)$, $t \in [0,2\pi]$ is the circle bordering the tool.

Avoiding Collisions by Rotating the Cutter

The necessary rotation can be determined as the solution of an expression similar to (7). We first note that the axis of the rotation of the tool includes the point $\vec{C}$ has the direction

$$(9) \qquad \vec{a} = \vec{m} \times \vec{n}$$

Here, $\vec{m}$ is the original tool axis. Let us further define the projection of a vector $\vec{X}$ onto the plane spanned by $\vec{n}$ and $\vec{m}$ as

$$(10) \qquad \vec{v}(\vec{X}) := \vec{X} - <\vec{X}, \vec{a}> \vec{a}$$

The condition for the rotation is that the angle between the projected vectors

$$(11) \qquad \vec{v}_1 = \frac{\vec{v}(\vec{K}(t) - \vec{C})}{\|\vec{v}(\vec{K}(t) - \vec{C})\|}$$

and

$$(12) \qquad \vec{v}_2 = \frac{\vec{v}(\vec{\mu}(\vec{K}(t)) - \vec{C})}{\|\vec{v}(\vec{\mu}(\vec{K}(t)) - \vec{C})\|}$$

is a maximum. In these expression $K(t)$ is the circular rim of the cutter. It is readily seen that the maximum condition may be written in the form

$$(13) \qquad d_0 = \max_{t \in [0, 2\pi]} <(\vec{v}_1(t) \times \vec{v}_2(t)), \vec{a}>$$

Since the vectors (11) and (12) are normalized, the corresponding angle is given by

$$(14) \qquad \alpha_0 = \arccos <\vec{v}_1(t_0), \vec{v}_2(t_0)>$$

where t_0 is the solution of (13).

Finding the Maximum

In order to determine the solution to the collision problem, we now have to find the parameter t_0 of the circle $\vec{K}(t)$, for which the expressions (8) and (13) reach their maximum values. This can be done by several standard methods. For a numerical solution the following procedure is quite efficient:

1. Chose n parameter values $t_i \in [0,2\pi]$, i=1,2,...n and determine the points $\vec{K}(t_i)$ of the circle.

2. Next determine the projection points $\vec{\mu}(\vec{K}(t_i))$ for these parameters. Evaluate the corresponding values l_i and d_i , respectively.

3. Find the maximum correction for the subset t_i .

4. If necessary, chose a finer distribution of parameter values around the obtained maximum and repeat the procedure.

Obviously, for our definition of the vectors under consideration, a collision only occurs for positive values of l_0 or α_0. If $\vec{C}$ is the only point of contact found within the first approximation, (8) and (14) yield $l_0 = 0$ and $\alpha_0 = 0$, respectively. Hence, no collision in the above sense occurs and repeating the listed steps may not be necessary if the initial distribution of parameters was fine enough.

5. Conclusions

In the preceding sections we have presented the main ideas necessary to extract milling courses from surface data. Applying these methods within a CAD/CAM system offers considerable advantages. To start with, we get a graphic simulation of the entire milling process which is of great help in the detection of possible problems (before they result in the damage of an expensive model). Further, many of the described steps always occur in a certain order when solving a problem. Hence, CAM commands can be linked together in macros and entire parts can be processed in the batch mode. Manpower remains necessary only for the much shorter timespan required to check the results.

Of course, the brief review we have given is far from a complete discussion of the problems involved. For instance, the milling across edges requires special attention. Especially, when dealing with larger parts special methods involving arrays of surfaces (so-called multisurfs) may be appropriate.

For technological reasons, the distribution of points on the MDIs cannot be fixed by only looking at the spatial tolerances. Rather, the spacing between neighboring points and the changes in orientation must not exceed certain limits.

Another question we have not covered is how to further process the milling courses we have obtainded. NC-machines do not directly understand MDIs and, more stringent, they require more information than just trajectories. The missing link to the machine is called a part program. There, the CAM curves are translated into the CLDATA which is the international standard for the programming of NC-machines. In addition to the preparation of the trajectories, necessary technological parameters such as the type of tool, number of revolutions and so on are supplied. As we already noted, a number of tool parameters already enter the problem solving within the CAM system, but there they are not stored explicitly in the results. The part program then gathers all required information. Part programs can be stored in a form which is easy to understand for the user and they can also be edited for use on a different machine.

Let us conclude with a brief outlook on current trends in the field of NC-milling. The use of personal computers is definitely on the rise and with continually improving hardware we can expect systems able to process even complicated problems in reasonable time for the near future.

The use of splines to describe milling courses is also being reconsidered. Of course, they have to be generalized in order to carry information about the tool axis. However, the amount of data needed for splines can be considerably smaller than for the corresponding polygons. Also the powerful methods to model and modify splines allow for a comfortable handling.

Another important development is the use of triangulation algorithms [7,8]. The milling courses are not directly computed from surface data. Rather, the freeform surfaces are first approximated by a set of planar triangles. The determination of milling courses and the collision check for these triangulations are both fast and stable.

Acknowledgement

We would like to thank B.Baruschke, K.H.Erlenmayer, H.J.Fetzer, P.Hertkorn, and J.Krause for valuable discussions and some help in the preparation of the manuscript.

References

1. G.Farin: Curves and surfaces for CAGD; Academic Press, Boston (1988) and references therein
2. J.Hoschek and D.Lasser: Grundlagen der geometrischen Datenverarbeitung; B.G.Teubner, Stuttgart (1989) and references therein
3. K.Damm and P.Spahn: CAD-CAM Einsatz in der Umformtechnik; Zeitschrift für Maschinenbau, Konstruktion, Fertigung, Vol.10, 665 (1985)
4. H.Henning: Fünfachsiges NC-Fräsen gekrümmter Flächen, ein Beitrag zur numerischen Flächendarstellung, Programmierung und Fertigung; Thesis, Universität Stuttgart (1976)
5. J.P.Duncan and S.G.Mair: 'The anti-interference features of polyhedral machining' Advances in Computer-Aided Manufacture; McPherson (ed.), North-Holland, Amsterdam (1977)
6. B.K.Choi and C.S.Jun: Ball-end cutter interference avoidance in NC machining of sculptured surfaces; CAD Vol.21, 371 (1989)
7. W.H.Frey and J.C.Cavendish: Fast planar mesh triangulations using the Delaunay triangulation; General Motors Research Laboratories, GMR-4555, 1 (1983)
8. N.Sapidis and R.Perucchio: Advanced techniques for automatic finite element meshing from solid models; CAD Vol.21, 248 (1989)

Aspects of Form Feature Modelling

Michael J. Pratt
Cranfield Institute of Technology
Dept. of Mathematics
Bedford, MK43 OAL
United Kingdom

1. Introduction

The topic of form features is arousing great interest in the field of computer-aided design and manufacture. These entities are used to represent certain shape configurations, usually fairly localised, on engineering parts or sub-assemblies. The applications envisaged for them are essentially of an engineering nature, but it is quite possible that other uses for them will emerge in the future. Although they are being studied with very practical ends in mind, they raise many interesting issues of a theoretical nature, concerned with representation and communication in geometric modelling.

It has been found difficult to define precisely what is meant by a 'form feature'. It is easy to provide examples, and many will be given later, but for the time being the rather vague definition first proposed by Pratt and Wilson (1985) will be used:

'Form feature:' A region of interest on the surface of a part.

On the whole this raises more questions than it answers. In particular, we may well ask 'What kind of region ?' The answer is unsatisfactory: 'Anything from a single point to a collection of many entire or partial faces of the part, not necessarily all connected to each other'. We may also ask 'Of interest to whom ?' Once again the answer is not very clear-cut: 'Anybody who is concerned with any aspect of the life-cycle of the part concerned'.

Some examples of form features will be given in the following paragraphs, which will serve to show that although the definition given above is not very enlightening it is nevertheless a valid one. Later, in Section 4, a more precise definition of a form feature will be given, formulated from the point of view of CAD modelling.

Before proceeding further we should note that other types of feature have also been identified for engineering purposes. For example, Shah & Rogers (1988a,b) define the additional concepts of *material features*, concerned with material composition and treatment in engineering parts, and *precision features*, concerned with engineering tolerances and surface finish requirements. The topic of tolerance relationships is closely allied to that of form features, however, as pointed out in Section 1.4, while the remainder of these topics is less interesting from the modelling point of view and will henceforth be ignored in this paper. In what follows, therefore, the word 'feature' should be taken to mean 'form feature'.

1.1 Manufacturing Features of Machined Parts

It was in the manufacturing field that the term 'form features' was first used. Machined parts typically exhibit through holes, blind holes, pockets, slots and grooves, all of which are examples of form features. Their importance for manufacturing planning is that each feature type has associated with it a comparatively small set of possible manufacturing options. The choice between these can be made on consideration of associated technological information (the surface finish required, for example) and of available production resources. A knowledge of the manufacturing features of a part is therefore a very useful start in the automatic generation of a process plan for its manufacture. This has recently been demonstrated by various workers, including Pavey et al. (1986), van't Erve & Kals (1986), Cutkosky et al. (1988) and Anderson & Chang (1989). A glossary of form features of machined and sheet metal parts was recently compiled by CAM-I (1986a).

This use of feature data leads to process planning of the generative type, in which each plan is worked out *ab initio*. An alternative would be to employ feature information in the determination of a part classification code. This has in fact been demonstrated by Kyprianou (1980) for certain classes of machined parts and by Bond & Jain (1988) for sheet metal parts. The existence of such a code allows the use of the variant rather than the generative approach to process planning; this is usually more efficient in companies which manufacture one or more ranges of similar parts. The code is used to retrieve process plans from a database of parts which are similar to the new one from the manufacturing point of view. The generation of the new process plan is then simply a matter of editing the old one appropriately. It may well assist in the generation of a part classification code if individual features are themselves classified in terms of some taxonomy which operates at a lower level (Pratt & Wilson 1985, Wilson & Pratt 1988).

It is important to note that each feature type defines a family of specific features; for example, 'flat-bottomed cylindrical blind hole' features may have any combination of radius and depth parameters. It should be noted also that (if we think for the moment in boundary representation terms) such features may be represented at least from the geometric point of view as sets of part faces. For example, a pocket feature comprises a base face and a set of side-wall faces. In this case the faces form a connected set, but features may also be composed of unconnected faces. Consider, for example, a cylindrical shaft with a circumferential groove. The groove separates two cylindrical faces which form a single feature from the manufacturing point of view, since they will normally both be produced by a single manufacturing operation.

Combinations of features are also of great importance. A countersunk hole provides an illustration, in which the cylindrical hole and the conical countersink may be regarded as features in their own right but their combination may be regarded as a 'compound feature'. A pattern of cylindrical holes is an example of another type of compound feature, in which the individual holes are all features of the same type but the pattern has some additional special significance.

All the examples so far given are of form features composed of entire part faces. Incomplete faces may also sometimes occur in a feature. Consider, for example, a small cube positioned with respect to a large one in such a way that one face of each lies in the same plane. Suppose that they are then unioned together. The two faces sharing the same geometry will coalesce to form a single non-rectangular face, and the small cube will form a projection feature defined by four of its original faces but only part of the non-rectangular face.

It is also possible that a feature may be defined in terms of one or more edges or vertices of a part. In a boundary representation modeller an edge may be labelled in the datastructure as 'chamfered', and in this case the chamfer feature is associated with the single edge. The modeller may have the capability of *evaluating* the label, in which case a new angled face will be interpolated in the model to replace the edge. Then the chamfer feature will be associated with the new face. This example shows that there may be more than one way of representing a single feature in a solid modeller. Corners may also be chamfered or bevelled, of course, and this illustrates the possibility mentioned earlier that a feature may be associated with a single point or vertex of a part.

1.2 Design Features

From the designer's point of view features are often concerned with the functionality of a part. A cylindrical hole feature may be a bearing housing, while a compound feature consisting of a set of fins on an engine may be

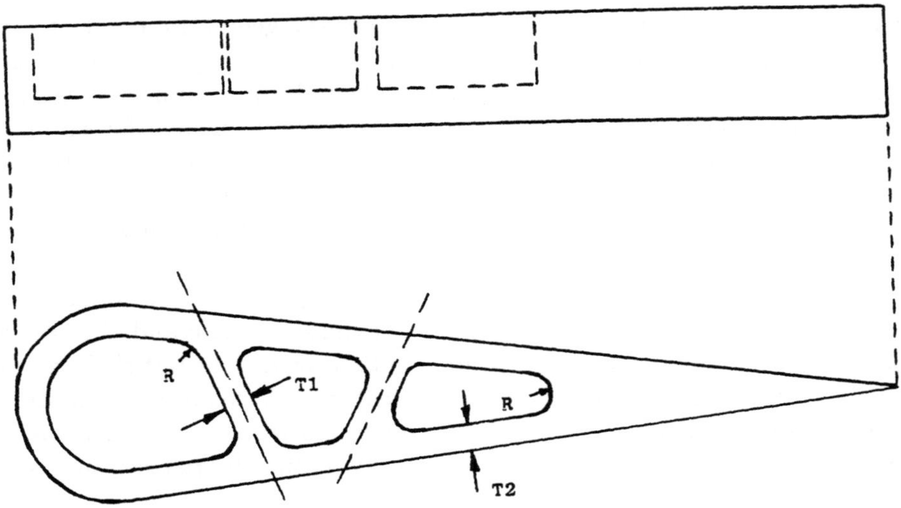

Figure 1 Machined Wing Rib

for cooling purposes, to give but two illustrations. The designer's features may sometimes correspond with the manufacturing features, but this will not always be so. A standard example is a wing rib which has a peripheral flange and several transverse stiffeners, as shown in Figure 1.

The designer may work with an initial shape prescribed by aerodynamic considerations; he will specify a thickness for the flange, the centrelines and thicknesses of the stiffeners and the blend radii where they meet the flange. When it comes to manufacturing, however, supposing the part is to be machined from the solid, the process planner will see a set of pocket features rather than a flange and some stiffeners. Matters are yet further complicated by the fact that the thickness of the stiffeners may in fact be of importance in a manufacturing context if they are likely to deform under the cutting forces applied.

1.3 Analysis Features

Analysis is an integral part of the overall design process, and we will consider a case where the use of finite element analysis is appropriate. The part to be analysed may have certain form features such as mounting lugs whose strength is particularly important, and these will be the salient features for the analysis. But there are also more subtle kinds of feature in this context, since the finite element mesh is usually generated on an *idealisation* of the original part model rather than on the part model itself. To give a few examples, parts with a high level of symmetry can frequently be analysed in terms of a small section of the original, rotational parts can be analysed in terms of a 2D profile, certain regions of the part may be approximated in terms of beam, plate or shell elements and certain design features of the original part (such as small holes in lightly stressed regions) may be neglected altogether. From the analyst's point of view, then, a part may exhibit symmetry features, rotational features, plate or shell features, and so on.

1.4 Tolerance and Inspection Features

The designer's dimensioning scheme for a part is usually related to the intended functionality of the part. The attachment of tolerance specifications to dimensions is aimed at ensuring that parts will fulfil their desired role despite the inevitability of process inaccuracies during manufacture. However, the imposition of tolerances lacks real meaning unless the part can be inspected, subsequent to manufacture, for compliance with those tolerances. There are certain standard types of inspection process, many of which are concerned with checking the dimensions of features such as those previously mentioned, i.e. diameters of holes or shafts, widths and depths of slots and so forth. Other types of feature also occur, however, for example a 'pair of parallel plane faces' feature, which may arise in connection with a tolerance either on distance between two planar faces or on their parallelism.

In modern tolerancing practice tolerances are specified with respect to datum planes, which may or may not be coincident with planar faces of the part. To handle non-coincident cases it is desirable to associate auxiliary geometry with the part model. This capability is provided by few solid modelling systems at present.

1.5 Assembly Features

These are features which are significant in the (possibly automated) assembly of some product from its component parts. Normally a set of faces on one part will mate with a corresponding set of faces on another. The sets of faces concerned must be grouped into features of the individual parts, and furthermore the logical relationship between these features must somehow be indicated in the assembly datastructure. Appropriate information concerning tolerances and fits must also be associated with these inter-feature relationships. Special methods will be needed for the representation of frequently occurring situations involving, for example, standard fastener types such as nuts and bolts or rivets used in the construction of assemblies (Nieminen et al. 1989).

1.6 Robotics Features

Under this heading may be grouped those features which are important from the point of view of robot manipulation. For example, a pair of parallel plane faces of a part might be labelled as a feature since they may be gripped by the end-effector of a robot involved in assembling this part with others.

1.7 Overall Shape Features

Although most of the features so far discussed have been localised in nature, the global shape characteristics of engineering parts can also have great importance. For example, parts to be machined from solid material are usually classified as either *rotational* or *prismatic*. In the former case the major manufacturing operation involved is turning on a lathe, while in the latter case the basic operation is milling. The distinction has immediate significance in determining whether the part can most conveniently be manufactured from cylindrical or rectangular-section stock material.

2. Form Features in a Solid Modelling Context

One of the main virtues of the solid modeller in engineering is that it generates product representations which, unlike the traditional engineering drawing, can in principle be interpreted by the computer. The potential therefore exists for the automation of various downstream processes once the initial model has been created. Moreover, we have seen that several applications such as the generation of finite element models, the planning of manufacturing methods and the assessment of manufacturing quality, are concerned with form features. It is therefore appropriate to examine how form features can be dealt with in the context of solid modelling.

The two major approaches to solid modelling are (Woodwark 1986, Rooney & Steadman 1987) constructive solid geometry (CSG) and bound-

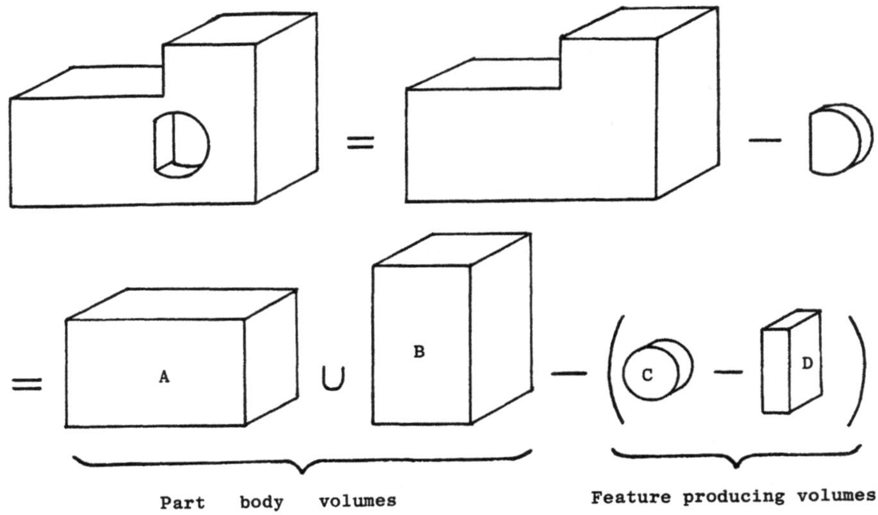

Figure 2 D-pocket as a CSG Feature

ary representation (Brep). The first provides a high-level geometric description of parts in terms of their volumetric constituents, while the second provides a low-level description in terms of a structured and connected set of faces, edges and vertices. A representation of a part in terms of its form features in fact comes midway between the two in terms of the level of information. In the CSG case a feature is not identical with a single primitive, but is generated by the interaction of two or more primitives, as illustrated in Figure 2.

While the overall object shape is represented by a global set of boolean operations on the primitives, the features are represented by more localised subsidiary boolean relations on them. In the Brep case a feature is generally represented in terms of a set of faces in the model, as mentioned in Section 1.1 above. On the assumption that any given face will normally be involved in no more than one feature, the number of features on a given part will usually be considerably smaller than the number of basic topological elements. In this case, then, the feature model is less localised than the individual elements in the basic datastructure, whereas in the CSG case it is more localised. It should be emphasized that the foregoing comments apply in the case of a single feature model for use in a specific application. For multiple applications it may be necessary to associate several different feature models with a single underlying solid model.

The point to be drawn from the last two paragraphs is that although feature information is required for the automation of certain applications, the fundamental datastructures of existing solid modellers provide information at a level which is either not sufficiently detailed (CSG) or which is too detailed (Brep). There are three possible ways of rectifying this situation. Firstly, the existing datastructures may be augmented to permit the representation of features. Secondly, new datastructures may be devised which permit the convenient representation of features. And thirdly, new datastructures may be devised in which feature entities are the basic components.

These three possibilities are examined in some detail in Section 4 of the paper, but first we will examine the mechanisms by which the feature information can be initially generated in the model.

3. Generation of Form Feature Information

Three approaches are available for the generation of feature data. Firstly, if the solid model has already been created then the features existing on it may be recognised *a posteriori*. Secondly, since an individual modelling operation often gives rise to a feature, all such features could be automatically recognised at their time of creation. Thirdly, the designer could actually be allowed to work in terms of features, in which case his intent with regard to features could be captured at the outset and the need for subsequent recognition to some extent avoided. From the discussion given below it will be seen that they all have their advantages and disadvantages; possibly each of them will have a place in the product modelling technology of the future.

3.1 A posteriori Feature Recognition

Feature recognition on a completed geometric model may be either manual or automatic. The manual method simply requires the operator to indicate those entities in the model which he wishes to group together to define a particular feature. It is comparatively easy to implement; existing systems providing this facility include the Cranfield Testbed Modeller (Hailstone 1985, Pavey et al. 1986) and a small number of commercially available systems.

The implementation of *automatic* feature recognition is a much more complex undertaking, but historically it was the first approach to be tried. Kyprianou, in a Cambridge PhD thesis (1980), describes a method which has been implemented with the Brep modeller BUILD (see also Parkinson 1983, Jared 1984). Features are identified by finding certain patterns of convex and concave edges occurring in the part datastructure. To give two simple examples, a square pocket in one face of an object is bounded by a loop of edges which are all convex, while a square boss is bounded by a set of concave edges at its base. These properties are characteristic of depression and protrusion features in general, and a recognition strategy was developed, based on these and related observations, for the detection of bosses and of blind and through holes. A related but distinct approach to automatic feature recognition from a Brep datastructure has been described by a group of Italian researchers (see Falcidieno & Giannini 1989 and the references therein).

Other methods have been devised which use *syntactic pattern recognition.* This requires the specification of a set of rules which must be satisfied by the elements forming a particular type of feature, e.g. opposite pairs of walls of a simple rectangular pocket must be parallel and all adjacent faces (including the floor) must be mutually perpendicular. Henderson (1984), for example, translates the information in a ROMULUS model into a set of PROLOG statements concerning the part model and then looks for connected groups of faces which conform to the rules defining various feature types (see also Staley et al. 1983, Henderson and Anderson 1984, Choi et al. 1984).

Both the above methods are implemented with boundary representation systems, but Woo (1982) has published an algorithm for the recognition of features in a CSG context. This has not been implemented in practice because no existing CSG modeller has all the required geometric capabilities, but the approach is interesting in several respects and is certainly worthy of study.

3.2 Feature Recognition during Model Creation

Some modellers provide the capability for defining profiles on planar faces of the model and sweeping them in, out or through to create depressions, protrusions or through holes respectively. In all these cases each sweep operation creates a feature, which may be classified at its time of creation by analysis of the geometry of the profile and the nature of the sweep. Other related modelling operations permit similar possibilities for building up a feature model as the design proceeds. However, this approach is not a universal answer to feature data generation. Firstly the features created are purely geometric or 'shape' features which may or may not correspond with the functional features required for application purposes. Secondly, the universally implemented boolean operations often do not give rise to features of any commonly recognised type, or lead to ambiguous situations.

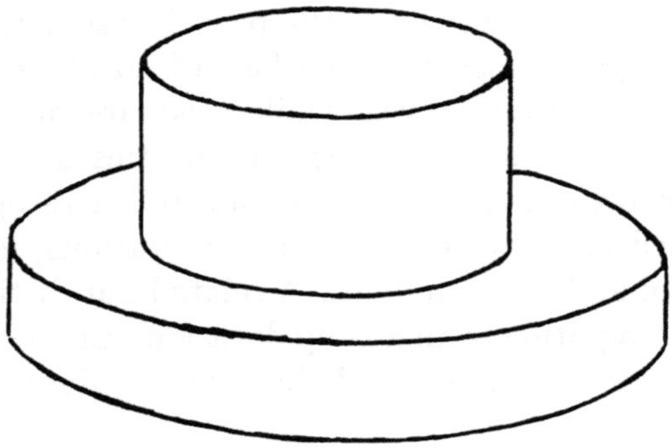

Figure 3 A boss on a disc, or flange on a cylinder ?

For example, if two coincident cubes are created and one is moved or rotated slightly before the two are unioned it is difficult to decide what is the basic body of the resulting object and what are the natures of the features arising on it. Alternatively, suppose a cylinder is unioned with a flat disc as shown in Figure 3; have we created a boss on the disc or a flange around the base of the cylinder ? Either interpretation may be valid for certain applications.

3.3 Design by Features

There are manifest advantages in allowing the designer actually to create his model in terms of features. Then, instead of subtracting a block or performing a sweep operation to create a rectangular pocket feature, he may simply pick an option 'RECTANGULAR POCKET' from a feature menu. The system might then call up a generic description of such a pocket and prompt him for details of size, location and so on, before performing the necessary computations and modifying the model appropriately. With this mode of operation the system would know the nature of the feature created, at least as far as the designer's view of it was concerned, and could set up the necessary information in the datastructure immediately. The designer's features will in many cases correspond to the features required by subsequent applications. When this is not so automatic recognition will still be required, however, though the cumbersome methods mentioned in Section 3.1 can be simplified if advantage is taken of the design feature information, as will be shown later. Design by features is therefore likely to lead to greater efficiency in the modeller/application interface. Another advantage is that the designer can specify not only the shape of the feature but also its intended functionality, e.g. a hole could be labelled as a bearing housing. It is not possible using automatic feature recognition to recapture designer intent regarding function; this must be recorded at the outset. The information will subsequently prove useful when the design is subsequently modified, possibly by some designer other then the one who originally created it.

This method of design is not without problems, unfortunately. Inevitably, extensive geometric checking will have to be provided to ensure that each feature created is of the stated type when the system interprets the designer's input. For example, he may call for a cylindrical blind hole feature but position it with respect to the base object so that its cylindrical face breaks through the boundary as shown in Figure 4. The feature thus created falls into some class which is not the one originally called for, and the user must be informed of an error. Similar problems arise when the creation of a new feature leads to a modification in some previously existing feature which changes its class. In fact it appears that the class of any feature may have to be continually monitored, using techniques akin to automatic recognition, throughout the design process.

The idea of design by features is not new, and a few commercial modelling systems possess limited implementations. However, a great deal of research remains to be done before the full benefits of this approach can be exploited. Some recent work in this area is described in Shah & Rogers (1988a,b).

4. Representation of Features in Solid Modelling

Various studies have established that a general-purpose feature-oriented design system must be extremely flexible in two major respects. Firstly, different organisations will wish to design in terms of different classes of features, and it is important that the system can be configured to meet the precise requirements of each designer as to feature definitions. Secondly, since the features required by downstream applications will not always correspond to those created by the designer, as previously mentioned, some form of recognition process must be provided for the identification of application feature types. These will vary not only between organisations but also between applications, and so a further requirement for system configurability

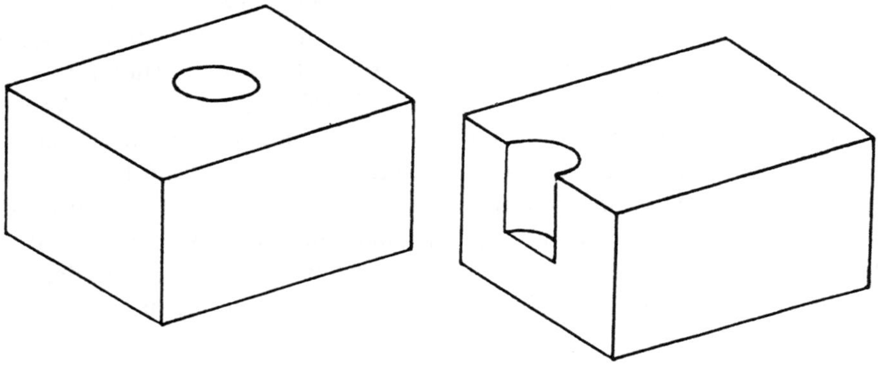

Figure 4 A 'circular blind hole' feature validly and invalidly positioned with respect to a block.

is indicated. Our two requirements pose questions regarding the manner in which feature definitions are implemented; in particular, can feature representations be found which are equally efficient both for the creation and for the automatic recognition of instances of standard or user-defined features?

Further considerations regarding feature representations stem from the operations which must be provided in a practical system. These include feature creation, deletion, modification, transformation, replication, interrogation and the attachment of properties or attributes. It may be necessary to provide a separate operation for the attachment of a newly created feature to a previously existing model, and if more than one type of feature representation is implemented (this possibility is discussed in some detail below) some means will be needed for conversion between representations.

There will also be a requirement for defining and manipulating compound features such as pattern features or primary features with one or more associated secondary features.

The issues mentioned above are complicated by the fact that two different basic approaches to feature representation have been identified, referred to by Pratt and Wilson (1985) as *implicit* and *explicit.* These are respectively descriptive (but concise) and detailed (but less succinct). To take a simple example, a cylindrical blind hole may be defined implicitly in terms of its radius and depth or explicitly in terms of the set of faces which compose it in a boundary representation model. Each type of representation has virtues in different situations. For automated process planning knowledge of the radius and depth are fundamental, while the equations of the side and bottom faces of the hole are of little interest. On the other hand, the generation of a shaded surface rendering of the model containing the feature does require precise surface information. It appears therefore that a general-purpose system may benefit from the use of feature representations having a hybrid nature.

The obvious way to represent a feature explicitly in a Brep context is as a collection of topological elements of the model, usually faces. The natural approach in a CSG system is to model features as interactions between volumetric primitives, but it is significant that this method was rejected at the University of Rochester, the home of the CSG philosophy, in favour of a boundary-oriented representation (Requicha & Chan 1985). We therefore concentrate on this approach.

An example of an explicit feature representation is provided by a list of the four walls and the bottom face of a rectangular pocket. Previous work by several authors recently summarised by the present writer (Pratt 1988) has suggested that these could with advantage be supplemented by an additional notional 'closure face' which converts the open set of part faces composing the feature into a closed feature volume. The presence of closure faces creates a non-manifold part/feature model having a cellular nature, in which closure faces may lie internal or external to the part boundary shell and in which three or more faces may meet at a single edge. Most boundary representation systems cannot cope with this situation, although at least three experimental systems now exist which do have the necessary capability (Weiler 1986, 1987, Gursoz & Prinz 1989, Masuda et al. 1989). However, Henderson (1984) and Pratt (1989) have shown that a volume-based part/feature model of the type described can be achieved using a relatively conventional commercially available manifold boundary representation system (ROMULUS).

An implicit feature representation consists of a reference to a feature type having some parametrised description, a list of parameter values for the instantiation of the feature and some information as to the position and orientation of the feature in the overall model. There will normally also be a reference to some procedure for evaluating the feature in the explicit form described earlier. Implicit representations have the virtue of conciseness, and are useful in creating idealised models in which fully explicit representation of all features would give rise to a mass of detail obscuring essential aspects of the design. As pointed out earlier, they also contain precisely the information needed for certain applications, in a conveniently accessible form. Furthermore, the feature descriptions in a feature library as required in any system implementing design by features will almost certainly be of the parametrised type on which this kind of implicit feature representation is based.

Since both types of representation have their virtues, the present writer has suggested the use of a hybrid approach (Pratt 1989). Essentially, two versions of each feature may be stored. The first is an 'ideal' version, which is of the implicit type and which describes the feature in terms of a few dimensional parameters together with some positioning and orientation information. This is referred to as the 'canonical feature volume (CFV)'. The second representation is explicit, and describes the feature as a closed volume in boundary representation form. This representation takes into account any non-standard interactions of the ideal feature with the part boundary or with other features, and is referred to as the 'attached feature volume (AFV)'. The volumes defined by the CFV and the AFV will often be identical, in which case the feature is said to be **regular**. Irregularity of a feature implies some volumetric interaction between the CFV and the part exterior volume or with some other feature volume, resulting in non-null intersection or difference volume for the cases of protrusion and depression features respectively. The approach described by Masuda et al. (1989) to the non-manifold modelling of parts and features achieves a similar aim. Many workers, including those mentioned, have advocated the use of a 'feature graph' to allow the representation of geometric adjacencies between features. If features are modelled as self-contained volumes, the result is a network structure linking a set of such volumes, having significant similarities to the classical CSG tree structure. Some of the advantages of such non-manifold approaches to part/feature modelling are as follows:

1) They give all the advantages of boundary representation modelling together with several of the advantages of the CSG philosophy, notably those relating to the constructional history of the model; this allows easier implementation of operations for editing the model, e.g. feature modification and deletion.

2) The presence of the closure face is important for manufacturing applications involving NC machining. Firstly it provides information regarding

the tool contact surface prior to the machining of a feature at any intermediate stage of manufacture. Secondly, it allows the ready generalisation of the feature volume concept to the *delta-volume* concept (CAM-I, 1982). A delta-volume is a volume of material to be removed which does not correspond to a feature of the part being machined; its form is determined in terms not only of the geometry of the part but also of the geometry of the stock material from which it is being made.

3) It permits the determination of various types of interactions between features which may be important for applications purposes. Since features are represented as volumes there is the added advantage that the actual volumes of interaction may be evaluated using standard boolean operations.

For certain applications the data describing the form feature itself may need considerable enhancement. For example, a tolerancing feature may make reference to a set of datum planes, some procedure for setting up the appropriate set of dimensions for the particular feature type (if these are not explicitly available in the model) and the associated tolerance values (Johnson et al. 1985). It is worth noting that design by features may allow much of this information to be generated automatically at the time of feature creation.

It is clearly desirable that the collection of faces making up an explicit feature description is structured in some way. For example, in a pocket feature the base and the side faces may be treated in different ways from a manufacturing point of view, e.g. using area clearance and profiling respectively. A depression feature such as this has a loop of edges which form its boundary; they are thereby distinguished from the other edges of the feature, and form the boundary of the closure face (or set of closure faces) of the feature volume.

The explicit and implicit types of representation have been discussed, but a further type of feature representation also exists, taking the form of an attribute attached to one or more topological entities. A common example is a thread attribute associated with the shank of a bolt model; in most cases it would not be desirable to represent the helical geometry of the thread explicitly. Chamfered or rounded edges may conveniently be represented in a similar way for certain purposes.

4.1 Feature Substructuring

Recent research carried out for CAM-I (Faux 1986, Ranyak & Fridshal 1988) has shown that the explicit representation of many manufacturing features can be decomposed into components which are more complex than the basic topological elements of a boundary representation model. A rectangular pocket, for example, may be thought of as built up of three components, each comprising a pair of parallel faces. Two of these pairs consist of opposing wall faces, and the third comprises the pocket floor and the 'closure face'. The first of the reports referred to is concerned with dimensions and tolerances, and it was found that from this point of view each of these pairs of faces is a feature, since it may be regarded as uncoupled from the remainder of the manufacturing feature definition for dimensioning, tolerancing and inspection purposes. Faux (1986) carried this idea further, suggesting that features in general should be thought of as built up from one or more such components. This will have advantages in the generation of application feature models from the designer's original feature model, as discussed in the next section. Faux further suggests that feature components (which he calls *feature primitives*) are defined not in terms of faces but rather of surfaces. Thus the three components of the rectangular pocket feature are now pairs of parallel planes. Other feature components defined by Faux are the single plane (which may be used to define the plane of symmetry of symmetric features such as slots and grooves), the cylindrical surface (used in the definition of cylindrical holes, for example) and a closed profile primitive used for defining general prismatic features.

The idea of using surfaces rather than faces for feature components has several advantages. For example, different boundary representation modellers may represent a complete cylinder in terms of any number of faces from one to four (assuming that the geometric representation is exact; faceting systems may use a much larger number of planar facets). By referring to the underlying surface rather than the individual faces lying on it the problems caused by such differences in approach are avoided. Further, as Faux points out, faces lying on the same surface are likely to be functionally related. For example, a part may have a planar face containing several pockets all having the same depth. Then the bottom faces of these pockets are logically related in the model since they all lie on the same surface; this logical relationship becomes a practical one when the pockets of the real part are machined, since they can all be dealt with in the same setup, often using identical machining strategies.

The requirement that feature components involving the same surface definitions are logically related implies that surface definitions should not be repeated in the model datastructure. For example, if two pockets are created as in the last paragraph then the floor faces of both should have pointers to the same surface entity in the model. In current modellers this will not always be the case; in some of them the creation of the second pocket will lead to the generation of a separate surface entity associated with the floor face, since the modeller will not check whether this face lies on a surface which already exists in the database. Redundancy of geometric definition will thus occur, and the desirable logical relationship between the floor faces of the two pockets will not be achieved. It therefore appears that new geometry should not be created until the system has checked whether newly created topological entities (e.g. faces, edges, vertices) lie on previously existing geometric entities (respectively surfaces, curves, points). Such checking will be time-consuming unless the geometry of surfaces and curves is stored in a suitable standard format, as discussed below.

Currently, a planar surface is frequently represented in terms of a reference point, lying on the surface, and a direction vector specifying the surface normal. Clearly there are infinitely many possible choices of reference point for any plane, though the normal direction is unique. Faux suggests that the choice of reference point should also be made unique, by requiring it to be the point on the plane closest to the origin of coordinates. To be more specific, the plane is specified by its perpendicular distance from the origin and its normal direction. This allows checking for correspondence of two planes simply by comparison of one scalar and one vector. Furthermore, sets of parallel planes are logically associated in the model since they all point to the same direction vector, while the distance between any pair of parallel planes can be ascertained by a simple subtraction of two scalars. It is possible to define similarly convenient standard representations for the other types of geometric entity commonly implemented in solid modellers.

In certain circumstances, however, it will be more convenient to define geometry in local rather than the global terms discussed above. Consider, for example, a compound feature whose primary component is a cylindrical boss in the top face of which a secondary coaxial cylindrical hole feature is defined. The question arises as to what happens to the hole if the height of the boss is edited. It may be desired that the base of the hole remains in the same position, so that the hole deepens as the boss extends. This will be achieved automatically if all geometry is represented in global terms. On the other hand, it may be desired for the depth of the hole to remain constant, and this result may be achieved by defining the plane of the base of the hole locally, as a fixed distance parallel offset from the top plane of the boss. The configuration interface of the modeller should allow either of these options to be adopted in the definition of compound features such as that described above.

4.2 Feature Rules

This is another considerable subject in its own right, which can here only be given a cursory glance. Firstly, rules can be used for the automatic recognition of features. Henderson (1984), for example, gives the following rules for recognising a cylindrical blind hole in a planar face:

If a depression feature has a circular closure face
and connected to this there is a cylindrical face
and this face is terminated by a valid bottom face
    (rotationally symmetric, sharing its axis with the cylinder),
then the feature is a cylindrical blind hole.

This represents a bottom-up approach to feature recognition. It appears to suffer from being too specific; for example, if the hole enters the face obliquely or the face is non-planar the quoted rules fail. From this point of view there seems to be virtue in the top-down approach based on the analysis of topological graphs (Kyprianou 1980, Falcidieno & Giannini 1989), which permitt the initial recognition of general depression features and the subsequent refinement of feature class by further geometric interrogations. Such an approach is probably more flexible and easily extensible.

Other types of rules may apply to the individual components within features. For example, if the faces composing a pocket feature are labelled as 'wall', 'bottom', 'closure', as suggested earlier, then when every pocket feature is installed on the part model the following rule could be invoked:

If the wall faces of the feature canonical
    volume intersect the part surface
and neither the bottom nor the closure
    face intersects the part surface
then this is a valid instance of a pocket feature.

If these rules are not satisfied, then what arises is some other kind of feature (unless, of course, some particular user decrees that for his purposes a pocket is allowed to break out of the side of an object - if he does desire this, then he must be allowed to amend the rules accordingly).

Reverting to the use of rules for feature recognition, now consider the case where a design has been created using design by features, and it is subsequently desired to identify another set of features for some downstream application. The labelling of feature elements at their time of creation can simplify this process considerably. Take as an example the machined wing rib of Figure 1. The designer's features are the peripheral flange and the stiffening members, as explained earlier, and so these will be the features represented in the model as first created. If these have their top and wall faces labelled automatically as they are generated, the recognition of the pocket features for manufacturing reduces simply to a search for closed strips of wall faces. Shah (1988) refers to the process of identifying a set of application features from an initial set of design features as 'feature transformation'. However, the word 'transformation' has several other important uses in the field of geometric modelling, and the present writer prefers the term 'feature transmutation' in the current context.

Still further types of rules may be concerned with such matters as dimensions of features. For example, an organisation may wish to make a distinction between a cylindrical hole and a circular pocket in terms of diameter, the rationale being that one will be drilled but the other (lacking the availability of a large enough drill) will be milled. Also, as pointed out by Faux (1986), some organisations may wish to restrict sizes of certain features such as bolt holes to a range of discrete sizes, to encourage the use of standard parts in design.

4.3 Features Redefined

At this stage, having acquired some useful hindsight, it is possible to give an improved definition of a form feature, at least as it relates to product modelling:

'Form feature:' A related set of elements of a product model, conforming to characteristic rules enabling its recognition and classification, which, regarded as an entity in its own right, has some significance during the life cycle of the product.

Note that this applies equally well to CSG models (where the elements referred to will be volumetric primitives) or to boundary representation models (where they will be topological or possibly geometrical entities). Note also that the definition implies that a product model may be dynamic or time-dependent, since for example a clamping feature at some stage of the manufacture of a machined part, consisting of a pair of parallel faces, may later be itself machined away in a subsequent operation. This opens up a new dimension (literally) on feature modelling, but one which cannot be pursued further here.

5. Feature Manipulation

Once feature datastructures have been set up in the solid model many convenient facilities are easily made available for geometric modification. Since a feature can be treated as a self-contained entity it can in principle be deleted, moved, scaled, copied or instanced. Most of these results are readily achieved in CSG systems by editing the underlying textual description of the model and re-evaluating, but in a Brep system which does not retain a history of the construction of the model such operations are either cumbersome or impossible unless features are implemented. The hybrid nonmanifold Brep/CSG approach to feature modelling described earlier appears to provide sufficient historical information to allow the relatively straightforward implementation of feature-based editing operations, as discussed by Pratt (1988, 1989). Such operations will be essential components in the advanced feature-based user interfaces of future CAD/CAM systems.

6. Concluding Remarks

For some applications, feature information must be supplemented by the full solid modelling data of the part. This will be necessary, for example, in checking that access to machined features is not blocked by other regions of the part, or in the detection of collisions of the tool holder with the part body in the simulation of machining. This indicates a need for a dynamic *applications* or *programming interface*, allowing applications programs to access the full functionality of the product modeller not only as regards features but also as regards the standard operations and interrogations conventionally available in solid modelling systems. A standardised interface of this kind has been proposed by CAM-I (1986b); Pratt & Wilson (1985) suggested extensions to this interface to enable the handling of feature information, and these extensions were found to be useful in practice by Pavey et al. (1986) in an experiment in automated process planning. However, in view of the current rapid development of features technology it is too early in the present writer's view to standardise the interface in this area. The same opinion holds for the 'form features model' included in the recently published Draft Proposal for an International Standard STEP for neutral file CAD data transfer (ISO 1989).

To conclude, in view of the potential importance of form features for automated applications of product modellers it is hardly surprising that commercial systems are now coming onto the market which provide feature facilities to a greater or lesser degree. Most of them are still very limited at present in what they offer, but clearly features technology has enormous potential for major developments in the integration of CAD with downstream applications. It is hoped that the present necessarily superficial survey of some of the concepts and problems may provide a stimulus to further work in the area.

7. References

Anderson, D.C. & Chang T.C. (1989) Automated Process Planning using Object-Oriented Feature-Based Design. In: F.-L. Krause & H. Jansen (eds.) Advanced Geometric Modelling for Engineering Applications (Proc. GI/IFIP International Symposium, West Berlin, 8-10 Nov. 1989). North-Holland Publ. Co., in preparation.

Bond, A.H. & Jain, R. (1988) The Formal Definition and Automatic Extraction of Group Technology Codes. In: Proc. ASME Computers in Engineering Conf., San Francisco, July/Aug. 1988.

CAM-I (1982) Design of an Advanced Numerical Control Processor. Report No. R-82-ANC-01, CAM-I Inc., Arlington, Texas.

CAM-I (1986a) Part Features for Process Planning. Report No. R-86-PPP-01, CAM-I Inc., Arlington, Texas.

CAM-I (1986b) The CAM-I Applications Interface Specification: Consolidated and Restructured Version. Report No. R-86-GM-01.1, CAM-I Inc., Arlington, Texas.

Choi, B.K., Barash, M.M. & Anderson, D.C. (1984) Automatic Recognition of Machined Surfaces from a 3D Solid Model. Computer Aided Design 16 (2), 81-86.

Cutkosky, M. & Tenenbaum, J. (1988) Features in Process Based Design. In: Proc. ASME Computers in Engineering Conf., San Francisco, July/Aug. 1988.

Falcidieno, B. & Giannini, F. (1989) Automatic Recognition and Representation of Shape-based Features in a Geometric Modelling System. Computer Vision, Graphics & Image Processing 48, 93-123.

Faux, I.D. (1986) Reconciliation of Design and Manufacturing Requirements for Product Description Data using Functional Primitive Part Features. Report No. R-86-ANC/GM/PP-01.1. CAM-I Inc., Arlington, Texas.

Gursoz, E.L. & Prinz, F.B. (1989) Corner-based Representation of Non-manifold Surface Boundaries in Geometric Modeling. Technical Report, Engineering Design Research Center, Carnegie Mellon University.

Hailstone, S.R. (1985) Explicit Form Features in Solid Modelling. MSc Thesis, Cranfield Institute of Technology, England.

Henderson, M.R. (1984) Extraction of Feature Information from Three Dimensional CAD Data. PhD Dissertation, Purdue University, May 1984.

Henderson, M.R. & Anderson, D.C. (1984) Computer Recognition and Extraction of Form Features: A CAD/CAM Link. Computers in Industry 5 (4), 315-325.

ISO (1989) Draft Proposal for STEP. Draft Proposal DP10303, International Standards Organisation.

Jared, G.E.M. (1984) Shape Features in Geometric Modelling'. In Solid Modelling by Computers (Proc. General Motors Solid Modeling Symposium, Warren, Michigan, Sept. 1983). Plenum Press, New York.

Johnson, R.H. & Associates (1985) Dimensioning and Tolerancing Final Report. Report No. R-85-GM-02.2, CAM- I Inc., Arlington, Texas, May 1985.

Kyprianou, L.K. (1980) Shape Classification in Computer Aided Design. PhD Dissertation, University of Cambridge, England.

Masuda, H., Shimada, K., Numao, M. & Kawabe, S. (1989) A Mathematical Theory and Applications of Non-Manifold Geometric Modelling. In: F.-L. Krause & H. Jansen (eds.) Advanced Geometric Modelling for Engineering Applications (Proc. GI/IFIP International Symposium, West Berlin, 8-10 Nov. 1989). North-Holland Publ. Co., in preparation.

Nieminen, J., Kanerva, J. & Mantyla, M. (1990) Feature-based Design of Joints. In: F.-L. Krause & H. Jansen (eds.) Advanced Geometric Modelling for Engineering Applications (Proc. GI/IFIP International Symposium, West Berlin, 8-10 Nov. 1989). North-Holland Publ. Co., in preparation.

Parkinson, A. (1983) Feature Recognition and Parts Classification in BUILD. Cambridge University CAD Group Document No. 112, available from DACAM, Cranfield Institute of Technology, England.

Pavey, S.G., Hailstone, S.R. & Pratt, M.J. (1986) An Automated Interface between CAD and Process Planning. In: Computer Aided Production Engineering (Proc. International Conf., Edinburgh, April 1986). Mechanical Engineering Publications Ltd., Bury St. Edmunds, England.

Pratt, M.J. (1988) Synthesis of an Optimal Approach to Form Feature Modelling. In: Proc. ASME Computers in Engineering Conf., San Francisco, July/Aug. 1988.

Pratt, M.J. (1989) A Hybrid Feature-based Modelling System. In: F.-L. Krause & H. Jansen (eds.) Advanced Geometric Modelling for Engineering Applications (Proc. GI/IFIP International Symposium, West Berlin, 8-10 Nov. 1989). North-Holland Publ. Co., in preparation.

Pratt, M.J. & Wilson, P.R. (1985) Requirements for the Support of Form Features in a Solid Modelling System. Report No. R-85-ASPP-01, CAM-I Inc., Arlington, Texas.

Ranyak, P. & Fridshal, R. (1988) Features for Tolerancing a Solid Model. In: Proc. ASME Computers in Engineering Conf., San Francisco, July/Aug. 1988.

Requicha, A.A.G. & Chan, S. (1985) Representation of Geometric Features, Tolerances and Attributes in Solid Modelers based on CSG. Tech. Memo. No. 48, Production Automation Project, University of Rochester, N.Y.

Rooney, J. & Steadman, P. (1987) Principles of Computer Aided Design. Open University & Pitman Press.

Shah, J.J. (1988) Feature Transformations between Application-specific Feature Spaces. Computer Aided Engineering Journal, 5, 6, 247-255.

Shah, J.J. & Rogers, M.T. (1988a) Functional Requirements and Conceptual Design of the Feature-based Modelling System. Computer Aided Engineering Journal 5 (1), 9-15.

Shah, J.J. & Rogers, M.T. (1988b) Feature-based Modeling Shell: Design and Implementation. In: Proc. ASME Computers in Engineering Conf., San Franciso, July/Aug 1988. American Society of Mechanical Engineers.

Staley, S.M., Henderson, M.R. & Anderson, D.C. (1983) Using Syntactic Pattern Recognition to Extract Feature Information from a Solid Geometric Data Base. Computers in Mechanical Engineering, Sept. 1983, 61-66.

van't Erve, A.H. & Kals, H.J.J. (1986) XPLANE: A Knowledge Base Driven Process Planning Expert System. In: Computer Aided Production Engineering (Proc. International Conf., Edinburgh, April 1986). Mechanical Engineering Publications Ltd., Bury St. Edmunds, England.

Weiler, K.J. (1986) Topological Structures for Geometric Modeling. PhD Thesis, Rensselaer Polytechnic Institute, Troy, NY.

Weiler, K.J. (1987) Non-manifold Geometric Boundary Modeling. Notes for Tutorial No. 26, (Advanced Topics in Solid Modeling), ACM SIGGRAPH Conf., July 1987, Anaheim, CA.

Wilson, P.R. & Pratt, M.J. (1988) A Taxonomy of Form Features for Solid Modeling. In: M.J. Wozny, H.W. McLaughlin & J. Encarnacao (eds.) Geometric Modeling for CAD Applications. North-Holland Publ. Co.

Woo, T.C. (1982) Feature Extraction by Volume Decomposition. In: Proc. Conf. on CAD/CAM Technology in Mechanical Engineering, Massachusetts Institute of Technology, March 1982. MIT Press.

Woodwark, J.R. (1986) Computing Shape. Butterworths, London.

Advanced Methods for Parametric Design

Dieter Roller

Abstract

In this paper first a short introduction into parametric design versus conventional rigid design is provided. Examples of application areas for parametric modelling techniques are given. In the main part, different methods are presented that have been developed for parametric modelling in CAD. These methods are variants programming, resolution of a system of constraint equations, rule based variation evaluation and a constructive approach to interactive parametric design. To illustrate the operation of each method, examples are given that demonstrate the main concepts. Eventually, the most important characteristics of the presented methods are summarized.

1. Introduction

Computer aided design and drafting systems have proven to be viable tools in industrial environments. However, in the product development cycle today still several design changes are typically needed before the full requirements for functionality, manufacturability and quality of a design are met. In many cases those changes only affect some changes in geometric dimensions of the design. Contemporary CAD systems support design changes with special modification commands like STRETCH, MOVE or ROTATE [14]. In this kind of modification philosophy the geometry is directly manipulated in order to accomplish dimensional design changes. Figure 1 demonstrates this principle by a simple example. In systems where the annotation is associated to the geometry, the dimensioning, hatching and texts and symbols are updated automatically in a subsequent process [13].

Another principle method for accomplishing dimensional design changes is parametric design [15]. In this technique the geometry of a design is being stored with variable coordinates and dimension parameters. After the assignment of concrete dimension values the corresponding instance of the design is evaluated (cf. Figure 2).

There are several design tasks, where dimensional variations are of immediate relevance. A very important example is the design of families of parts. Here the members of the parts family just differ in geometric dimensions. Due to economic reasons there is even an ever-increasing requirement for re-used parts and parts families in the development and design of technical products. With the parametric design method the design needs to be entered in the system for one member of the

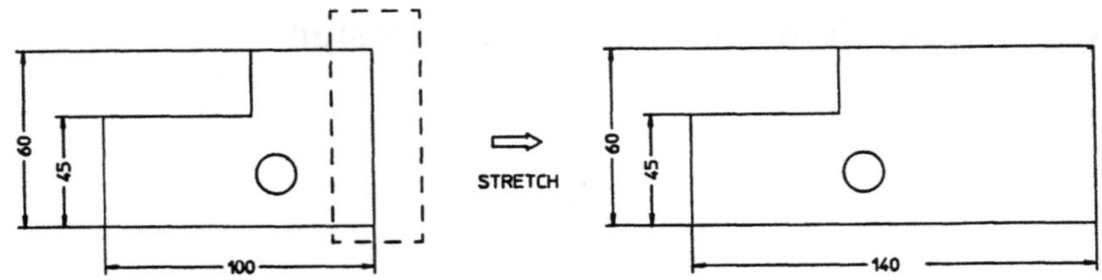

Figure 1: Dimensional modification in a conventional CAD system

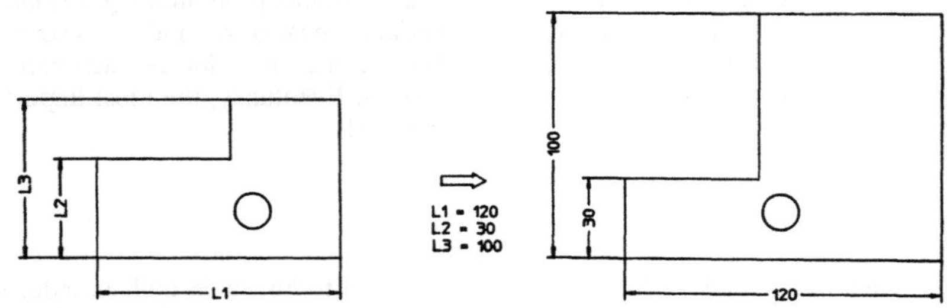

Figure 2: Dimensional modification in a parametric CAD system

family only. All other instances according to the special dimension values can be generated automatically. For this application it is common to provide the sets of dimension values in tabular form (cf. Figure 3).

Parametric design can also overcome the serious limitation of the application of CAD systems during the conceptual design. In this design phase, the designer does not yet specify all the dimensional details of the design. Rather, the principal shape of the design is fixed and functional constraints are transformed into geometric constraints. Parametric design allows the user to model the geometry and geometric constraints without specifying the concrete dimensions. Even sketch inputs or scanned and vectorized paper drawings can be converted into correctly sized geometry through a parametric approach.

In tolerance analysis parametric design also can be a powerful tool. For example, maximum or minimum conditions can be reflected in the geometry by applying the corresponding dimension values. Further obvious applications of parametric design are simulation and kinematics.

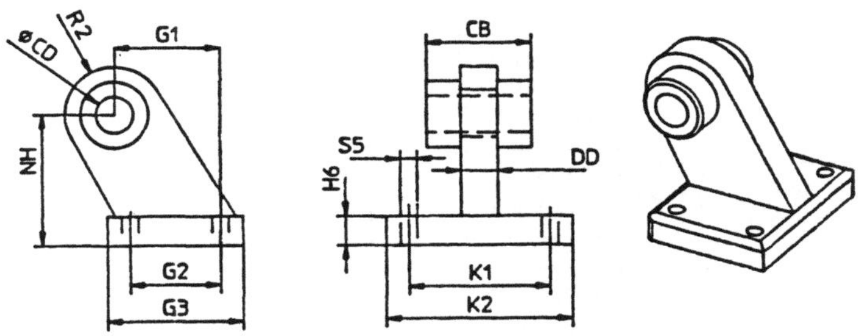

PISTON DIAMETER	K2	G3	S5	K1	G2	G1	CB	CD	R2	NH	H6	DD
⌀ 320	234	186	26	170	122	150	120	45	54	200	40	25
⌀ 250	200	160	20	150	110	128	110	40	43	165	35	25
⌀ 200	162	130	16	122	90	105	90	30	36	135	30	25
⌀ 160	156	126	14	118	88	97	90	30	36	115	25	25
⌀ 125	124	90	11	94	60	70	70	25	33	90	20	25

Figure 3: Example of a table for parameter value input

There are specific design applications where the general shape of the solution is known and the problem is to evaluate the needed dimensions and to subsequently create the corresponding variant. In these cases, the integration of parametric design with design rules evaluation can be the bases for a knowledge based design automation [17].

Concluding from the above outlined applications, parametric design is a fundamental technology for

- fast dimensional design changes,
- efficient design of family of parts,
- representation of standard parts in libraries,
- conceptual design,
- sketch input and automatic interpretation of paper drawings,
- tolerance analysis,
- simulation and kinematics,
- knowledge based design automation.

2. Basic technical considerations

Several methods have been developed to support parametric design in a CAD system. Figure 4 shows the principle that is common to all the methods that will be presented in the following sections.

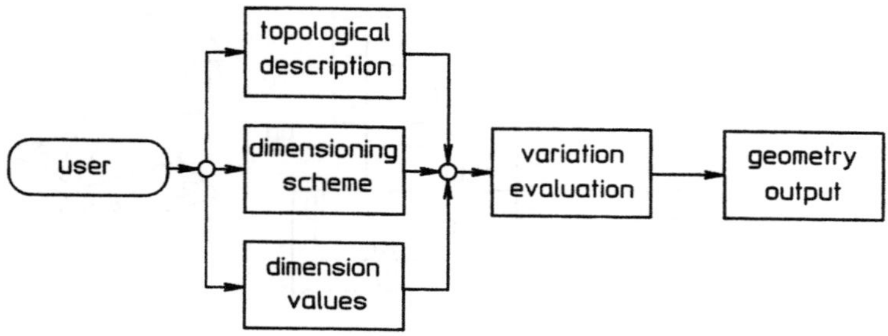

Figure 4: Schematic of the parametric design principle

The user first enters the topology and the dimensioning scheme of the design. The topological description essentially defines the connectivity of the geometric elements, while the dimensioning scheme specifies the dimensional constraints. This information needs to be entered only initially and is called the *primary design.* All dimensional variants are generated by a variation evaluation method, based on the topological description, the dimensioning scheme and the corresponding dimension values. The various methods for parametric design differ in the representation of the topological description and the dimensioning scheme as well as in the approach to evaluate the variants.

In engineering drawings and designs normally not all dimensional constraints of the geometry are specified explicitly. Examples of constraints that are not usually included in the dimensioning scheme of a drawing are

- parallel lines,
- tangential lines,
- colinear lines,
- orthogonal lines,
- horizontal lines,
- vertical lines.

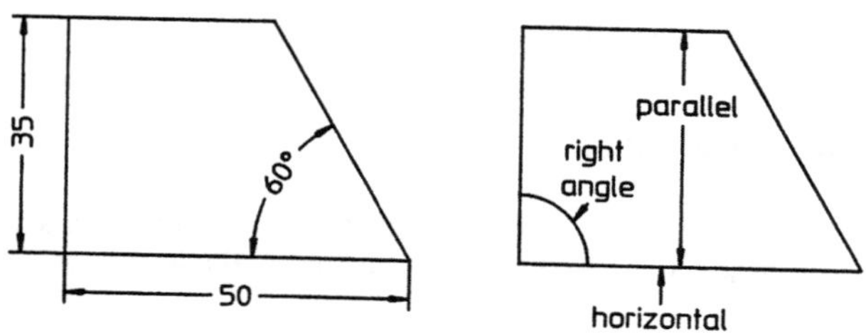

Figure 5: Left: Dimensioning scheme. Right: Underlying implicit dimensional constraints

In this paper we refer to these as *implicit dimensional constraints.* Figure 5 shows explicit dimensions given by a dimensioning scheme, and implicit dimensional constraints in a simple example. The integrity of a set of dimensional constraints depends on the explicit dimensioning as well as the underlying implicit dimensional constraints.

In a parametric design system the complete constraint set needs to be maintained. It is therefore important to provide the user with a clear constraint feedback during the design steps. Figure 6 shows an example of a design with attached constraint annotation in form of dimension labels and constraint icons.

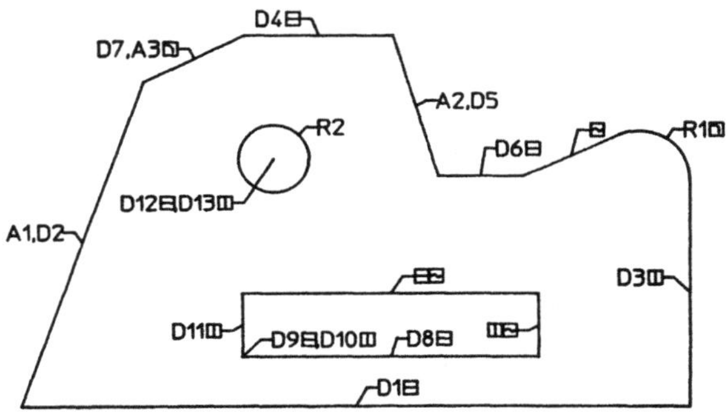

Figure 6: Example with dimension labels and constraint icons

Table 1 lists the meaning of the dimension labels and constraint icons that are used in Figure 6.

D1⊟	distance D1, horizontal line
D11◫	distance D11, vertical line
D9⊟D10◫	distance D9 along horizontal, distance D10 along vertical
⧄	flexible line
⊟⧄	horizontal, flexible line
◫⧄	vertical, flexible line
R1◳	fillet radius R1
D7,A3⧅	chamfer size D7, chamfer angle A3
A1,D2	angle A1, distance D2 along this line
R2	radius R2

Table 1: Meaning of the dimension labels and constraint icons in Figure 6

In the following sections we will present a programming method and three different graphically interactive methods for parametric design. We concentrate on the principles of those methods rather than on concrete commercial implementations.

3. Variants programming

Variants programming is one of the most wide-spread methods, particularly for the representation of families of parts and elements in standard parts libraries. This method is based on a programming language that supports the generation of models in a CAD data structure. A CAD macro language is particularly well suited for this task, if it supports

- geometry and annotation creation commmands,
- display commands,
- references to elements in the data structure,
- value assignment from keyboard and files,
- basic arithmetic and vector calculations,
- program control mechanisms, (LOOPS, IF THEN, WHILE etc.).

In this method the primary design is being done by writing a program that generates variants of a design, based on specific dimension value inputs. We are going to explain this for a simple example that is shown in Figure 7. This example represents a slot with a variable width, length and orientation. The programmming language that was chosen for this example is the macro language of the CAD system Hewlett Packard Mechanical Engineering Series 10 [12].

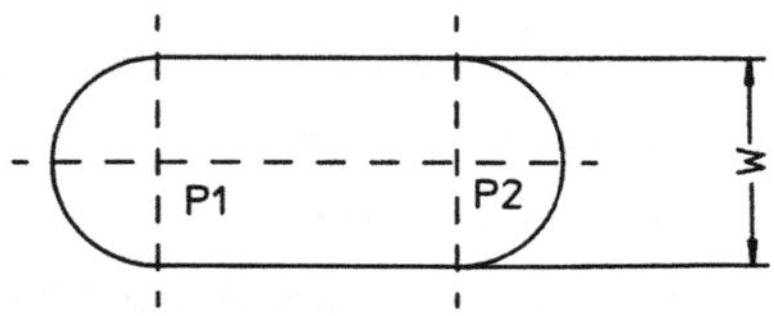

Figure 7: Macro example *parametric slot*

The task for the parametric slot macro to be programmed is to generate an arbitrary number of slots. These slots shall have a width W that the user can enter as a parameter. In order to fix the location, orientation and lenght of each individual slot, the user is asked to enter two reference points P1, P2 that represent the center points for the rounding circles. The generation of slots shall continue until the user enters the command END. Figure 8 shows a macro program for this parametric design example.

```
DEFINE SLOT
  LOCAL W
  LOCAL P1
  LOCAL P2
  LOCAL V
  READ NUMBER 'Enter slot width' W
  LOOP
    FOLLOW OFF
    COLOR WHITE
    LINETYPE SOLID
    READ PNT 'Enter first center point' P1
    READ PNT 'Enter second center point' P2
    LET V ( NORMAL ( P2 - P1 ) * ( W / 2 ) )
    ARC CEN_BEG_END P1 ( P1 + V ) ( P1 - V )
    ARC CEN_BEG_END P2 ( P2 - V ) ( P2 + V )
    LINE POLYGON ( P1 - V ) ( P2 - V )
    LINE POLYGON ( P1 + V ) ( P2 + V )
    LET V ( V * 1.5 )
    COLOR YELLOW
    LINETYPE DOT_CENTER
    LINE POLYGON ( P1 - V ) ( P1 + V )
    LINE POLYGON ( P2 - V ) ( P2 + V )
    LET V ( ROT V 90 )
    LINE POLYGON ( P1 + V ) ( P2 - V )
  END_LOOP
  COLOR WHITE
  LINETYPE SOLID
END_DEFINE
```

Figure 8: Macro program for the *parametric slot*

This method works for three-dimensional designs in an analogous way. Of course, then the set of macro commands need to include the 3D design commands of the CAD system. Another interesting feature of this apporach is the fact that also logic constraints (e.g. *IF L1 100 THEN L2 = 25*) can easily be built into the program.

However, for this method obviously special programming know-how is required from the user. Variants programming therefore is only acceptable in special cases, where this investment can be justified.

4. System of constraint equations

This approach is based on translating all dimensional constraints into equations, where the unknowns are the coordinates of the characteristic points of the geometric model. Subsequently this system of equations is solved by an iterative numerical method. Several versions of this approach have been proposed [6 ,7, 8, 9, 10, 11].

For example the distance D between two points

$$P_1 = (x_1, y_1) \text{ and } P_2 = (x_2, y_2)$$

can be translated into the constraint equation

(1) $$f = (x_1 - x_2) + (y_1 - y_2) - D = 0.$$

Obviously constraint equations in general are non-linear. Table 2 shows some further dimensional constraints and their corresponding constraint equations.

Dimension name	Entities constrained	Equation
Horizontal distance	P_1, P_2	$X_1 - X_2 - D = 0$
Vertical distance	P_1, P_2	$Y_1 - Y_2 - D = 0$
Linear distance	P_1, P_2	$(X_1 - X_2)^2 + (Y_1 - Y_2)^2 - D^2 = 0$
Distance from point to line	$P_1, \overrightarrow{P_2P_3}$	$\hat{U} \times V - D = 0$
		$\hat{U} = \frac{(X_3 - X_2)}{\lvert\overrightarrow{P_2P_3}\rvert}\hat{\imath} + \frac{(Y_3 - Y_2)}{\lvert\overrightarrow{P_2P_3}\rvert}\hat{\jmath}$
		$V = (X_2 - X_1)\,\hat{\imath} + (Y_2 - Y_1)\,\hat{\jmath}$
Angular dimension	$\overrightarrow{P_1P_2}, \overrightarrow{P_3P_4}$	$\lvert\overrightarrow{P_1P_2} \times \overrightarrow{P_3P_4}\rvert \,/\, (\overrightarrow{P_1P_2} \bullet \overrightarrow{P_3P_4})$ $- \mathrm{Tan}(A) = 0$

Table 2: Selected dimensional constraint equations

A system of constraint equations can be resolved by the Newton-Raphson iterative method as follows:

Let $F(X) = 0$ be the system, with $X = (x_i)_{i \leq m}$, $F = (f_j)_{j \leq m}$.

The sequence (X^n), with X^0 chosen arbitrarily is then computed by

(2) $$X^{n+1} = X^n - [F'(X^n)]^{-1} F(X^n)$$

where

(3) $$F'(X^n) = \left(\frac{\partial f_i}{\partial x_j}(X^n)\right)_{i,j \leq m}$$

is the Jacobian matrix.

We are now going to show the transformation of explicit and implicit dimensional constraints into constraint equations on a simple geometric example (cf. Figure 9). This example represents a triangle characterized by the three points

$$P_1 = (x_1, y_1), P_2 = (x_2, y_2) \text{ and } P_3 = (x_3, y_3).$$

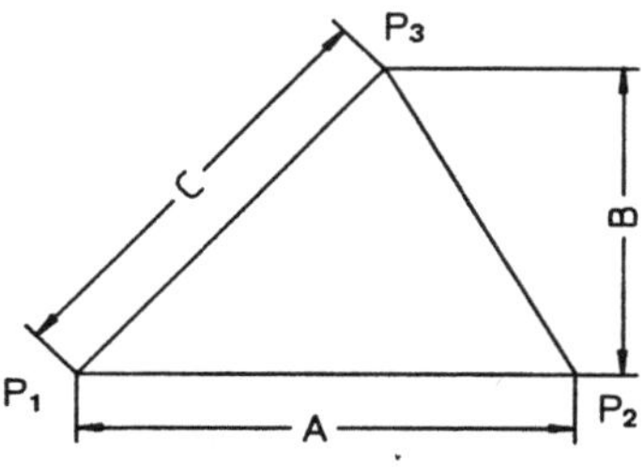

Figure 9: Geometric shape example

The triangle is dimensioned by two given side lengths A and C and the height B. Further constraints are that P_1 is at the origin of the coordinate system and the side $\overrightarrow{P_1 P_2}$ is horizontal. Thus, for the 6 unknowns x_1, x_2, x_3, y_1, y_2, y_3, we get the following constraint equations:

$$f_1 = (x_1 - x_2)^2 + (y_1 - y_2)^2 - A = 0$$

$$f_2 = y_3 - y_2 - B = 0$$

$$f_3 = (x_1 - x_3)^2 + (y_1 - y_3)^2 - C = 0$$

$$f_4 = x_1 = 0$$

$$f_5 = y_1 = 0$$

$$f_6 = y_2 - y_1 = 0$$

This is a general and powerful method and can handle variations of two- as well as three-dimensional models. In fact, all constraints that can be expressed by equations can be supported. If a new constraint is inserted by the user, only the corresponding constraint equation needs to be inserted into the system of equations. Also cyclic constraint situations, where constraints depend on each other, can be resolved due to the simultaneous solution of the equations.

If the dimensioning scheme has not been specified properly, the Jacobian may be singular. For this case Light [9] has proposed a solution that is based on a reduction of the number of equations and setting unconstrained variables to zero. In practical applications, it is normally requested to provide a completely dimensioned geometry. Therefore, methods have been developed for checking the consistency and completeness of a constraint scheme before the numerical method is applied.

Fitzgerald [4] proposed such a method for axial dimensions. A more comprehensive graph based method has been developed by Chyz [2].

A limitation of the constraint equations method is that in cases of multiple solutions only one of them is computed by the numerical method. Moreover, this solution is not unique and depends on the start vector. Also it should be noticed that the evaluation of a variant, i. e. the resolution of the system of constraint equations, is very computation intensive.

5. Rule based variation evaluation

More recently artificial intelligence based methods have been developed for calculating the actual shape of the geometry for a given set of dimensional constraints. These approaches employ an expert system to sequentially propagate point coordinates that can be evaluated immediately from given constraints and known coordinates. Several versions of this type of approach have been proposed [1, 18, 19, 20], Here we will follow Sunde [19].

Dimensional constraints, including tangency constraints between circles or circles and lines, as well as radius constraints, can be expressed using only distance and angle constraints. E.g., a tangency constraint between a line and a circle is represented by a right angle constraint between the line and the radius line from the center of the circle to the tangent point.

We assume that these translations have been made for a design before this method is being applied. As a formal notation to express constraints we use so-called *CA-sets* and *CD-sets*

- A CA-set is a set of pairs of points with mutually constrained angles.
- A CD-set is a set of points with mutually constrained distances.

These sets are created when constraints are added to the design as follows:

- When an angle constraint between two pairs of points is introduced then the two CA-sets of the referred pairs of points are united. Initially all pairs of points are stored in disjunct CA-sets.
- When a distance constraint between two points is given, a new CD-set is created. This CD-set contains these two points.

The evaluation of a variant that satisfies all constraints for given dimension values is performed by an expert system. The philosophy hereby is to successively evaluate point coordinates by applying geometric rules. A design is completely determined if all points belong to the same CD-set. As a foundation the geometric knowledge comprises facts, production rules and verification rules.

Facts are specified by the user. They denote the existence of geometric entities and constraints. The production rules are used to generated higher level information. They are performed in situations where the conditions for combining two or more CA- or CD-sets into larger CA- or CD-sets are fulfilled.

Examples for production rules are:

1. If a new distance constraint is introduced then a CD-set is generated that contains the two referred points.
2. If an angle constraint is introduced then the CA-sets of the referred two pairs of points are united.
3. If two CD-sets have a point in common and an angle is specified between these CD-sets then the CD-sets are united.

Figure 10 and Figure 11 show examples for situations where the conditions for the above listed production rules 2 and 3 are fulfilled. It can be shown that rules about triangles, quadrilaterals and parallelograms support the evaluation of a wide class of geometric models.

Some elements of the conditions for the production rules are not stored explicitly in the database. In these cases verification rules are used for testing the truth of fact elements.

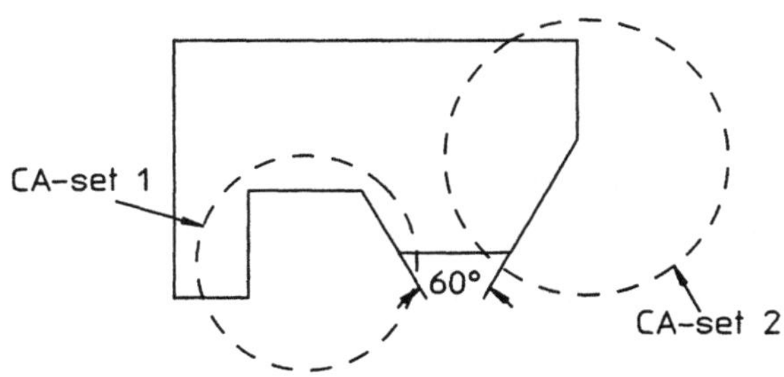

Figure 10: Condition for union of two CA-sets

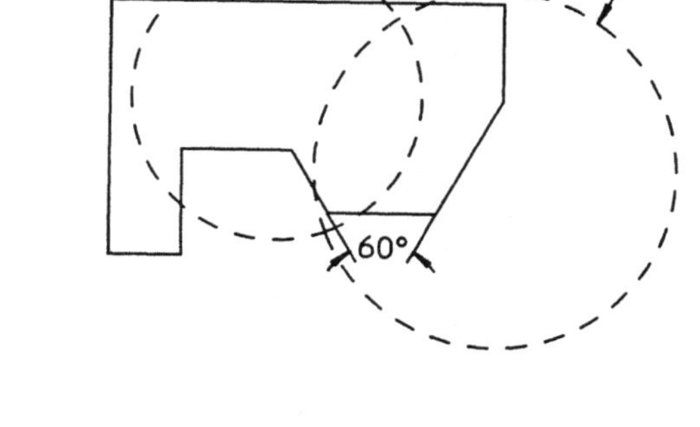

Figure 11: Condition for union of two CA-sets

Potential benefits of this method are the following. Ill defined constraint situations may be explained to the user by using capabilities of the expert system. Incorporation of design rules in respect to dimension values can be achieved in a way where the design knowledge is maintained separately from the geometric algorithms.

Besides the current performance limitation and memory requirement for the rule based approaches, another severe drawback is that for cyclic constraint situations, that may occur in practice, the deducing process ends up in a loop and no solution is found. Not much practical experience, particularly for three-dimensional designs, has been published about this method yet.

6. Constructive approach

This approach utilizes the construction sequence of a design. Different parametric design methods based on this approach can be found in [3, 5, 16]. Here, we focus on [16] and describe an advanced version that has been proven to be very effective in many industrial installations. This method is based on the following key points:

- introduction of three classes of geometry: FIX, VARIABLE and FLEXIBLE,
- automatic build up of implicit constraints through inference from command semantics,
- capturing construction principles by tracking the usage of construction geometry,
- storing the construction sequence,
- generating a construction plan consisting of conventional CAD commands,
- processing the construction plan for given sets of dimension parameters.

Frequently variation is only needed for a subset of the dimensions of a part. The likelihood of inconsistencies in dimensioning can be minimized, if the variational capability can be limited to the relevant dimensions. In order to support this principle, three different classes of geometry are provided for the user. More precisely, the geometry input commands can be executed in three different modes:

- *FIX:* Geometric elements that are to have fixed dimensions are constructed in FIX mode. This means, for example, that lines created in this mode will keep their length and angle when performing a variation of the design. Their relative position, however, will depend on connected elements, if there are any.
- *VARIABLE:* For elements that require variable dimension parameters (length, radius, angle) the corresponding creation command is used in VARIABLE mode. For the users convenience and efficiency of the design input, dimension variables will be generated with an automatic incrementing of an index. In order to constantly provide a graphical feedback of the current design, these variable elements will be displayed with the arbitrary dimensions that have been chosen for their initial creation.
- *FLEXIBLE:* Frequently the length of an element is not explicitly known to the user, but it has to fit between two existing points, e.g. endpoints of other lines. In these cases the FLEXIBLE mode will be used. A FLEXIBLE line can be thought of as an expandable rubber line which stretches between its constrained points.

The consequent selection of the appropriate creation command mode reduces the complexity for the user in the later stage of producing design variants.

In order to relieve the user from explicitly specifying all of the implicit constraints in a tedious post-process of the construction, the implicit dimensional constraints are determined directly from the used construction commands. Typical construction commands are LINE_TWO_POINTS, LINE_HORIZONTAL, LINE_VERTICAL, LINE_PERPENDICULAR, CIRCLE_CONCENTRIC. For example, the construction of a straight line using the command LINE_HORIZONTAL will be reflected in the constraint, that this particular line has to be horizontal in all variants. Analogous constraints are built up with the other above mentioned command examples. If a line is to be constructed with no particular implicit constraints, then the command LINE_TWO_POINTS is used, which generates a line from one arbitrary point to another.

Many designs can be accomplished by creation of the geometry using the type of commands explained above. However, there are cases where the coordinates of points can more easily be determined using so-called construction geometry. Construction geometry lines are unbounded help lines and are usually supported in CAD systems. In order to support this type of construction in a parametric environment, construction geometry is enriched with dimensional constraints similar to the ones described for the regular geometry.

After the primary design has been completed, a construction plan is being built. The construction plan consists of a series of geometry creation commands. These commands are stored with variables as parameters. For all variables formula expressions are maintained that represent the relation of the variables and the dimension parameters.

In order to perform the generation of variants, new values may be assigned to the dimension parameters. The construction plan is then executed with the new set of dimension values. In order to perform the commands in the construction plan, the expressions for the referred variables are evaluated.

We will now discuss this method step by step on the example of a plate with a tilted slot as shown in Figures 12 and 13.

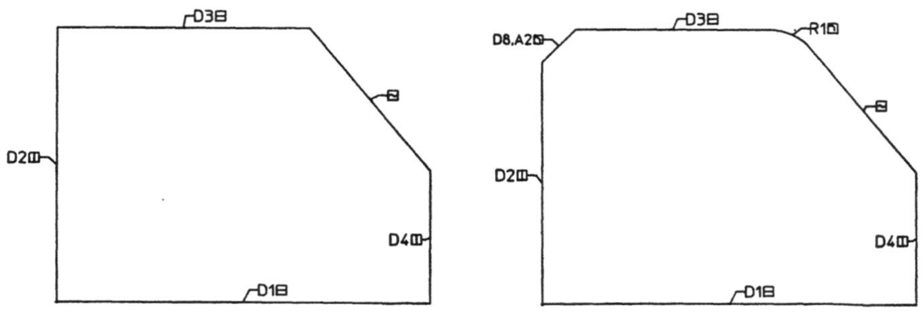

Figure 12: Left: First design step. Right: Modification using the CHAMFER and FILLET functions

For the annotation of the created constraints the dimension labels and constraint icons are used, as presented in the introduction. First the outer contour lines are generated (cf. Figure 12 left side). In the next step CHAMFER and FILLET commands are used in VARIABLE mode to create the chamfer and the fillet radius (cf. Figure 12, right side).

Then construction geometry is used to design the tilted slot (cf. Figure 13, left side). In the last step the contour of the slot is overdrawn with an OVERDRAW function. This function automatically generates FLEXIBLE lines based on the intersection points of the construction geometry.

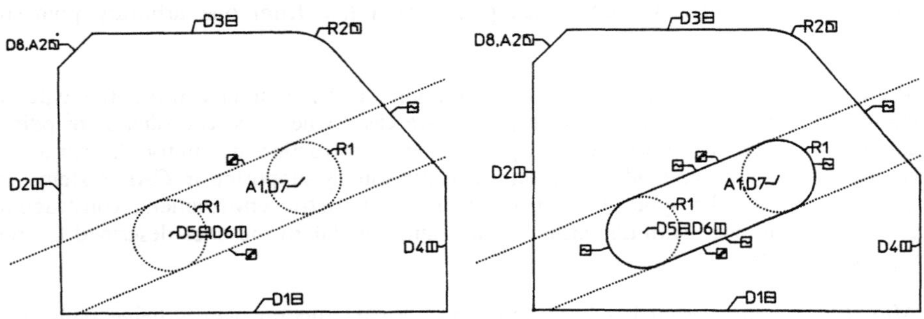

Figure 13: Left: Use of variable construction geometry. Right: Final part

This method has proven to be very fast. The processing for variants of the example in Figure 13 takes only a fraction of a second. Therefore it becomes practical to only store dimension tables together with the primary design instead of storing each variant separately. This can save a significant amount of storage space. Another important feature of the constructive approach is the efficient way of entering the constraints.

While the utilization of information about the design sequence results in the fast speed, the main drawback is that this method cannot be applied directly to create variations of conventionally generated designs.

7. Conclusion

We have discussed several different methods for parametric design. There is obviously no best solution to the parametric design problem. Rather, for particular applications the most appropriate methods should be considered. Table 3 summarizes the main characteristics and includes some general comments to the presented methods.

Variants programming	in most commercial systems possible, general method for analytical and logical constraints, power depends on available language, programming knowledge required
System of constraint equations	for all analytical constraint descriptions, robust algorithms available, combination with constraints checker recommended, provides in unambiguous cases only one solution, computation intensive,
Rule based variation evaluation	potentially well suited for integration of design rules, little experience in industrial use, not for cyclic constraints, time and memory intensive
Constructive approach	very fast variants generation, easy to use, not applicable for conventional data files, not for cyclic constraints

Table 3: Characteristic features and notes to the presented methods

A challenge for future system developments will be to couple parametric design with a knowledge based support. As a goal, during the design the user should be intelligently assisted in selecting parameter values that meet design rules in respect to functionality, manufacturability, quality and costs.

References

1. Aldefeld, B.: Variation of geometries based on a geometric- reasoning method. CAD, vol. 20, no. 3, April 1988, pp. 117-126
2. Chyz W.: Constraint Management for CSG. Master Thesis, MIT, June 1985
3. Cugini, U., Folini, F. and Vincini, I.: A procedural system for the definition and storage of technical drawings in parametric form. Eurographics 88, D.A. Duce and P. Jancene (eds.), North Holland, 1988, pp. 183-196
4. Fizgerald W.: Using axial dimensions to determine the proportions of line drawings in computer graphics. CAD, vol. 13, no. 6, November 1981, pp. 377-382
5. Gossard, D., Zuffante, R. and Sakurai, H.: Representing dimensions, tolerances and features in MCAE systems. IEEE Computer Graphics and Applications, March 1988, pp. 51-59
6. Hillyard, R. and Braid, I.: Analysis of dimensions and tolerances in computer-aided mechanical design. CAD, vol. 10, no. 3, May 1978, pp. 161-166

7. Hillyard, R. and Braid I.: Characterizing non ideal shapes in terms of dimensions and tolerances. Computer Graphics, vol. 12 no. 3, August 1978, pp. 234-238

8. Lee, K. and Andrews, G.: Inference of the positions of components in an assembly: Part 2. CAD, vol. 17, no. 1, January 1985, pp. 20-24

9. Light, R.: Symbolic Dimensioning in Computer-Aided Design. Master Thesis, MIT, May 1979

10. Light, R., Lin, V. and Gossard D.: Variational geometry in CAD. Computer Graphics, vol. 15, no. 3, August 1981, pp. 171-177

11. Light, R. and Gossard, D.: Modification of geometric models through variational geometry. CAD, vol. 14, no. 4, July 1982, pp. 209-214

12. ME 10d Mechanical Engineering CAD System Writing Macros Manual. Hewlett Packard, Böblingen, Edition 1, March 1989

13. Roller, D., Mainguy, J.-P., Kurz, W.: Internal design of design automation software and its consequences for the User. MICAD 86, Proceedings of the Fifth European Conference on CAD/CAM and Computer Graphics, Hermes-Verlag, Paris, 1986, pp. 765-783

14. Roller, D.: Effiziente Modellierung und Modellmodifikation von mechanischen Teilen. CAD und Computergraphik, Nr. 3/4, Okt. 1989, Wien, pp. 115-123

15. Roller, D., Schonek, F., Verroust, A.: Dimension-driven geometry in CAD: a survey. Theory and Practice of Geometric Modelling, W. Strasser, H.-P. Seidel (eds.), Springer-Verlag, 1989, pp. 509-523

16. Roller, D.: A system for interactive variation design". Geometric Modelling for Product Engineering. M. Wozny, J. Turner and K. Preiss (eds.), North Holland, 1989, pp. 207-220

17. Roller, D.: Parametrische Formelemente als Basis für intelligentes CAD. Procedings of GI-Fachgespräch Graphik und KI, Königswinter bei Bonn, 3.-4. April 1990, Springer-Verlag, 1990

18. Sunde, G.: Specification of shape by dimensions and other geometric constraints. IFIP WG. 5.2 on Geometric Modeling, Rensselaerville, NY, May 1986

19. Sunde, G.: A CAD system with declarative specification of shape. Eurographics Workshop on Intelligent CAD Systems, April 21-24, 1987, Noorwijkerhout, The Netherlands (1987), pp. 90-104

20. Sunde, G. and Kallevik, V.: A dimension-driven CAD system - utilizing AI techniques in CAD. Report no. 860216-1, Senter for Industriforskning, December 1987

A Tutorial Introduction to Blossoming

Tony deRose
Michael Lounsbery
Department of Computer Science & Engineering FR-35
University of Washington
Seattle, WA 98195

Ronald Goldman
Department of Computer Science
University of Waterloo
Waterloo, Ontario
Canada N2L 3GI

Abstract

A powerful new technique for analyzing Bézier and B-spline curves and surfaces has been developed recently by Ramshaw and de Casteljau. The method, called *blossoming* or *polarization*, is based on an old idea from multilinear algebra where polynomials are studied by replacing them with simple multivariable functions. The purpose of these notes is to provide a brief tutorial introduction to the basic concepts and uses of the technique. Several implementation consequences are also identified.

1 Introduction

The theory of *blossoms*, or *polar forms*, is a highly geometric way to view much of the field of computer aided geometric design (CAGD)[11, 12]. Blossoming provides a simple yet powerful tool for deriving many of the fundamental properties of common curve and surface paradigms. The theory also unifies these paradigms, and provides a solid base upon which to build a geometric modeling system.

Our approach will be fairly informal, stressing intuition rather than rigor; more rigorous treatments of this material can be found elsewhere [5, 6, 11, 12, 14]. In the following, we assume that the reader is familiar with the standard properties of Bézier curves and B-splines. An introduction to this material can be found in [2] and [9]. We begin by reviewing some of the necessary geometric concepts in Section 2. In Section 3, we motivate the definition of blossoms by examining in detail the case of quadratic Bézier curves. The extension of these ideas to higher order curves is presented in Sections 4.1-4.4. B-splines are treated in Sections 4.5-4.7, and a short discussion of surfaces appears in Section 5. We conclude in Section 6 with some remarks on how blossoming can be used as an effective programming tool.

2 Geometric Preliminaries

We outline here the basic geometric facts necessary for manipulating blossoms. For a more complete introduction, consult [7] or [9].

We will be working in the context of *affine spaces*. Intuitively, affine spaces are a slight generalization of the more familiar Euclidean spaces. For our current purposes, it is sufficient to think of an affine space as a collection of points that is closed under *affine combinations*. An affine combination of points $\mathbf{p}_1, ..., \mathbf{p}_n$ is an expression of the form

$$\alpha_1 \mathbf{p}_1 + \cdots + \alpha_n \mathbf{p}_n, \tag{1}$$

where $\alpha_1, ..., \alpha_n$ are real numbers that sum to one.

Affine combinations have nice geometric interpretations. For example, if $\mathbf{p} = \alpha_1 \mathbf{p}_1 + \alpha_2 \mathbf{p}_2$, then $\mathbf{p}$ lies on the line segment $\mathbf{p}_1 \mathbf{p}_2$ so as to break the segment into subsegments of relative lengths $\alpha_2 : \alpha_1$, as shown in Figure 1.

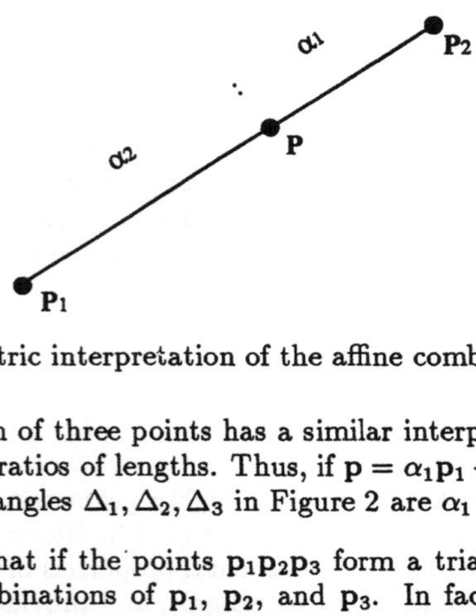

Figure 1: The geometric interpretation of the affine combination $\mathbf{p} = \alpha_1 \mathbf{p}_1 + \alpha_2 \mathbf{p}_2$

An affine combination of three points has a similar interpretation, except that ratios of areas are used instead of ratios of lengths. Thus, if $\mathbf{p} = \alpha_1 \mathbf{p}_1 + \alpha_2 \mathbf{p}_2 + \alpha_3 \mathbf{p}_3$, then the ratios of the areas of the subtriangles $\Delta_1, \Delta_2, \Delta_3$ in Figure 2 are $\alpha_1 : \alpha_2 : \alpha_3$.

Figure 2 illustrates that if the points $\mathbf{p}_1 \mathbf{p}_2 \mathbf{p}_3$ form a triangle, then other points $\mathbf{p}$ can be written as affine combinations of $\mathbf{p}_1$, $\mathbf{p}_2$, and $\mathbf{p}_3$. In fact, every other point $\mathbf{p}$ in the plane of $\mathbf{p}_1 \mathbf{p}_2 \mathbf{p}_3$ can be written *uniquely* as an affine combination of these points. If $\mathbf{p} = \alpha_1 \mathbf{p}_1 + \alpha_2 \mathbf{p}_2 + \alpha_3 \mathbf{p}_3$, the coefficients $\alpha_1, \alpha_2, \alpha_3$ are the *barycentric coordinates* of $\mathbf{p}$ relative to the *domain triangle* $\mathbf{p}_1 \mathbf{p}_2 \mathbf{p}_3$. The notion of barycentric coordinates extends naturally to affine spaces of arbitrary dimension, with the generalization of domain triangles being domain simplexes.

In addition to affine combinations, *affine maps* play a central role in the theory of blossoms. An affine map is defined in much the same way as a linear map. In particular, a map $T : X \to Y$ between affine spaces X and Y is said to be an affine map if it preserves affine combinations. That is, if $\mathbf{x}_1, ..., \mathbf{x}_k$ are points in X, then

$$T(\alpha_1 \mathbf{x}_1 + \cdots + \alpha_k \mathbf{x}_k) = \alpha_1 T(\mathbf{x}_1) + \cdots + \alpha_k T(\mathbf{x}_k)$$

must hold for all k, for all choices of $\mathbf{x}_1, ..., \mathbf{x}_k$, and for all sets of coefficients $\alpha_1, ..., \alpha_k$ that sum to one.

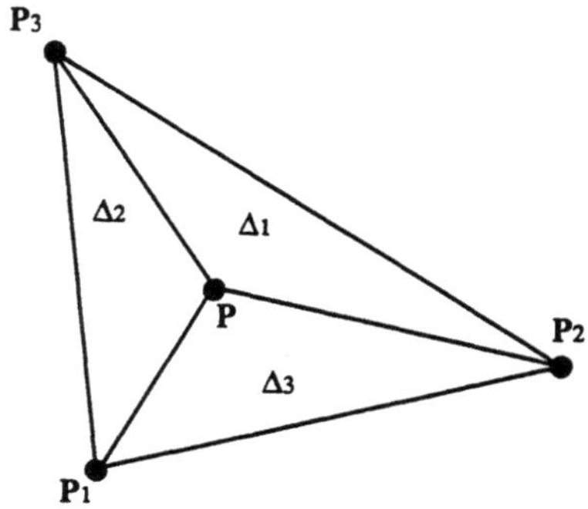

Figure 2: The geometric interpretation of the affine combination $\mathbf{p} = \alpha_1\mathbf{p}_1 + \alpha_2\mathbf{p}_2 + \alpha_3\mathbf{p}_3$.

3 Motivation

To motivate some of the ideas underlying blossoming, let us begin by examining quadratic Bézier curves, an example of which is shown in Figure 3. Recall that they are defined by the formula

$$Q(u) = \sum_{i=0}^{2} \mathbf{v}_i \cdot B_i^2(u),$$

where $B_i^2(u)$ are the quadratic Bernstein polynomials

$$B_i^2(u) = \binom{2}{i} u^i (1-u)^{2-i}.$$

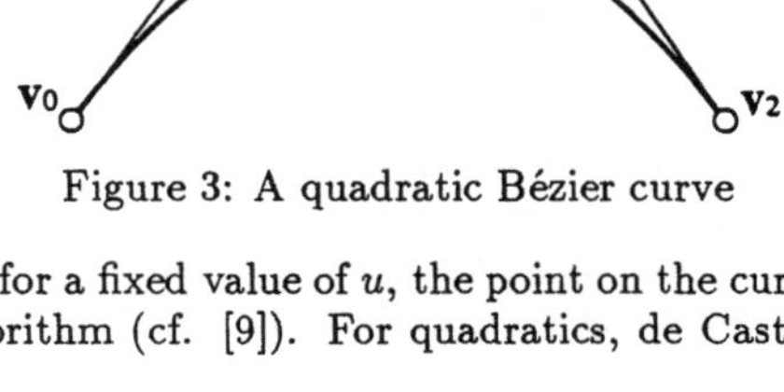

Figure 3: A quadratic Bézier curve

It is well known that for a fixed value of u, the point on the curve $Q(u)$ can be computed using de Casteljau's algorithm (cf. [9]). For quadratics, de Casteljau's algorithm may be stated as:

$$\begin{aligned} \mathbf{v}_0^1 &\Leftarrow (1-u)\mathbf{v}_0 + u\mathbf{v}_1 \\ \mathbf{v}_1^1 &\Leftarrow (1-u)\mathbf{v}_1 + u\mathbf{v}_2 \\ \mathbf{v}_0^2 &\Leftarrow (1-u)\mathbf{v}_0^1 + u\mathbf{v}_1^1 \\ Q(u) &\Leftarrow \mathbf{v}_0^2 \end{aligned}$$

The standard proof for de Casteljau's algorithm proceeds by induction by establishing and exploiting a recurrence relation for the Bernstein polynomials. As is often the case with inductive proofs, this derivation of de Casteljau's algorithm suffers from a lack of intuition. Blossoming provides an alternative proof that yields substantially more insight. To see this, let us look at a somewhat nonintuitive bivariate function $q(u_1, u_2)$ defined by

$$q(u_1,u_2) = \mathbf{v}_0(1-u_1)(1-u_2) + \mathbf{v}_1\{u_1(1-u_2)+u_2(1-u_1)\} + \mathbf{v}_2 u_1 u_2.$$

Notice that this function q has some interesting properties, namely:

1. q is *symmetric* with respect to its 2 arguments: $q(u_1,u_2) = q(u_2,u_1)$.
2. q is *bi-affine* or *2-affine*: When $a+b=1$,
   (a) $q(at+bv, u_2) = a\,q(t,u_2) + b\,q(v,u_2)$.
   (b) $q(u_1, at+bv) = a\,q(u_1,t) + b\,q(u_1,v)$.
3. q agrees with Q on its *diagonal*: $q(u,u) = Q(u)$.
4. $q(0,0) = \mathbf{v}_0$
5. $q(0,1) = q(1,0) = \mathbf{v}_1$
6. $q(1,1) = \mathbf{v}_2$

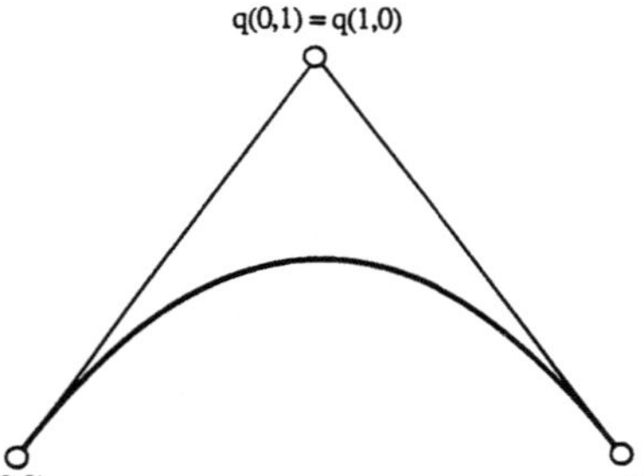

Figure 4: Relabeling of control points.

With observations 4, 5, and 6, we can relabel our diagram of the quadratic Bézier curve as shown in Figure 4. From observations 3-6, we can see that the function q is closely related to the curve Q. q is called a *blossom* (or *polar form*), and is the *polarization* of Q. By working with the blossom of Q instead of with Q itself, we exploit a powerful tool for manipulating Q. The blossom q lets us work singly with small, simple, affine components of Q.

As an example of a simple affine blend, we hold one of the 2 arguments in q constant, and allow the other to vary. Then

$$\begin{aligned} q(u,0) &= q((1-u)\cdot 0 + u\cdot 1, 0) \\ &= (1-u)\,q(0,0) + u\,q(1,0), \end{aligned}$$

showing that the point $q(u,0)$ is an affine blend of the points $q(0,0)$ and $q(1,0)$. As we can see in Figure 5, $q(u,0)$ is the unique point on the line between $q(0,0)$ and $q(1,0)$ that breaks

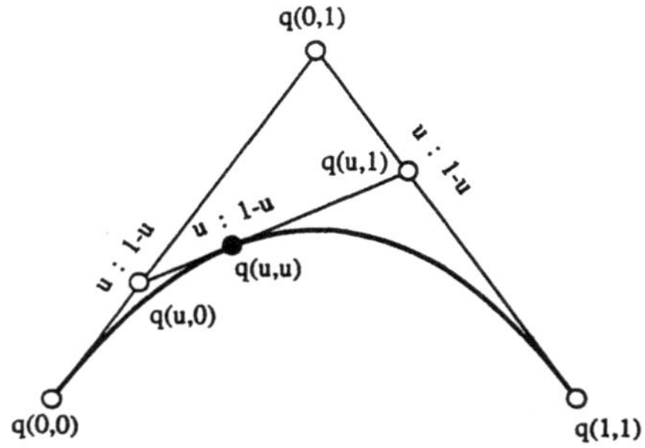

Figure 5: De Casteljau's algorithm for quadratics as blossom evaluation.

the intervals into the ratio $u : 1-u$. Similarly, $q(u,1)$ is on the line between $q(0,1)$ and $q(1,1)$.

Now that we have $q(u,1)$ and $q(u,0)$, we can derive $q(u,u)$ in a similar manner:

$$\begin{aligned} q(u,u) &= q(u,(1-u)\cdot 0 + u\cdot 1) \\ &= (1-u)\,q(u,0) + u\,q(u,1). \end{aligned}$$

This tells us that $q(u,u)$ is at position u on the line between $q(u,0)$ and $q(u,1)$. From property 3, $q(u,u)$ is exactly equivalent to $Q(u)$. Repeated affine combinations of blossom values evaluated with u has led us to the point $Q(u)$ on the original Bézier curve. This is simply a rederivation, using blossom notation, of the de Casteljau algorithm for evaluating quadratic Bézier curves.

Building on simple affine combinations such as these, we will see how to manipulate arbitrarily complex polynomials, both curves and surfaces.

4 Curves

4.1 Blossom Definition

Blossoms for arbitrary degree polynomials are based on the following theorem from multilinear algebra:

Theorem 1 Let Q be a polynomial of degree d. If $D \geq d$, then there is a unique symmetric D-affine function $q(u_1, \ldots, u_D)$ such that $q(u, \ldots, u) = Q(u)$.

To be more constructive, let us determine the blossom of the extremely simple polynomial $Q(u) = u^2$, and let us choose D in Theorem 1 to be 3. Thus, we seek a tri-affine function $q(u_1, u_2, u_3)$ such that $q(u,u,u) = u^2$. As an initial attempt, consider the following as a candidate solution:

$$q(u_1, u_2, u_3) = u_1 u_2.$$

This function is affine in each argument and agrees with Q on its diagonal, but is not symmetric. In particular, $q(u_1, u_2, u_3) \neq q(u_3, u_2, u_1)$. We can solve this problem by symmetrizing over the three variables, taken two at a time, to yield

$$q(u_1, u_2, u_3) = \frac{u_1u_2 + u_2u_3 + u_1u_3}{3}.$$

This new function meets all criteria, so by uniqueness it must be the function we seek. Reviewing what we have done, if $Blossom_D()$ represents the mapping from a polynomial to its D-affine blossom, then we have discovered that

$$Blossom_3(u^2) = \frac{u_1u_2 + u_2u_3 + u_1u_3}{3}.$$

Generalizing slightly, it is not difficult to show that

$$Blossom_D(u^k) = \frac{\sum_{i_1,\ldots,i_k} u_{i_1} \cdot u_{i_2} \cdots u_{i_k}}{\binom{D}{k}}, \tag{2}$$

where the summation is taken over all indices $i_1, \ldots, i_k$ such that each index is chosen from the set $\{1, \ldots, D\}$, and such that no two indices are equal.

The operator $Blossom_D()$ is *linear*, meaning that if a and b are real numbers, or, more generally, vectors, and if $P(u)$ and $Q(u)$ are two polynomials, then

$$Blossom_D(aP(u) + bQ(u)) = a\ Blossom_D(P(u)) + b\ Blossom_D(Q(u)).$$

The linearity of $Blossom_D()$, together with Equation 2 implies that the D-affine blossom of an arbitrary polynomial $Q(u) = \sum_k \mathbf{c}_k u^k$ can be expressed as

$$Blossom_D(Q(u)) = Blossom_D(\sum_k \mathbf{c}_k u^k) = \sum_k \mathbf{c}_k \sum_{i_1,\ldots,i_k} \frac{u_{i_1} \cdots u_{i_k}}{\binom{D}{k}}. \tag{3}$$

For example, if $Q(u) = 1 + 2u + 4u^2 - u^3$, then

$$Blossom_3(Q(u)) = 1 + 2\frac{u_1 + u_2 + u_3}{3} + 4\frac{u_1u_2 + u_1u_3 + u_2u_3}{3} - u_1u_2u_3.$$

4.2 Blossoms and Bézier Representations

Given a blossom $q(u_1, \ldots, u_d)$, suppose we want to find the Bézier control points $\mathbf{v}_0, \ldots, \mathbf{v}_d$ of the curve $Q(u) = q(u, \ldots, u)$. To solve this problem, we can expand the first argument of q using the identity $u = (1 - u) \cdot 0 + u \cdot 1$, then use the fact that q is affine in its first argument. This gives us

$$Q(u) = (1 - u)q(0, u, \ldots, u) + uq(1, u, \ldots, u).$$

Continue this process, recursively expanding the terms $q(0, u, \ldots, u)$ and $q(1, u, \ldots, u)$. When this is done and terms are collected, the resulting expression is

$$Q(u) = \sum_{i=0}^{d} q(0, \ldots, 0, \underbrace{1, \ldots, 1}_{i}) \binom{d}{i} u^i (1 - u)^{d-i}.$$

Since a polynomial has a unique set of Bézier control points, we deduce that

$$\mathbf{v}_i = q(0, \ldots, 0, \underbrace{1, \ldots, 1}_{i}).$$

To summarize, given a blossom $q(u_1, ..., u_d)$, the ith Bézier control point of its diagonal polynomial can be extracted by evaluating q at 1 a total of i times, and at 0 a total of $d - i$ times.

This observation can be generalized for Bézier curves parametrized on intervals other than $[0, 1]$. More specifically, the relationship between Bézier control points parametrized on an arbitrary interval $[s, t]$ to blossom values is $\mathbf{v}_i = q(s, ..., s, \underbrace{t, ..., t}_{i})$.

The above discussion shows that if the blossom is known in the sense that an arbitrary value can be computed, then the Bézier control points arise by evaluating the blossom at simply described values. Consider now the converse problem: given the Bézier control points of a polynomial $Q(u)$, compute an arbitrary value of its blossom $q(u_1, ..., u_d)$. In principle, one could write down an expression similar to Equation 3 in terms of the Bézier control points instead of the power basis coefficients $\mathbf{c}_k$. A simpler approach, however, is to provide an algorithm to compute arbitrary values. Such an algorithm is remarkably simple, as shown in Figure 6. De Casteljau's algorithm can be seen to be a special case of the algorithm of Figure 6, occurring when $u_1 = \cdots = u_d = u$; that is, for the computation of a point on the diagonal.

EvaluateBlossom($\mathbf{v}_0, ..., \mathbf{v}_d, u_1, ..., u_d$)
for $i = 1$ to d do
    for $j = 0$ to $d - i$ do
        $\mathbf{v}_j = (1 - u_i)\mathbf{v}_j + u_i \mathbf{v}_{j+1}$
        { $\mathbf{v}_j$ *now equals* $q(u_1, ..., u_i, \underbrace{1, ..., 1}_{j}, 0, ..., 0)$ }
    end
end
return $\mathbf{v}_0$

Figure 6: Evaluation algorithm for an arbitrary blossom value $q(u_1, ..., u_d)$ given the Bézier control points $\mathbf{v}_0, ..., \mathbf{v}_d$.

4.3 Subdivision

Referring to the de Casteljau algorithm depicted in Figure 7, notice that the points

$$q(0, 0, \ldots, 0, u), q(0, \ldots, 0, u, u), \ldots, q(u, u, \ldots, u)$$

are all computed by the algorithm. These are all of the form $q(s, \ldots, s, \underbrace{t, \ldots, t}_{i})$, where $s = 0$ and $t = u$. The results of Section 4.2 therefore imply that these values are exactly the Bézier control points of Q defined on the interval $[0, u]$. Similarly, the points $q(u, \ldots, u, 1, \ldots, 1)$ were computed as well. These are precisely the control points of Q defined on the interval $[u, 1]$. De Casteljau's algorithm therefore effectively subdivides the original Bézier curve into two subcurves by deriving as a side effect the Bézier control points that characterize each subinterval.

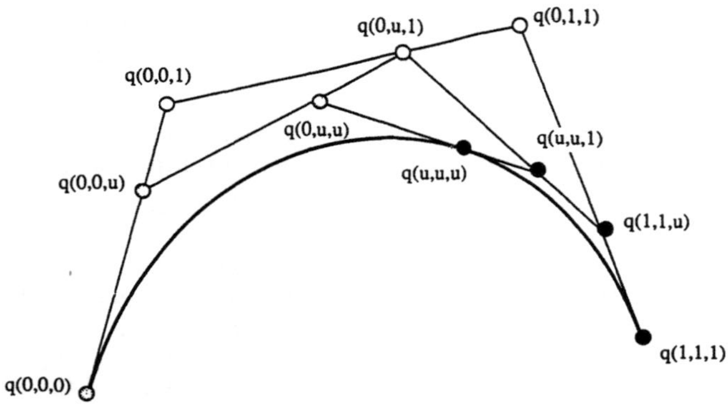

Figure 7: De Casteljau's algorithm for cubics as blossom evaluation.

4.4 Degree Raising

Suppose we have a curve $Q(u) = \sum_{i=0}^{2} \mathbf{v}_i B_i^2(u)$ and we want to find a curve $\tilde{Q}(u) = \sum_{i=0}^{3} \tilde{\mathbf{v}}_i B_i^3(u)$ such that $Q(u) = \tilde{Q}(u)$.

Our method for accomplishing this will be:

1. Find the blossom q of Q.
2. Build the blossom $\tilde{q}$ of Q from q.
3. Evaluate $\tilde{q}$ to extract the $\tilde{\mathbf{v}}_i$.

This process is summarized in Figure 8.

We start with a bi-affine blossom q, fully characterized by $\mathbf{v}_0$, $\mathbf{v}_1$, and $\mathbf{v}_2$. Our goal in degree raising is to find a tri-affine blossom $\tilde{q}$ such that

$$\tilde{Q}(u) = \tilde{q}(u, u, u) = q(u, u) = Q(u).$$

The expression

$$\tilde{q}(u_1, u_2, u_3) = \frac{q(u_1, u_2) + q(u_2, u_3) + q(u_1, u_3)}{3}$$

defines a symmetric tri-affine function that has the proper behavior on the diagonal — by uniqueness, it must therefore be $\tilde{q}$.

We still need to find the control points $\tilde{\mathbf{v}}_i$ that characterize $\tilde{q}$. Now that we have $\tilde{q}$, this is accomplished by simple evaluation:

$$\begin{aligned}
\tilde{\mathbf{v}}_0 &= \tilde{q}(0,0,0) &= \tfrac{q(0,0)+q(0,0)+q(0,0)}{3} &= \mathbf{v}_0 \\
\tilde{\mathbf{v}}_1 &= \tilde{q}(0,0,1) &= \tfrac{q(0,0)+q(0,1)+q(0,1)}{3} &= \tfrac{\mathbf{v}_0+2\mathbf{v}_1}{3} \\
\tilde{\mathbf{v}}_2 &= \tilde{q}(0,1,1) &= \tfrac{q(0,1)+q(1,1)+q(0,1)}{3} &= \tfrac{2\mathbf{v}_1+\mathbf{v}_2}{3} \\
\tilde{\mathbf{v}}_3 &= \tilde{q}(1,1,1) &= \tfrac{q(1,1)+q(1,1)+q(1,1)}{3} &= \mathbf{v}_2
\end{aligned}$$

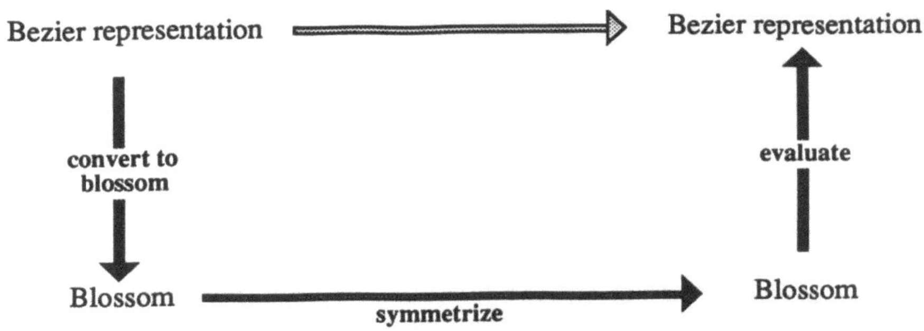

Figure 8: Schematic diagram of how blossoms are typically used to solve change of representation problems.

The above procedure for degree raising extends to curves of any degree. In general, raising a degree d Bézier curve Q whose blossom is q to a degree $d+1$ curve $\tilde{Q}$ with blossom $\tilde{q}$ is based upon the symmetric, multiaffine relation

$$\tilde{q}(u_1, \ldots, u_{d+1}) = \frac{q(u_1, u_2, \ldots, u_d) + \cdots + q(u_1, \ldots, u_{i-1}, u_{i+1}, \ldots u_{d+1}) + \cdots + q(u_2, \ldots u_{d+1})}{d+1}.$$

4.5 B-splines

The previous sections have indicated that blossoms are useful for analyzing individual polynomial curve segments. In this section, we explore the use of blossoms to study piecewise curves, i.e., B-splines. As a prerequisite, we must first understand the relationship between derivatives and blossoms.

4.5.1 Blossoms, Derivatives and C^k Continuity

Two curve segments $F : [r, s] \to X$ and $G : [s, t] \to X$, where X is an affine space, are said to meet with C^k continuity at s if they have matching derivatives up to order k at s:

$$F^{(i)}(s) = G^{(i)}(s), \quad i = 0 \ldots k,$$

where superscript (i) refers to the ith derivative.

Suppose two segments F and G meet with C^k continuity at s. What can be said about the relationship between their blossoms? To address this question, let us examine the connection between derivatives and blossoms.

Let f denote F's blossom, and let g denote G's. To simplify the discussion, we will also assume that F and G are of common degree d. The definition of $F^{(1)}(s)$ that is most convenient here is

$$\begin{aligned} F^{(1)}(s) &:= \frac{d}{du} F(s+u)|_{u=0} \\ &= \frac{d}{du} f(s+u, s+u, \ldots, s+u)|_{u=0}. \end{aligned}$$

To continue, we need to invoke the chain rule, which in this instance takes the form

$$\frac{d}{du}f(\underbrace{s+u}_{=u_1},\ldots,\underbrace{s+u}_{=u_d}) = \sum_{i=1}^{d}\frac{\partial f}{\partial u_i}(u_1,\ldots,u_d)|_{u_1=s,\ldots,u_d=s}\,\frac{du_i}{du}|_{u=0}.$$

The terms $\frac{du_i}{du}|_{u=0}$ are simply disposed of, since $u_i = s+u$ implies that $\frac{du_i}{du}|_{u=0} = 1$. Concentrating now on the terms containing a partial derivative of f, we get:

$$\begin{aligned}\frac{\partial f}{\partial u_i}|_{u_1,\ldots,u_d=s}(u_1,\ldots,u_d) &= \frac{\partial}{\partial u_i}((1-u_i)f(u_1,\ldots,u_{i-1},0,u_{i+1},\ldots,u_n)\\ &\qquad +u_i f(u_1,\ldots,u_{i-1},1,u_{i+1},\ldots u_d))|_{u_i,\ldots,u_d=s}\\ &= f(u_1,\ldots,u_{i-1},1,u_{i+1},\ldots,u_n)|_{u_i,\ldots,u_d=s}\\ &\qquad -f(u_1,\ldots,u_{i-1},0,u_{i+1},u_d)|_{u_i,\ldots,u_d=s}\\ &= f(1,s,\ldots,s)-f(0,s,\ldots,s).\end{aligned}$$

A similar result holds for other terms containing partial derivatives of f. Putting this together, and using the symmetry of f, we find that:

$$F^{(1)}(s) = n\,(f(1,s,\ldots,s)-f(0,s,\ldots,s))\,.$$

Applying the process once again reveals that

$$F^{(2)}(s) = n(n-1)\,(f(1,1,s,\ldots,s)-2f(1,0,s,\ldots,s)+f(0,0,s,\ldots,s))\,.$$

By iterating this procedure, the general form for the ith derivative can be shown to be

$$F^{(i)}(s) = \frac{n!}{(n-i)!}\sum_{j=0}^{i}(-1)^j\binom{i}{j}f(\underbrace{1,\ldots,1}_{i-j},\underbrace{0,\ldots,0}_{j},s,\ldots s). \tag{4}$$

Coming back to the question of continuity, if F and G meet with C^1 continuity at s, then we have

$$\begin{aligned}F(s) &= G(s)\\ F^{(1)}(s) &= G^{(1)}(s).\end{aligned}$$

Hence, their blossoms satisfy:

$$\begin{aligned}f(s,\ldots,s) &= g(s,\ldots,s)\\ f(1,s,\ldots,s)-f(0,s,\ldots,s) &= g(1,s,\ldots,s)-g(0,s,\ldots,s).\end{aligned} \tag{5}$$

Adding the first equation in Equation 5 to the second, we find that the left side becomes

$$f(1,s,\ldots,s)-f(0,s,\ldots,s)+f(s,s,\ldots,s),$$

which is an affine combination of three points. Since f and g are multiaffine, the sum of the two equations can be written as

$$f(1-0+s,s,\ldots,s) = g(1-0+s,s,\ldots,s).$$

This shows that if F and G meet with C^1 continuity, then their blossoms satisfy

$$\begin{aligned} f(s,\ldots,s) &= g(s,\ldots,s) \\ f(s+1,s,\ldots,s) &= g(s+1,s,\ldots,s). \end{aligned} \tag{6}$$

Moreover, since every value x_1 can be written as an affine combination of s and $s+1$, Equation 6 implies that

$$f(x_1,s,\ldots,s) = g(x_1,s,\ldots s), \tag{7}$$

for all x_1. This shows that the conditions of Equation 5 imply Equation 7. Since the converse clearly holds as well, we deduce that Equation 7 is a necessary and sufficient condition for C^1 continuity. The generalization for C^k continuity follows quickly from Equation 4. As a theorem, we have:

Theorem 2 Two polynomial segments $F:[r,s]\to X$ and $G:[r,s]\to X$ of degree n, having blossoms f and g respectively, meet with C^k continuity at s if and only if

$$f(x_1,\ldots,x_k,s,\ldots,s) = g(x_1,\ldots,x_k,s,\ldots,s)$$

for all $x_1\ldots x_k$.

4.6 B-splines and Piecewise Blossoms

We saw in Section 4.2 that a blossom $f_i(u_1,u_2,u_3)$ is completely determined by the progressive values $f_i(s,s,s)$, $f_i(s,s,t)$, $f_i(s,t,t)$, $f_i(t,t,t)$, since these are the Bézier control points for the diagonal polynomial $F_i(u) = f_i(u,u,u)$ on the interval $[s,t]$. More generally, if t_i, t_{i+1}, t_{i+2}, t_{i+3}, t_{i+4}, t_{i+5} are distinct non-decreasing values called *knots*, then f_i is determined by the *progressive* blossom values $f_i(t_i,t_{i+1},t_{i+2})$, $f_i(t_{i+1},t_{i+2},t_{i+3})$, $f_i(t_{i+2},t_{i+3},t_{i+4})$, $f_i(t_{i+3},t_{i+4},t_{i+5})$. That is, once these values have been specified, any other value of f_i can be computed. To see that this is so, notice that these values can be used to compute the values $f_i(t_{i+2},t_{i+2},t_{i+2})$, $f_i(t_{i+2},t_{i+2},t_{i+3})$, $f_i(t_{i+2},t_{i+3},t_{i+3})$, $f_i(t_{i+3},t_{i+3},t_{i+3})$, as shown in Figure 9. This latter set of values are the Bézier control points for the segment parametrized on the interval $[t_{i+2},t_{i+3}]$. Since the Bézier control points uniquely characterize a blossom, an arbitrary blossom value can then be computed using the algorithm of Figure 6. A more direct approach proceeds by constructing an arbitrary blossom value for a segment of degree d directly from the progressive values

$$\mathbf{v}_{i+j} = f_i(t_{i+j},t_{i+j+1},\ldots,t_{i+j+d-1}), \qquad j=0,\ldots,d,$$

using the algorithm of Figure 10.

Returning to cubic curves, the values $f_i(t_i,t_{i+1},t_{i+2})$, $f_i(t_{i+1},t_{i+2},t_{i+3})$, $f_i(t_{i+2},t_{i+3},t_{i+4})$, $f_i(t_{i+3},t_{i+4},t_{i+5})$, then, characterize a polynomial $F_i(u) := f_i(u,u,u)$ on the interval $[t_{i+2},t_{i+3}]$. The idea now is to build a piecewise C^2 curve F such that $F(u) := F_i(u) = f_i(u,u,u)$ when $u\in[t_{i+2},t_{i+3}]$. We will do this by building a *piecewise* blossom f. Recall from Theorem 2 that F_{i-1} and F_i meet with C^2 continuity at t_{i+2} if and only if their blossoms satisfy $f_{i-1}(t_{i+2},u,v) = f_i(t_{i+2},u,v)$ for all u,v.

Now we can characterize f_i by the progressive blossom values given above. We can similarly characterize f_{i-1} by the progressive values of three knots, beginning at $t_{i-1}\ldots t_{i+4}$.

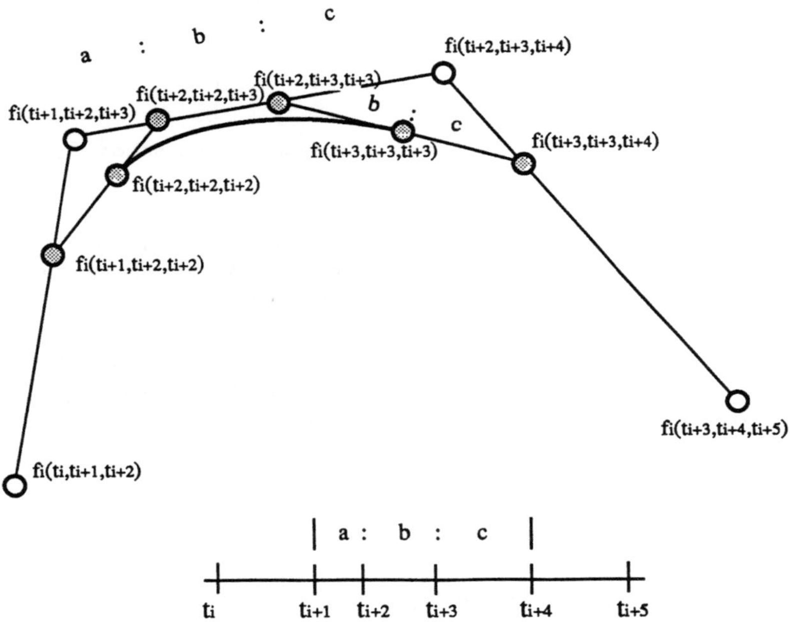

Figure 9: Another way to characterize a curve segment.

We require that f_{i-1} and f_i have values that agree whenever certain knot arguments agree, as given by the following conditions:

$$\begin{aligned} f_{i-1}(t_i, t_{i+1}, t_{i+2}) &= f_i(t_i, t_{i+1}, t_{i+2}) \\ f_{i-1}(t_{i+1}, t_{i+2}, t_{i+3}) &= f_i(t_{i+1}, t_{i+2}, t_{i+3}) \\ f_{i-1}(t_{i+2}, t_{i+3}, t_{i+4}) &= f_i(t_{i+2}, t_{i+3}, t_{i+4}). \end{aligned}$$

With these conditions, the C^2 constraints will be satisfied (this is easily verified). Thus, there are only 5 points that determine f_{i-1} and f_i, and hence F_{i-1} and F_i. These points are:

$$\begin{gathered} f_{i-1}(t_{i-1}, t_i, t_{i+1}) \\ f_{i-1}(t_i, t_{i+1}, t_{i+2}) = f_i(t_i, t_{i+1}, t_{i+2}) \\ f_{i-1}(t_{i+1}, t_{i+2}, t_{i+3}) = f_i(t_{i+1}, t_{i+2}, t_{i+3}) \\ f_{i-1}(t_{i+2}, t_{i+3}, t_{i+4}) = f_i(t_{i+2}, t_{i+3}, t_{i+4}) \\ f_i(t_{i+3}, t_{i+4}, t_{i+5}). \end{gathered}$$

Moreover, no matter where these 5 points are placed, f_{i-1} and f_i will meet with C^2 continuity at t_i. We may continue by placing another segment f_{i+1}, and another knot t_{i+6}. The result of this construction is seen in Figure 11.

We can improve our notation by dropping the subscripts on the functions and defining

```
EvaluateBlossomProgressive( v_i, ..., v_{i+d}, t_i, ..., t_{2d-1+i}, u_1, ..., u_d)
for k = 0 to d - 1 do
    for ℓ = 0 to d - k - 1 do
        { α, β are barycentric coords of u_{k+1} in [t_{i+k+ℓ}, t_{d+i+k+ℓ}]. }
        β = (u_{k+1} - t_{i+k+ℓ}) / (t_{d+i+k+ℓ} - t_{i+k+ℓ})
        α = 1 - β
        v_{i+ℓ} = α v_{i+ℓ} + β v_{i+ℓ+1}
        { v_{i+ℓ} now equals q(u_1, ..., u_{k+1}, t_{i+k+l}, ... t_{i+d+l-1}) }
    end
end
return v_i
```

Figure 10: Evaluation algorithm for an arbitrary blossom value $q(u_1, ..., u_d)$ given progressive blossom values and knots.

a piecewise blossom $f(u_1, u_2, u_3)$ using the rule:

$$f(u_1, u_2, u_3) = \begin{cases} f_i(u_1, u_2, u_3) & \text{if } u_i\text{'s are not all equal and } u_1, u_2, u_3 \in [t_i, t_{i+5}] \\ f_j(u_1, u_2, u_3) & \text{if } u_i\text{'s are all equal and } u_i \in [t_{j+2}, t_{j+3}]. \end{cases}$$

(Remark: strictly speaking, this isn't strong enough; see Ramshaw[11, 12].) The labeling becomes much simpler now, since we can drop subscripts of f's in Figure 11.

The values $f(t_{i-1}, t_i, t_{i+1}), \dots, f(t_{i+2}, t_{i+3}, t_{i+4}), \dots, f(t_{i+4}, t_{i+5}, t_{i+6})$ are the cubic B-spline control points for the *knot vector* $t_{i-1}, \dots, t_{i+6}$. In fact, when $u_1 = u_2 = \cdots = u_d = u$ the algorithm of Figure 10 is exactly the de Boor algorithm for computing a point on a B-spline (cf. Bartels et al [2] or Farin [9]).

These ideas can be used to define the cubic B-spline design scheme:

Input: A sequence of control points $\mathbf{v}_0 \dots \mathbf{v}_m$, a sequence of knot values $t_0 \dots t_{m+2}$, and a point $u \in [t_2, t_m]$.

Output: A point $F(u)$.

Such that: The curve $F(u)$, $u \in [t_2, t_m]$ is C^2.

Method: Let i be such that $u \in [t_{i+2}, t_{i+3}]$, and set $\mathbf{v}_{i+j} = f(t_{i+j}, t_{i+j+1}, t_{i+j+2})$, $j = 0, 1, 2, 3$, then use the de Boor algorithm of Figure 10 to compute the point $f_i(u, u, u) = F(u)$.

We now generalize this construction to an arbitrary degree d: For a general degree d, C^{d-1} continuity at t_{i+d-1} requires that adjacent blossoms must satisfy

$$f_{i-1}(t_{i+d-1}, u_2, \dots, u_d) = f_i(t_{i+d-1}, u_2, \dots, u_d)$$

for all $u_2, \dots, u_d$. This is guaranteed if f_{i-1} and f_i have blossom values that agree at their

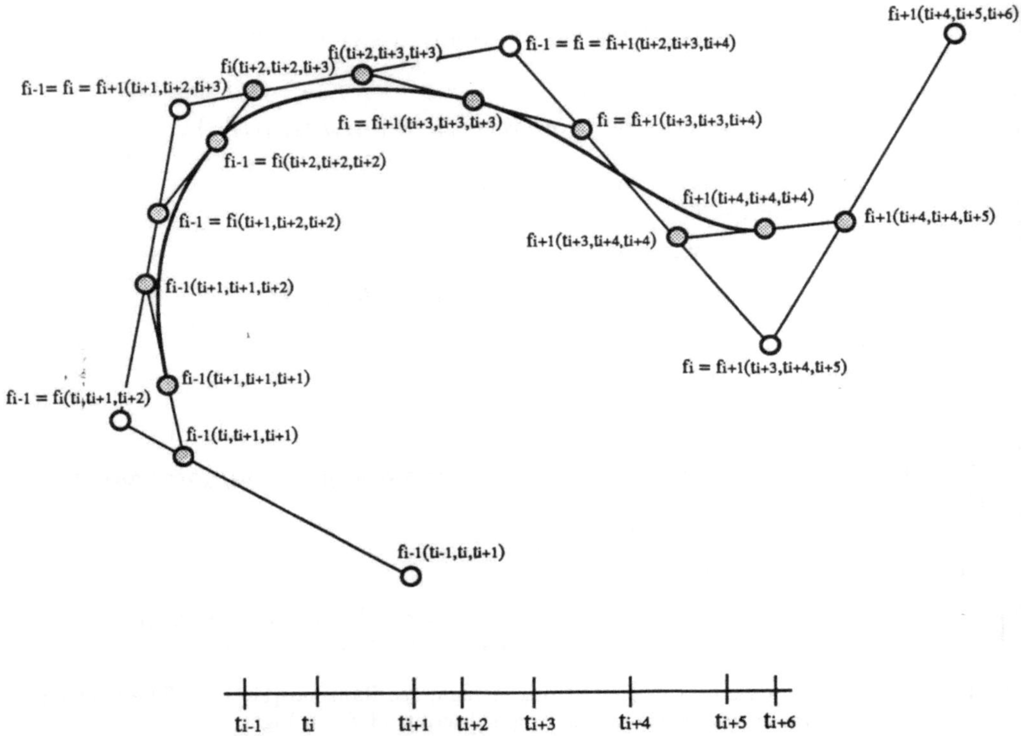

Figure 11: Three segments joining with C^2 continuity.

overlapping knots:

$$\begin{array}{ccc} f_{i-1}(t_i, \ldots, t_{i+d-1}) & = & f_i(t_i, \ldots, t_{i+d-1}) \\ \vdots & & \vdots \\ f_{i-1}(t_{i+d-1}, \ldots, t_{i+2d-2}) & = & f_i(t_{i+d-1}, \ldots, t_{i+2d-2}) \end{array}$$

We can therefore define a piecewise blossom $f(u_1, \ldots, u_d)$ according to:

$$f(u_1, \ldots, u_d) = \begin{cases} f_i(u_1, \ldots, u_d) & \text{if } u\text{'s are not all equal and } u_1, \ldots, u_d\text{'s} \in [t_i, t_{i+2d-1}] \\ f_j(u_1, \ldots, u_d) & \text{if } u\text{'s are all equal and } u \in [t_{j+d-1}, t_{j+d}]. \end{cases}$$

4.7 Knot Insertion

Knot insertion refers to the problem of finding control points for a curve over a knot vector $\hat{T}$, given the control points for the curve over a knot vector T, where $\hat{T}$ is a *refinement* of T; that is, where $T \subset \hat{T}$.

The CAGD literature describes two major knot insertion algorithms: Boehm's algorithm [3], and the Oslo algorithm [4]. The Oslo algorithm is capable of solving the problem no matter how many knots are added to T to obtain $\hat{T}$, whereas Boehm's algorithm is restricted to the case where T and $\hat{T}$ differ by a single knot. (Boehm's algorithm can be used to insert any number of knots by successively inserting one knot at a time.) Boehm's algorithm was originally proved using properties of B-spline basis functions; here we show how Boehm's algorithm is interpreted from the perspective of blossoming. For concreteness we shall restrict the discussion to cubic curves. The general case presents no conceptual difficulties, but the additional notation is sufficiently cumbersome that we omit it here.

Let $F(u)$ be the cubic B-spline curve in question, let $T = \{t_i\}$ be the original knot vector, let $\{\mathbf{v}_i\}$ be the control points of Q over T, and let $\hat{t} \in [t_i, t_{i+1}]$ be the knot to be inserted. From the blossoming point of view, we know that the B-spline control points are obtained by evaluating the blossom f of F at consecutive triples of knots. After $\hat{t}$ is included, the control points

$$\begin{aligned} \mathbf{v}_{i-1} &= f(t_{i-1}, t_i, t_{i+1}) \\ \mathbf{v}_i &= f(t_i, t_i, t_{i+1}) \end{aligned}$$

are no longer valid for $\hat{T}$. Instead, these must be replaced with the new control points

$$\begin{aligned} \hat{\mathbf{v}}_{i-1} &= f(t_{i-1}, t_i, \hat{t}) \\ \hat{\mathbf{v}} &= f(t_i, \hat{t}, t_{i+1}) \\ \hat{\mathbf{v}}_{i+1} &= f(\hat{t}, t_{i+1}, t_{i+2}). \end{aligned}$$

This procedure is summarized in Figure 12.

The Oslo algorithm, while somewhat more complicated due to its generality, can be shown to be equivalent to the evaluation of off-diagonal blossom values using the algorithm of Figure 10.

5 Surfaces

We now turn to the theory of blossoms for surfaces. The generalization to surfaces can be done in two different ways, yielding either tensor product or non-tensor product descriptions. We first examine the tensor product construction.

A bi-d-c tensor product surface $F(u, v)$ takes the form

$$F(u, v) = \sum_{i,j} \mathbf{w}_{i,j} b_i(u) b_j(v),$$

where $\mathbf{w}$'s denote the (rectangular array of) control points, and where b's denote a collection of univariate blending functions of degree d. Linearity of the operator $Blossom_D()$ implies that blossoming can occur independently in u and v, creating a function $f(u_1, ..., u_D; v_1, ..., v_D)$ with the properties

1. f is affine in each of its $2D$ argument;
2. f is symmetric with respect to interchange of u's;
3. f is symmetric with respect to interchange of v's;

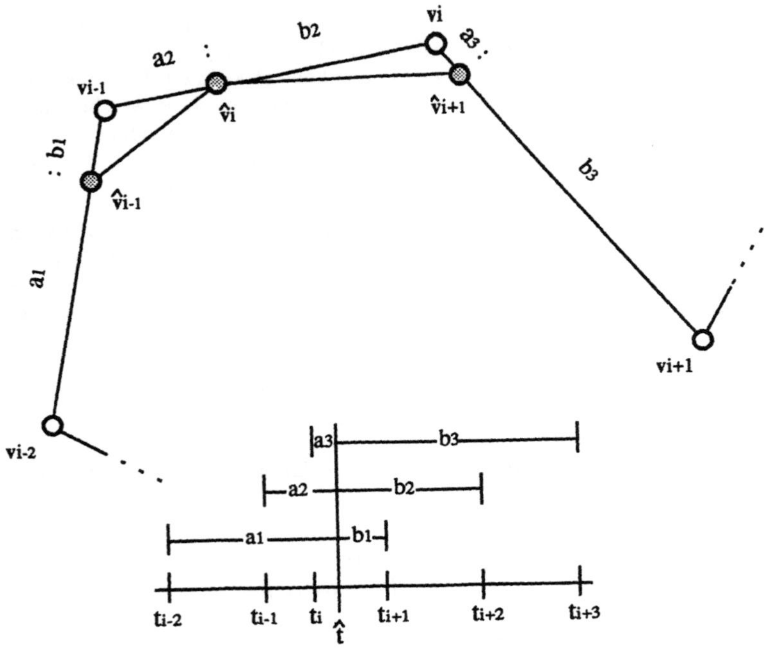

Figure 12: Boehm's knot insertion algorithm.

4. $F(u,v) = f(u,...,u; v,...,v)$.

Note that f is not guaranteed to be symmetric if one of the u arguments is interchanged with one of the v's. For instance, in general

$$f(u_1, u_2, ..., u_D; v_1, v_2, ..., v_D) \neq f(v_1, u_2, ..., u_D; u_1, v_2, ..., v_D).$$

All results of the blossoming theory for curves can be applied to $f(u_1, ..., u_d; v_1, ..., v_D)$, including extraction of Bézier and B-spline points by evaluation, degree raising in u or v, etc.

Non-tensor product surface forms arise when the domain parameter is allowed to range over $\Re^2$ instead of $\Re$. (The following readily generalizes to higher dimensions.) We will map u into a point $Q(u)$ on a surface using the following uniqueness theorem for surfaces:

Theorem 3 Let Q be a bivariate polynomial of degree d. If $D \geq d$, there is a unique symmetric D-affine function q such that $Q(u) = q(\underbrace{u,\ldots,u}_{D})$.

Similar to our treatment of curves, q is the blossom of Q. The Bézier points of Q can be extracted directly by evaluating q at simply described values. Figure 13 shows what these values are for quadratic surfaces. In general, if q is a blossom of Q, then Q's Bézier control

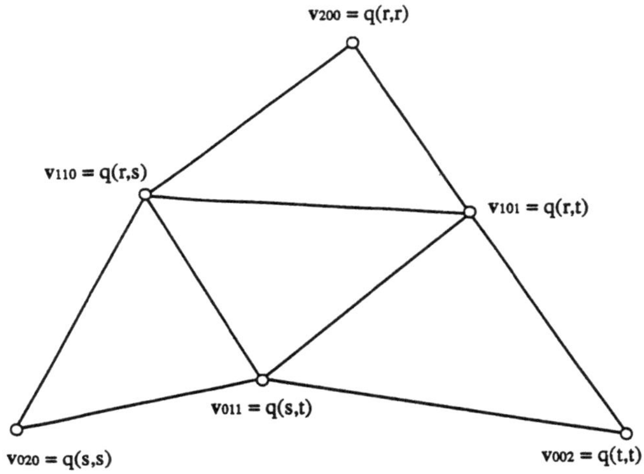

Figure 13: Blossom labels for quadratic surfaces.

points relative to a domain triangle rst are given by

$$\mathbf{v}_{i_1,i_2,i_3} = q(\underbrace{r,\ldots,r}_{i_1},\underbrace{s,\ldots,s}_{i_2},\underbrace{t,\ldots,t}_{i_3}).$$

5.1 Evaluation

Given a point $u = \alpha_1 r + \alpha_2 s + \alpha_3 t$, we would like to find the point $Q(u) = q(u,u)$. Just as for curves, we can find it by a series of affine combinations. First, we have

$$\begin{aligned} q(u,u) &= q(\alpha_1 r + \alpha_2 s + \alpha_3 t, u) \\ &= \alpha_1 q(r,u) + \alpha_2 q(s,u) + \alpha_3 q(t,u). \end{aligned}$$

Also,

$$\begin{aligned} q(r,u) &= q(r, \alpha_1 r + \alpha_2 s + \alpha_3 t) \\ &= \alpha_1 q(r,r) + \alpha_2 q(r,s) + \alpha_3 q(r,t), \end{aligned}$$

and similar expressions can be found for $q(s,u)$ and $q(t,u)$. Geometrically, this is very simple, as shown in Figure 14.

5.2 Subdivision

Subdivision for Bézier surfaces comes directly from the points evaluated in the de Casteljau algorithm. The subdivision given by the algorithm breaks the surface into 3 pieces about the

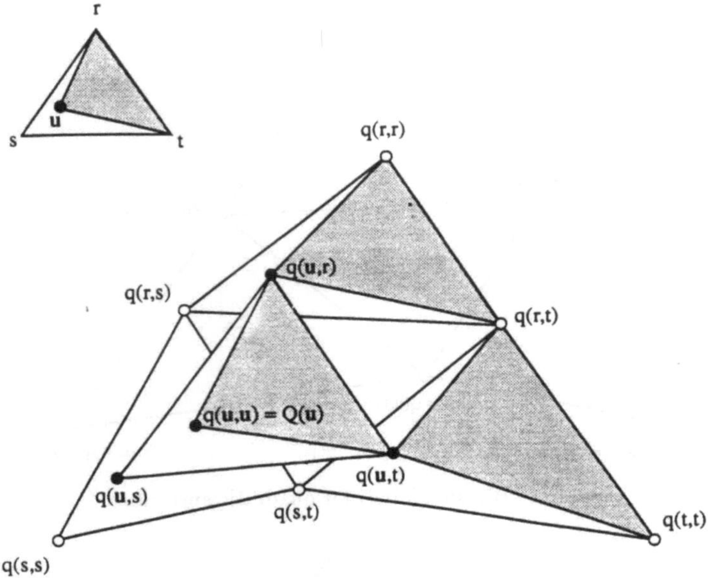

Figure 14: De Casteljau's algorithm for surfaces using blossom labeling.

point of evaluation. Using blossoming, the proof is again fairly simple. We need only observe that the parameters of the intermediate points found by the algorithm have the proper form to be control points for a degree d polynomial. By uniqueness, they must be correct. For example, the subtriangle shown in gray in Figure 14 has as its control points $q(r,r)$, $q(u,r)$, $q(r,t)$, $q(u,u)$, $q(u,t)$, and $q(t,t)$. These blossom values are sufficient to describe Q when its domain is restricted to the triangle urt.

5.3 Degree Raising

Given a quadratic Q defined by its control points $\mathbf{v}_{i_1,i_2,i_3}$, suppose we want to raise its degree to a cubic $\tilde{Q}$ defined by the control points $\tilde{\mathbf{v}}_{i_1,i_2,i_3}$ such that $Q(u) = \tilde{Q}(u)$. The blossom $\tilde{q}$ of $\tilde{Q}$ in terms of q is

$$\tilde{q}(u_1, u_2, u_3) = \frac{q(u_1, u_2) + q(u_2, u_3) + q(u_1, u_3)}{3}.$$

Note that this is the same as the equation used in degree raising a Bézier curve! The domain is in a higher dimension, but the resulting equations are exactly the same.

As with degree raising of curves, the control points $\tilde{\mathbf{v}}_{i_1,i_2,i_3}$ can now be determined easily by evaluation. For example,

$$\begin{aligned}
\tilde{\mathbf{v}}_{300} &= \tilde{q}(r,r,r) = \tfrac{q(r,r)+q(r,r)+q(r,r)}{3} = \mathbf{v}_{200} \\
\tilde{\mathbf{v}}_{111} &= \tilde{q}(r,s,t) = \tfrac{q(r,s)+q(s,t)+q(r,t)}{3} = \tfrac{\mathbf{v}_{110}+\mathbf{v}_{011}+\mathbf{v}_{101}}{3}
\end{aligned}$$

Unlike the close relation between Bézier curves and Bézier surfaces, a similarly nice generalization of B-spline curves to surfaces is yet to be found.

6 Blossoms as Abstract Data Types

We have thus far been treating blossoming primarily as a theoretical tool for developing algorithms for the manipulation of Bézier and B-splines curves and Bézier surfaces. In this section, we briefly describe how blossoming can also be used as an effective tool for computer programming.

The key to exploiting blossoming for programming is to create a software library that supports blossoms as an *abstract data type* [1]. An abstract data type (ADT) is simply a collection of data types together with a collection of operations for manipulating them. An ADT for performing affine and Euclidean geometric programming has previously been detailed [7, 8]. Here we outline how the geometric ADT can be expanded to embrace blossoming, and hence Bézier and B-spline curves and surfaces.

The fundamental data types supported in the geometric ADT include `Space`, `Point`, `Vector`, and `AffineMap`. The `Space` data type is a model for arbitrary (finite) dimensional affine spaces. `AffineMaps` between spaces can be created by specifying how the map transforms the points of a domain simplex. For instance, if the points rst form a triangle in a two dimensional affine space X, and if RST are points in a space Y (Y need not be two dimensional), then an affine map $M : X \rightarrow Y$ can be created by pseudo-code similar to

```
M := AffineMapCreate( r, s, t, R, S, T)
```

Once `M` is created, executing a statement like

```
AffineMapEvaluate(M,p)
```

returns the point $M(p)$ in Y.

Blossoms can be introduced into the ADT by providing a `Blossom` data type. Since blossoms are nothing more than symmetric multiaffine maps, support for them is most easily achieved as a generalization of the `AffineMap` data type; that is, the `AffineMap` type can be considered as a univariate blossom. One method of creating blossoms would be to generalize the pseudo-code above by specifying an appropriate number of argument lists, together with their image under the blossom. This is not terribly convenient in practice, as there are a number of restrictions on the argument lists used. A somewhat more convenient way is to specify a blossom by specifying its diagonal polynomial in Bézier form. For instance, if `V` denotes the control net (a sequence of points in some space Y) for a degree d triangular Bézier surface Q, defined on a domain triangle rst in a two dimensional space X, then Q's blossom can be created using pseudo-code such as

```
q := BlossomCreateFromBezierNet( r, s, t, V).
```

Once created, an arbitrary value of `q` can be evaluated using pseudo-code such as

```
BlossomEvaluate( q, u1, ...  , ud)
```

where `u1, ..., ud` are arbitrary points in X. For curves, occurring when the domain space X is one dimensional, it is also convenient to create blossoms from the B-spline representation of their diagonal.

The advantage to this approach is that the fundamental operations of many algorithms are encapsulated in the blossom evaluation routine `BlossomEvaluate`. The implementation of this algorithm is essentially given in Figure 10. Indeed, as we've already seen, the following algorithms are all based on blossom evaluation:

- The de Casteljau and de Boor algorithms.
- Sablonniere's algorithm [13] for the conversion between Bézier and B-spline curves, illustrated in Figure 9.
- Boehm's knot insertion algorithm.
- The Oslo algorithm.
- Bézier subdivision.

References

[1] A. Aho, J. Hopcroft, and J. Ullman. *Data Structures and Algorithms*. Addison-Wesley, 1983.

[2] Richard H. Bartels, John C. Beatty, and Brian A. Barsky. *An introduction to splines for use in computer graphics & Geometric Modeling*. Morgan Kaufmann, Los Altos, CA, 1987.

[3] W. Boehm. Inserting new knots into B-spline curves. *CAD*, 12(4):199–201, 1980.

[4] E. Cohen, T. Lyche, and R. Riesenfeld. Discrete B-splines and subdivision techniques in computer aided geometric design and computer graphics. *Computer Graphics and Image Processing*, 14(2):87–111, 1980.

[5] Pierre de Casteljau. Formes à pôles. Hermes, Paris, 1985.

[6] Pierre de Casteljau. *Shape Mathematics and CAD*. Kogan Page, Ltd., London, 1986.

[7] Tony D. DeRose. A coordinate-free aproach to geometric programming. In W. Strasser and H.-P. Seidel, editors, *Theory and Practice of Geometric Modeling*, pages 291–306. Springer-Verlag, Berlin, 1989.

[8] Tony D. DeRose. Coordinate-free geometric programming. Technical Report 89-09-16, University of Washington, Seattle, WA 98195, September 1989.

[9] Gerald Farin. *Curves and Surfaces for Computer Aided Geometric Design*. Academic Press, Boston, 1988.

[10] Gary Herron. Techniques for visual continuity. In Gerald E. Farin, editor, *Geometric Modeling: Algorithms and New Trends*, pages 163–174. SIAM, 1987.

[11] Lyle Ramshaw. Blossoming: A connect-the-dots approach to splines. Technical Report 19, Digital Systems Research Center, Palo Alto, CA, 1987.

[12] Lyle Ramshaw. Béziers and B-splines as multiaffine maps. In *Theoretical Foundations of Computer Graphics and CAD*, pages 757–776. Springer, New York, 1988.

[13] P. Sablonniere. Spline and Bézier polygons associated with a polynomial spline curve. *CAD*, 10(4):257–261, 1978.

[14] Hans-Peter Seidel. Computing B-spline control points. In *Theory and Practice of Geometric Modeling*, pages 17–32. Springer-Verlag, Berlin, 1989.

Springer Series
Computer Graphics – Systems and Applications

Former Subseries of SYMBOLIC COMPUTATION

J. L. Encarnação, R. Schuster, E. Vöge (eds.):
Product Data Interfaces in CAD/CAM Applications.
Design, Implementation and Experiences.
IX, 270 pages, 147 figs., 1986

U. Rembold, R. Dillmann (eds.):
Computer-Aided Design and Manufacturing.
Methods and Tools. Second, revised and enlarged edition.
XIV, 458 pages, 304 figs., 1986

G. Enderle, K. Kansy, G. Pfaff:
Computer Graphics Programming. GKS – The Graphics
Standard. Second, revised and enlarged edition.
XXIII, 651 pages, 100 figs., 1987

Y. Shirai:
Three-Dimensional Computer Vision.
XII, 297 pages, 313 figs., 1987

D. B. Arnold, P. R. Bono:
CGM and CGI. Metafile and Interface Standards
for Computer Graphics.
XXIII, 279 pages, 103 figs., 1988

J. L. Encarnação, P. C. Lockemann (eds.):
Engineering Databases. XII, 229 pages, 152 figs., 1990

P. Wisskirchen:
Object-Oriented Graphics. From GKS and PHIGS
to Object-Oriented Systems.
XIII, 236 pages, 83 figs., 1990

J. L. Encarnação, R. Lindner, E. G. Schlechtendahl:
Computer Aided Design. Fundamentals and System
Architectures. Second, revised and extended edition.
XII, 432 pages, 240 figs., 1990

T. Yagiu:
Modeling Design Objects and Processes.
Approx. 300 pages, 82 figs., 1991

H. Hagen, D. Roller (eds.):
Geometric Modeling. Methods and Applications.
VII, 286 pages, 140 figs., 1991

MIX
Papier aus verantwortungsvollen Quellen
Paper from responsible sources
FSC® C105338

If you have any concerns about our products,
you can contact us on
ProductSafety@springernature.com

In case Publisher is established outside the EU,
the EU authorized representative is:
Springer Nature Customer Service Center GmbH
Europaplatz 3, 69115 Heidelberg, Germany

Printed by Libri Plureos GmbH
in Hamburg, Germany